SMART SENSOR SYSTEMS

SMART SENSOR SYSTEMS: EMERGING TECHNOLOGIES AND APPLICATIONS

Edited by

Gerard Meijer

Delft University of Technology and SensArt,
The Netherlands

Michiel Pertijs

Delft University of Technology, The Netherlands

Kofi Makinwa

Delft University of Technology, The Netherlands

IEEE PRESS

WILEY

Library of Congress Cataloging-in-Publication Data

Smart sensor systems : emerging technologies and applications / edited by Gerard C.M. Meijer, Michiel Pertijs, Kofi Makinwa.
        p. cm.
    Includes bibliographical references and index.
    ISBN 978-0-470-68600-3 (cloth)
    1. Detectors–Design and construction. 2. Detectors–Industrial applications. 3. Microcontrollers. I. Meijer, G. C. M. (Gerard C. M.)
    TA165.S55 2008
    681'.25–dc22

                                                                                    2008017675

A catalogue record for this book is available from the British Library.

ISBN: 9780470686003

Cover picture: © Martil Instruments

Typeset in 10/12pt TimesLTStd by Laserwords Private Limited, Chennai, India

1   2014

Contents

About the Editors

Gerard Meijer

Gerard Meijer received his M.Sc. and Ph.D. degrees in Electrical Engineering from Delft University of Technology, Delft, The Netherlands, in 1972 and 1982, respectively. Since 1972, he has been a member of the research and teaching staff of Delft University of Technology, and since 2001, an Antoni van Leeuwenhoek Professor. His major interests are in the field of Analog Electronics and Electronic Instrumentation. In 1984, and on a part-time basis from 1985–1987, he was seconded to Delft Instruments Company, Delft, The Netherlands, where he was involved in the development of industrial level gauges and temperature transducers. In 1996 he co-founded the company SENSART, where he is a consultant for the design and development of sensor systems. In 1999 the Dutch Foundation of Technical Sciences STW awarded Meijer with the honorary title of Simon Stevin Meester. Meijer is chairman of the National STW Platform on Sensor Technology.

Michiel Pertijs

Michiel Pertijs received the M.Sc. and Ph.D. degrees in electrical engineering (both *cum laude*) from Delft University of Technology, Delft, The Netherlands, in 2000 and 2005, respectively. From 2005 to 2008, he was with National Semiconductor, Delft, where he designed precision operational amplifiers and instrumentation amplifiers. From 2008 to 2009, he was a Senior Researcher with imec / Holst Centre, Eindhoven, The Netherlands. In 2009, he joined the Electronic Instrumentation Laboratory of Delft University of Technology, where he is now an Associate Professor. He heads a research group working on integrated circuits for medical ultrasound and energy-efficient smart sensors. Dr. Pertijs is an Associate Editor of the IEEE *Journal of Solid-State Circuits* (*JSSC*) and a member of the technical program committees of the International Solid-State Circuits Conference (ISSCC), the European Solid-State Circuits Conference (ESSCIRC), and the IEEE Sensors Conference. He received the ISSCC 2005 Jack Kilby Award for Outstanding Student Paper, the JSSC 2005 Best Paper Award, and the 2006 Simon Stevin Gezel Award from the Dutch Technology Foundation STW.

Kofi Makinwa

Kofi Makinwa received the B.Sc. (First class honors) and M.Sc. degrees from Obafemi Awolowo University, Ile-Ife, Nigeria in 1985 and 1988 respectively. In 1989, he received the M.E.E. degree (*cum laude*) from the Philips International Institute, Eindhoven, The Netherlands and in 2004, the Ph.D. degree from Delft University of Technology, Delft, The Netherlands. From 1989 to 1999, he was a Research Scientist with Philips Research Laboratories, Eindhoven, The Netherlands, where he worked on interactive displays and on optical and magnetic recording systems. In 1989 he joined the Electronic Instrumentation Laboratory of Delft University of Technology, where he is currently an Antoni van Leeuwenhoek Professor and head of the laboratory. His main interests are in the design of precision analog circuitry, smart sensors and sensor interfaces.

Dr. Makinwa is on the technical program committees of the European Solid-State Circuits Conference (ESSCIRC) and the Advances in Analog Circuit Design (AACD) workshop. He was on the Program Committee of the International Solid-State Circuits Conference (ISSCC) from 2006 to 2012. He has also been a three-time guest editor of the *Journal of Solid-State Circuits* (*JSSC*) and a two-term distinguished lecturer of the IEEE Solid-State Circuits Society. He is a co-recipient of the 2005 Simon Stevin Gezel Award from the Dutch Technology Foundation, as well as of several best paper awards: from the *JSSC* (2005, 2011), the ISSCC (2005, 2008, 2011) and the ESSCIRC (2006, 2009), among others. In 2013, at the 60th anniversary of ISSCC, he was recognized as one of its top ten contributors. He is an IEEE fellow, an alumnus of the Young Academy of the Royal Netherlands Academy of Arts and Sciences and an elected member of the IEEE Solid-State Circuits Society AdCom, the society's governing board.

List of Contributors

Pedram Afshar, Medtronic Neuromodulation, Minneapolis, USA

Bernhard Boser, Berkeley Sensor & Actuator Center, University of California, Berkeley, USA

Zu-Yao Chang, Electronic Instrumentation Laboratory, Delft University of Technology, Delft, The Netherlands

Peng Cong, Medtronic Neuromodulation, Minneapolis, USA

Tim Denison, Medtronic Neuromodulation, Minneapolis, USA

Chinwuba Ezekwe, Robert Bosch, Palo Alto, California, USA

Ali Heidari, Guilan University, Rasht, Iran

Johan Huijsing, Electronic Instrumentation Laboratory, Delft University of Technology, Delft, The Netherlands

Blagoy Iliev, Martil Instruments, Heiloo,The Netherlands

Xiujun Li, Exalon, Delft, The Netherlands and Sensytech, Delft, The Netherlands

Kofi Makinwa, Electronic Instrumentation Laboratory, Delft University of Technology, Delft, The Netherlands

Gerard Meijer, Electronic Instrumentation Laboratory, Delft University of Technology and SensArt, Delft, The Netherlands

Stoyan Nihtianov, Electronic Instrumentation Laboratory, Delft University of Technology, Delft, The Netherlands

Jos Oudenhoven, imec/Holst Centre, Eindhoven, The Netherlands

Michiel Pertijs, Electronic Instrumentation Laboratory, Delft University of Technology, Delft, The Netherlands

Gheorghe Pop, Martil Instruments, Heiloo, The Netherlands

Valer Pop, imec/Holst Centre, Eindhoven, The Netherlands

Michael Renaud, imec/Holst Centre, Eindhoven, The Netherlands

Zhichao Tan, Electronic Instrumentation Laboratory, Delft University of Technology, Delft, The Netherlands

Albert Theuwissen, Harvest Imaging, Bree, Belgium and Electronic Instrumentation Laboratory, Delft University of Technology, Delft, The Netherlands

Roland Thewes, Technische Universität Berlin (Berlin Institute of Technology), Berlin, Germany

Hubregt Visser, imec/Holst Centre, Eindhoven, The Netherlands

Ruud Vullers, imec/Holst Centre, Eindhoven, The Netherlands

Ziyang Wang, imec/Holst Centre, Eindhoven, The Netherlands

Preface

This book is intended as a reference for designers and users of sensors and sensor systems, and as a source of inspiration and a trigger for new ideas. For a major part, it is based on material used in the multidisciplinary "Smart Sensor Systems" course, which has been held annually at Delft University of Technology since 1995. The goals of this course are to present the basic principles of smart sensor systems to a broad, multidisciplinary audience, to develop a common language and scientific background to discuss the challenges associated with the design of such systems, and to facilitate mutual cooperation. In this way, we hope to contribute to the continuous expansion of the community of people advancing the exciting field of smart sensor systems.

As diverse and widespread as smart sensors may be today, research and development in this field is far from complete. It is driven by the continuous demand for lower cost, size and power consumption, and for higher performance and greater reliability. Moreover, new sensing principles and technologies are continuously emerging, and so significant effort is required to bring them to maturity. Often, this process involves more than just improving the performance of the transducer concerned. The *system* around the transducer plays an equally important, if not a more important, role. This system includes the electronics that interface with the transducer, the package that protects the transducer from the environment, and the calibration procedure that ensures that a certain performance specification is met.

This book focuses on these important system aspects, and, in particular, on the design of *smart* sensor systems, in which sensors and electronics are combined in a single package or even on a single chip to provide improved functionality, performance and reliability. In a previous book entitled "Smart Sensor Systems," the basics of such systems were covered. This book complements this prior publication by covering a number of emerging sensing technologies and applications, as well as discussing, in more detail, the system aspects of smart sensor design.

The book opens by discussing the exciting possibilities afforded by the combination of sensors and electronics: the accurate processing of small sensor signals (Chapter 1), the adoption of self-calibration techniques (Chapter 2), and the integration of precision instrumentation amplifiers (Chapter 3). This is followed by a discussion of a number of sensor systems in which system aspects play a key role: sensing of physical and chemical parameters by means of impedance measurement (Chapter 4); low-power angular-rate sensing using feedback and background-calibration techniques (Chapter 5); sensor systems for the detection of biomolecules, such as DNA (Chapter 6); optical sensor systems-on-a-chip in the form of CMOS image sensors (Chapter 7); and smart sensors capable of interfacing with the human nervous system (Chapter 8). Finally, the book also describes emerging technologies for the generation and storage of energy, since these are the key to realizing truly autonomous sensor systems (Chapter 9).

During the course of writing this text, we have been assisted by many people. We gratefully acknowledge the feedback and suggestions provided by our reviewers: Reinoud Wolffenbuttel of Delft University of Technology, Michael Kraft of the Fraunhofer Institute for Microelectronic Circuits and Systems, Michiel Vellekoop of the University of Bremen, Jan Bosiers of Teledyne DALSA, Firat Yazicioglu of imec, and the authors who also acted as reviewers. At our publisher, John Wiley & Sons, Ltd., we would like to acknowledge the Project Editors Richard Davies, Liz Wingett, and Laura Bell, for their support, encouragement and help in arranging agreements as well as to Production Editor Genna Manaog and Sangeetha Parthasarathy of Laserwords for help throughout the production of this book. Furthermore, we want to express our gratitude to the universities, research institutes and companies who permitted the use of figures and illustrations to make this book attractive for our readers. Finally, we would like to thank our spouses, Rumiana, Hannah and Abi, for their love and support.

Gerard Meijer, Michiel Pertijs and Kofi Makinwa
Delft, The Netherlands

1

Smart Sensor Design*

Kofi Makinwa

*Electronic Instrumentation Laboratory, Delft University of Technology,
Delft, The Netherlands*

1.1 Introduction

Sensors have become a ubiquitous part of today's world. Modern cars employ tens of sensors, ranging from simple position sensors to multi-axis MEMS accelerometers and gyroscopes. These sensors enhance engine performance and reliability, ensure compliance with environmental standards, and increase occupant comfort and safety. In another example, modern homes contain several sensors, ranging from simple thermostats to infrared motion sensors and thermal gas flow sensors. However, the best example of the ubiquity of sensors is probably the mobile phone, which has evolved from a simple communications device into a veritable sensor platform. A modern mobile phone will typically contain several sensors: a touch sensor, a microphone, one or two image sensors, inertial sensors, magnetic sensors, and environmental sensors for temperature, pressure and even humidity. Together with a GPS receiver for position location, these sensors greatly enhance ease of use and have extended the utility of mobile phones far beyond their original role as portable telephones.

Today, most of the sensors in a mobile phone, as well as most sensors intended for consumer applications, are made from silicon. This is mainly because silicon sensors can be mass-produced at low cost by exploiting the large manufacturing base established by the semiconductor industry. Another important motivation is the fact that the electronic circuitry required to bias a sensor and condition its output can be readily realized on the same substrate or, at least, in the same package. It also helps that semiconductor-grade silicon is a highly pure material with well-defined physical properties, some of which can be tuned by doping, and which can be precisely machined at the nanometer scale.

Silicon is a versatile material, one that exhibits a wide range of physical phenomena and so can be used to realize many different kinds of sensors [1]. For example, magnetic fields can

---

* This chapter is an expanded and updated version of [7].

Smart Sensor Systems: Emerging Technologies and Applications, First Edition.
Edited by Gerard Meijer, Michiel Pertijs and Kofi Makinwa.

be sensed via the Hall effect, temperature differences can be sensed via the Seebeck effect, mechanical strain can be sensed via the piezo-resistive effect and light can be sensed via the photo-electric effect. In addition, measurands that do not directly interact with silicon can often be indirectly sensed with the help of silicon-compatible materials. For example, humidity can be sensed by measuring the dielectric constant of a hygroscopic polymer [2], while gas concentration can be sensed by measuring the resistance of a suitably adsorbing metal oxide [3]. It should be noted that although silicon sensors may not achieve best-in-class performance, their utility and increasing popularity stems from their small size, low cost and the ease-of-use conferred by their co-integrated electronic circuitry.

Sensors are most useful when they are part of a larger system that is capable of processing and acting upon the information that they provide. This information must therefore be transmitted to the rest of the system in a robust and standardized manner. However, since sensors typically output weak analog signals, this task must be performed by additional electronic circuitry. Such *interface electronics* is best located close to the sensor, to minimize interference and avoid transmission losses. When they are both located in the same package, the combination of sensor and interface electronics is what we shall refer to as a *smart* sensor [4].

In addition to providing a robust signal to the outside world, the interface electronics of a smart sensor can be used to perform traditional signal processing functions such as filtering, linearization and compression. But it can also be used to increase the sensor's reliability by implementing self-test and even self-calibration functionality (as will be discussed in Chapter 2). A recent trend is towards sensor fusion, in which the outputs of multiple sensors in a package are combined to generate a more reliable output. For example, the outputs of gyroscopes, accelerometers and magnetic sensors can be combined to obtain robust position estimates, thus enabling mobile devices with indoor navigational capability.

This chapter discusses the design of smart sensor systems, in general, and the design of smart sensors in standard integrated circuit (CMOS) technology, in particular. Examples will be given of the design of state-of-the-art CMOS smart sensors for the measurement of temperature, wind velocity and magnetic field. Although the use of standard CMOS technology constrains the performance of the actual sensors, it minimizes cost, and as will be shown, the performance of the overall sensor *system* can often be significantly improved with the help of the co-integrated interface electronics.

1.2   Smart Sensors

A smart sensor is a system-in-package in which a sensor and dedicated interface electronics are realized. It may consist of a single chip, as is the case with smart temperature sensors, image sensors and magnetic field sensors. However, in cases when the sensor cannot be implemented in the same technology as the interface electronics, a two-chip solution is required. Since this also decouples the production yield of the circuit from that of the sensor, a two-chip solution is often more cost effective, even in cases where the sensor could be co-integrated with the electronics. Examples of two-chip sensors are mechanical sensors, such as MEMS accelerometers, gyroscopes and microphones, whose manufacture requires the use of micro-machining technology.

Since silicon chips, and especially their connections to the outside world, are rather fragile, smart sensors must be protected by some kind of packaging. The design of an appropriate package can be quite challenging since it must satisfy two conflicting requirements: allowing the sensor to interact with the measurand, while protecting it (and its interface electronics) from environmental damage. In the case of temperature and magnetic sensors, more or less standard integrated circuit packages can be employed. Standard packaging can also be used for inertial sensors, provided that a capping die or layer is used to protect their moving parts. In general, however, most sensors require custom packaging, which significantly increases their cost and usually involves a compromise between performance and robustness.

As has been noted earlier, silicon sensors are not necessarily best-in-class. However, the co-integrated interface electronics can be used to improve the performance of the overall system, either by operating the sensor in an optimal mode or by compensating for some of its non-idealities. This requires a good knowledge of the sensor's characteristics. For example, electronic circuitry can be used to incorporate MEMS inertial sensors in an electro-mechanical feedback loop, which, in general, results in improved linearity and wider bandwidth [5]. An example of such a system will be presented in Chapter 5, which describes the use of feedback and compensation circuits to enhance the performance of a MEMS gyroscope. Knowledge of the sensor's characteristics is also necessary to compensate for its cross-sensitivities, for example, to ambient temperature and packaging stress. The design of a smart sensor thus involves the optimization of an entire system and is, therefore, an exercise in system design.

1.2.1 Interface Electronics

To communicate with the outside world, the output of a smart sensor should preferably be a digital signal, although duty-cycle or frequency modulated signals are also microprocessor-compatible and so are sometimes used. The current trend in smart sensor design is to digitize the sensor's output as early as possible, and then to perform any additional signal conditioning, such as filtering, linearization, cross-sensitivity compensation and so on, in the digital domain. This approach facilitates the interconnection of several sensors via a digital bus, and takes advantage of the flexibility and ever-increasing digital signal processing capability of integrated circuitry. A similar trend can be observed in radio receivers, whose ADCs are moving closer and closer to the antenna, and which are thus employing more and more digital signal processing [6].

However, most sensors output low-level analog signals. This is especially true of silicon sensors such as thermopiles, Hall plates and piezo-resistive strain gauges, whose outputs contain information at the microvolt level. One reason for this is the nature of the transduction mechanisms available in silicon. Another is that their small size limits the amount of energy that they can extract from their environment. While this is a desirable feature in a sensor, which should not disturb, that is, extract energy from, the physical process that it observes, it makes the design of transparent interface electronics quite challenging. Great care must be taken to ensure that circuit non-idealities, such as thermal noise and offset, do not limit the performance of a smart sensor.

A further design challenge arises from the fact that the signal bandwidth of most sensors includes DC. As a result, the design of transparent interface electronics, especially in today's mainstream CMOS technology, involves a constant battle against random error sources such as drift and $1/f$ noise, as well as against systematic errors caused by component mismatch, charge injection and leakage currents.

Fortunately, most sensors are quite slow compared to the switching speed of transistors, and so dynamic error correction techniques, which essentially trade speed or bandwidth for precision, can be used to correct for systematic errors [7]. As the term "dynamic" implies, these techniques act continuously to reduce such errors, and so also mitigate the effects of low-frequency random errors due to drift and $1/f$ noise. In general, dynamic error correction techniques can be divided into two categories: sample-and-correct techniques and modulate-and-filter techniques.

An example of a sample-and-correct technique is auto-zeroing (Figure 1.1), in which the input of an amplifier is periodically shorted, while its output is fed to an offset-canceling integrator [8]. During normal operation the integrator's input is disconnected, thus freezing its output and canceling the amplifier's instantaneous offset (and $1/f$ noise). The main drawback of auto-zeroing is that the need to short-circuit the amplifier's input reduces its availability. However, this can be circumvented by using two, alternately auto-zeroed, amplifiers in a so-called ping-pong configuration [9].

An alternative way to reduce amplifier offset is known as chopping, and it is an example of a modulate-and-filter technique. The input signal is modulated by a square-wave, amplified and then demodulated [8]. As shown in Figure 1.2, this sequence of operations modulates the amplifier's offset (and $1/f$ noise) to the chopping frequency f_{ch}, which facilitates their removal by a low-pass (averaging) filter. However, the filter also limits the amplifier's useful bandwidth. This drawback can be circumvented by the use of chopper-stabilized amplifiers in which a chopper amplifier is used to improve the low frequency characteristics of a wide-band main

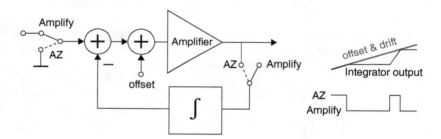

Figure 1.1   Simplified block diagram of an auto-zeroed amplifier

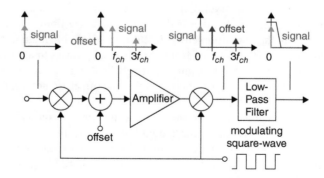

Figure 1.2   Simplified diagram of a chopper amplifier

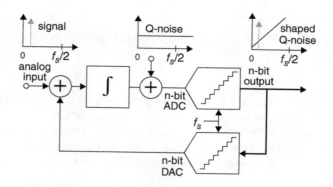

Figure 1.3 Simplified diagram of a sigma-delta modulator

amplifier [10]. The design of precision amplifiers based on various combinations of chopping and auto-zeroing will be discussed in more detail in Chapter 3.

High-resolution analog-to-digital conversion can be achieved by employing a technique known as sigma-delta (or delta-sigma) modulation, in which a low-pass filter, an ADC and a DAC are combined to form a feedback loop [11]. As shown in Figure 1.3, the ADC's quantization error (which can be usefully modeled as random noise) will then be high-pass filtered when it is referred to the input of the loop. This *noise-shaping* property of the loop allows a sigma-delta modulator to achieve very high resolution in a narrow bandwidth. The quantization noise outside this bandwidth can then be removed by a succeeding digital low-pass filter (not shown in Figure 1.3). By combining various dynamic error correction techniques with sigma-delta modulation, ADCs with more than 20-bit resolution and 18-bit linearity have been realized [12, 13].

1.2.2 Calibration and Trimming

Like all sensors, the accuracy of a smart sensor can only be evaluated by calibrating it against a known standard, after which its systematic inaccuracy is known. This can then be reduced in a subsequent trimming operation. The main limitation on the sensor's accuracy then becomes its stability over time. Trimming is a powerful technique, which can be used to correct for many errors resulting from manufacturing tolerances and process spread. However, in sensors intended for high-volume production, it should be seen as a method of last resort, since the associated calibration requires extra test equipment and takes up costly production time. These topics will be discussed in more detail in the following chapter.

1.3 A Smart Temperature Sensor

In this section, the design of a high-accuracy temperature sensor in standard CMOS will be described [14]. The sensing element is the substrate bipolar junction transistor that is available in all CMOS processes. However, it is a parasitic device, whose characteristics exhibit significant process spread. As a result, the resulting temperature sensor must be trimmed in order to achieve inaccuracies less than $\pm 2°C$.

1.3.1 Operating Principle

The base-emitter voltage V_{BE} of a bipolar junction transistor is given by:

$$V_{BE} = \frac{kT}{q} \ln \frac{I_C}{I_S} \tag{1.1}$$

where I_C is its collector current and I_S is a process-dependent parameter which also depends on the transistor's size. As shown in Figure 1.4, V_{BE} is a near-linear function of temperature with a slope of approximately $-2\,\text{mV}/°\text{C}$. A voltage that is proportional-to-absolute-temperature (PTAT) can then be obtained by measuring the difference between the base-emitter voltages of two, nominally identical, bipolar junction transistors $Q_{1,2}$ biased at a $1 : p$ current ratio:

$$\Delta V_{BE} = \frac{kT}{q} \ln p. \tag{1.2}$$

If the current ratio p is well defined, ΔV_{BE} will be an accurate function of absolute temperature, since it does not depend on I_S or any other process-dependent parameters. However, it is a small signal, with a sensitivity of about $140\,\mu\text{V/K}$ (for $p = 5$), which means that low-offset interface electronics is required.

In order to digitize ΔV_{BE}, a reference voltage is also required. As shown in Figures 1.4 and 1.5, a so-called band-gap reference voltage V_{REF} ($\sim 1.2\,\text{V}$) can be obtained by combining V_{BE} with a scaled version of ΔV_{BE}. Both voltages can then be applied to an ADC, which determines their temperature-dependent ratio μ:

$$\mu = \frac{\alpha \Delta V_{BE}}{V_{BE} + \alpha \Delta V_{BE}} = \frac{V_{PTAT}}{V_{REF}}. \tag{1.3}$$

Assuming that the interface electronics is ideal, the sensor's main source of error will be the effect of process spread on V_{BE}. As discussed in [4] in Chapter 7, [16] and shown in Figure 1.5, this only affects the slope of V_{BE}, while the extrapolated value of V_{BE} at 0 K, known as V_{BE0}, remains the same. This means that the effect of process spread can be corrected for by calibrating the sensor at room temperature and then adding a PTAT correction voltage to V_{BE}, for instance by trimming the current I_2 in the Figure 1.4 circuit. This will be discussed in more detail in Chapter 2.

1.3.2 Interface Electronics

A simplified block diagram of the sensor's interface electronics is shown in Figure 1.6. It is based on a second-order single-bit sigma-delta modulator, which converts ΔV_{BE} and V_{BE} into

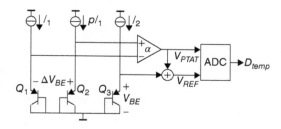

Figure 1.4   Simplified circuit diagram of the CMOS smart temperature sensor

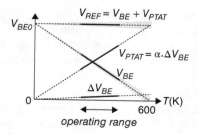

Figure 1.5 Temperature dependence of the voltages generated by the circuit in Figure 1.4; the shaded areas indicate the effect of process spread

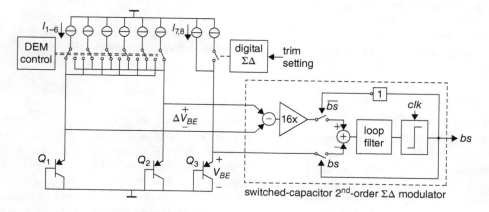

Figure 1.6 Simplified circuit diagram of the CMOS smart temperature sensor

a temperature-dependent bitstream bs. The modulator employs a charge-balancing scheme in which its input is either V_{BE} or $\alpha\Delta V_{BE}$ depending on the instantaneous value of the bitstream. It can be shown [14, 16], that the resulting bitstream average is exactly equal to μ, as given by (3). So in contrast to the scheme shown in Figure 1.4, an explicit reference voltage V_{ref} does not need to be generated, which simplifies the required circuitry. The scale factor $\alpha(= 16)$ is established by appropriately sizing the sampling capacitors at the input of the modulator.

To achieve the targeted inaccuracy of 0.1°C, the errors introduced by the interface electronics should all be reduced to the 0.01°C level. This means, for instance, that the modulator's offset should be less than 2 μV, while the bias current ratio $p = 5$ and the scale factor α should be accurate to within about 100 ppm. Dynamic error correction techniques were used to achieve this level of accuracy, since the manufacturing tolerance of a typical CMOS process mean that the best-case component mismatch is only about 0.1%.

Figure 1.6 also shows how a technique known as dynamic element matching (DEM) was used to obtain an accurate 1:5 bias current ratio. Via a set of switches, one of six nominally equal current sources I_{1-6} is connected to Q_1, while the others are connected to Q_2. This leads to six possible connections, each of which may result in an inaccurate ΔV_{BE} due to the mismatch between the current sources. The *average* value of ΔV_{BE}, however, is much more accurate, because the mismatch errors cancel out [4, Chapter 7; 15]. The required averaging can conveniently be performed by the same digital filter that suppresses the sigma-delta modulator's quantization noise. A similar DEM scheme was used to average out errors due to mismatch in the sampling capacitors of the modulator.

The modulator's input-referred offset was reduced by the use of correlated double sampling (a technique very similar to auto-zeroing) in the first integrator [8, 17]. Since this did not reduce the offset sufficiently, the entire modulator was also chopped, which ensures that its residual offset is well below the 2 µV level.

As shown in Figure 1.6, the sensor was trimmed by adjusting the bias current of transistor Q_3 with the help of a 10-bit current DAC, consisting of a digital first-order sigma-delta modulator whose output modulates one of the bias current sources. The DAC covers the expected trimming range with a resolution of 0.01°C. Since the bitstream output of the DAC may interfere with the bitstream output of the main modulator, the timing of the DAC, as well as that of the other DEM schemes, was synchronized to the main modulator's bit-stream [18].

The resulting sensor consumes 190 µW and achieves an inaccuracy of ±0.1°C over the military temperature range (−55°C to 125°C) after a single room-temperature trim. This level of accuracy still represents the state-of-the-art for CMOS temperature sensors [19].

1.3.3   Recent Work

Recent work has focused on simplifying the sensor's calibration, as well as reducing its power dissipation. Calibrating a temperature sensor with the help of a reference sensor can be a time-consuming and hence expensive process, since achieving the necessary thermal equilibrium between the sensors can take several minutes. By regarding ΔV_{BE} as a sufficiently accurate measure of temperature, and then digitizing it with respect to an accurate *external* voltage reference, the sensor can be *voltage-calibrated* in less than a second with an inaccuracy of less than 0.1°C [20]. Increasing the sensor's efficiency was achieved by using a more efficient two-step ADC that combines a coarse step, based on binary search, followed by a fine step, based on sigma-delta modulation [21]. Over the military temperature range (−55°C to 125°C), the resulting sensor achieves an inaccuracy of ±0.15°C after voltage calibration, which is only slightly less than the state-of-the-art. However, it dissipates only 5 µW, which is nearly 40x less than its predecessor [14].

1.4   A Smart Wind Sensor

In this section, the design of a smart wind sensor is described, that is, a solid-state sensor that measures wind speed and direction with *no* moving parts [22]. The sensor makes use of the fact that wind passing above a heated object will cool it asymmetrically. Wind speed and direction can then be determined by measuring the resulting temperature gradient. If the object is a chip, it can be heated by passing current through resistors, while the wind-induced temperature gradient can be sensed by integrated thermopiles.

1.4.1   Operating Principle

As shown in Figure 1.7, the flow of air over a heated disc will cool it non-uniformly. The result is a temperature gradient δT between any two points located symmetrically around the

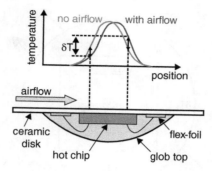

Figure 1.7 Operating principle of the wind sensor

center of the disk. The magnitude of δT is proportional to the square-root of flow speed, while its direction is aligned with that of the flow. Thus, by measuring δT, both wind speed and direction can be determined [23, 24].

Although heaters and temperature sensors can be readily integrated on a standard CMOS chip, the requirement that it must sense a flow-induced temperature gradient precludes the use of standard packages. Instead, as shown in Figure 1.7, the chip is glued to the underside of a thin ceramic disc, while the airflow is passed over the other side. This simple and robust packaging solution ensures that the chip is in good thermal contact with the flow. The disc is then mounted in an aerodynamic housing, which ensures that the wind sensor is only sensitive to the horizontal components of the wind [23].

As shown in Figure 1.8, four heaters and four p + /Al thermopiles are realized on the sensor chip. The thermopiles are configured to measure orthogonal components of the flow-induced temperature gradient. Since silicon is a good thermal conductor, these are quite small: in the order of a few tenths of a degree. As a result, the output of the thermopiles is at the microvolt level. In a first-generation sensor, these signals were digitized by precision *off-chip* electronics, and the results used to compute wind speed and direction. The resulting errors in the computed wind speed and direction are typically less than 5% and 3° respectively [25].

Due to manufacturing tolerances, the assembled wind sensor must be calibrated and trimmed. This is because, in general, the chip will not be located exactly in the center of the disc, and so the hot-spot on the disc may not be centered on the chip. The result is a flow-dependent thermal offset, which can be much larger than the actual flow-induced temperature differences. This offset can be cancelled by trimming the power dissipated in the four heaters, so as to center the heat distribution on the chip [26]. The sensor is then calibrated in a wind tunnel. The resulting data is stored in a non-volatile memory, and used to compensate for the effects of any residual offset and gain errors. However, the whole procedure is time consuming and adds significantly to the sensor's cost.

To circumvent these problems, the *smart* wind sensor was operated in an alternative mode: the temperature-balance mode [27, 28]. In this mode, the flow-induced temperature gradient is continuously *canceled* by dynamically adjusting the power dissipated in the heaters. This automatically centers the heat distribution on the chip, and as a result, any thermal offset becomes a well-defined function of flow speed. Moreover, the heater power does not need to be manually trimmed. Flow speed and direction can then be computed from the differential heat power required to *cancel* each component of the flow-induced temperature gradient [29].

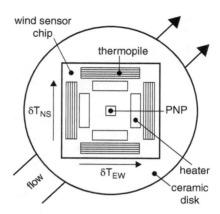

Figure 1.8   Schematic layout of the wind sensor

This approach also simplifies the interface electronics, which now has the much easier task of digitizing the relatively large signals (several tens of milliwatts) applied to the heaters instead of the microvolt-level outputs of the thermopiles.

1.4.2   Interface Electronics

The block diagram of the smart wind-sensor chip is shown in Figure 1.9. It consists of three thermal sigma-delta modulators, two of which are arranged to cancel the north-south δT_{ns} and east-west δT_{ew} components of any on-chip temperature gradient [22]. Their bitstream outputs are then a digital representation of the differential heating powers δP_{ns} and δP_{ew} required to cancel the two components δT_{ns} and δT_{ew} respectively. The heat pulses generated by the modulators are low-pass filtered by the sensor's thermal capacitance, and thus the sensor itself functions as the modulators' loop filter. So in addition to the flow sensor, each modulator only requires the implementation of a clocked comparator, which leads to a very compact architecture. Since the thermopile's output is at the microvolt level, the comparator was auto-zeroed to reduce its offset [22, 29].

A third thermal sigma-delta modulator maintains the sensor at a constant temperature (the overheat $\Delta T \sim 10°C$) above ambient temperature. In this mode, the magnitude of δP will be proportional to the square-root of wind speed [23]. The temperature of the chip T_{chip} is measured by a substrate PNP transistor at the center of the chip, while an external transistor measures ambient temperature T_{amb} (Figure 1.9). As in the smart temperature sensor, these transistors are biased at two different collector currents, in order to generate a voltage proportional to the overheat. By using an auto-zeroed comparator and well-matched current sources, the error in the overheat due to process spread will be limited to about $\pm 1°C$. Although this error will alter the sensor's sensitivity, its effect is also taken into account by the sensor's calibration.

After calibration, which will be discussed in more detail in Chapter 2, the smart sensor was tested in a wind tunnel at wind speeds ranging from 1 m/s to 25 m/s. The errors in the computed speed and direction were less than 4% and 2°, which are slightly less than those of an earlier wind sensor *without* on-chip interface electronics [23–25]. Due to the compact interface architecture, this was achieved with no increase in chip area.

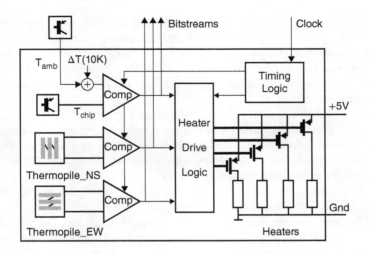

Figure 1.9   Block diagram of the smart wind sensor

1.4.3   Recent Work

Recent work has focused on simplifying the sensor's construction and reducing its power dissipation. In [30], the sensor was operated in a so-called constant power mode, in which its overheat was *not* regulated, thus eliminating the need for the external temperature sensor and the associated overheat control loop. As a result, the heater power can be drastically reduced, since no guard-band needs to be maintained to accommodate errors in the overheat control loop. To maintain resolution at such decreased heater power levels, the in-band quantization noise of the thermal sigma-delta modulators was reduced by connecting an electrical filter (an integrator) in series with the sensor's thermal filter. The lower bound on heater power dissipation was then found to be set by the integrator's residual offset. In [31], this was reduced by the application of system-level chopping. The resulting wind sensor dissipates only 25 mW, 16x less than that of [22], while achieving the same accuracy after calibration, that is, less than 4% and 2° error in wind speed and direction, respectively.

1.5   A Smart Hall Sensor

In this section, the design of a smart magnetic field sensor intended for compass applications is described [32]. The sensor is based on the Hall effect, that is, the fact that if a plate carrying current along one axis is exposed to a magnetic field, a voltage will be induced across its transverse axis. The magnitude of this Hall voltage is proportional to the current through the plate and to the normal component of the magnetic field [4, Chapter 9; 33].

1.5.1   Operating Principle

In a CMOS process, a Hall plate will usually consist of an n-well layer, resulting in a sensitivity of about 100 V/AT. At a typical bias current of 1 mA, the earth's 50 μT (max) magnetic field will then result in a Hall voltage of only 5 μV. Accurately digitizing such a small voltage presents a significant interfacing challenge.

Furthermore, silicon Hall plates exhibit considerable offset (5 mT to 50 mT) due to mechanical stress, doping variations and lithographic errors. Although this is much larger than the earth's magnetic field, this would not be a major problem in itself since magnetic compasses are usually calibrated to compensate for the presence of nearby ferromagnetic materials. The main problem in a compass application is offset *drift*, which will result in a time-varying angular error.

The offset of a Hall plate can be reduced to the 10 μT level by the spinning current method; in which the Hall plate's bias current is spatially rotated while its output is averaged in time [34]. This also reduces drift, but not sufficiently for a compass application, especially in the presence of the mechanical stress caused by low-cost plastic packaging. Offset and drift can also be reduced by orthogonally coupling two or more Hall plates [23]. The smart sensor described here uses the spinning current technique and *four* orthogonally-coupled Hall plates to achieve the smallest possible offset and drift.

Packaging presents another challenge, since standard IC packages typically contain trace amounts of ferromagnetic materials, which may distort the magnetic field around the sensor. To avoid such errors, a custom package which is free of ferromagnetic materials has been developed [35]. It is designed so that an electronic compass can be made by mounting three sensors both vertically (as shown in Figure 1.10) and horizontally on a PCB.

1.5.2   Interface Electronics

The block diagram of the sensor's interface electronics is shown in Figure 1.11. It consists of a voltage-to-current converter (VIC), whose output is digitized by a first-order sigma-delta modulator. The output of the modulator is averaged over an entire spinning-current cycle by an up/down counter, and the result is transmitted to the outside world via a RS-232/SPI/μWIRE compatible serial interface.

Due to their offset, the output of the four Hall plates can be as high as 50 mV during the various phases of a spinning cycle. The average value, however, is much smaller and is less than 50 nV under zero field conditions. The interface electronics should, therefore, have an input-referred offset of less than 50 nV and a linear dynamic range of about 120 dB, which is quite challenging.

To achieve this level of linearity, the VIC consists of two opamps (Figure 1.12), each with a DC gain of over 120 dB, which generate its output current by applying the output voltage of the Hall plates across a resistor. A so-called nested chopping scheme was used to reduce its offset to the desired 50 nV level [36]. As shown in Figure 1.9, the VIC is first chopped by

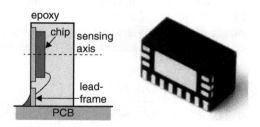

Figure 1.10   Custom-packaged smart Hall sensor

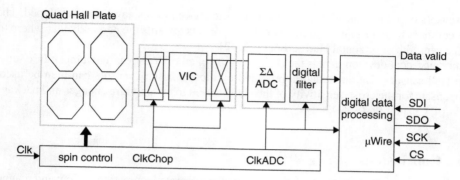

Figure 1.11   Block diagram of the smart Hall sensor

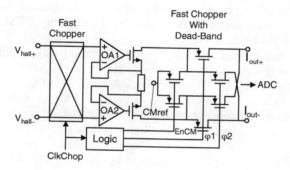

Figure 1.12   Schematic diagram of the chopped voltage-to-current converter

a pair of "fast" choppers driven by the 12.5 kHz ClkChop signal. The residual offset (due to spikes associated with the operation of the input choppers) was further reduced by creating a dead-band [37]. This was implemented via the EnCM signal (a 1 ms pulse), which connects the VIC output to a reference voltage CMref after every ClkChop transition, while simultaneously opening the output switches. To reduce the input-referred offset even further, the entire front-end is chopped at about 10 Hz by periodically inverting the polarity of the Hall plate's bias current and simultaneously inverting the sign of the modulator's bitstream.

The result is a sensor with an offset of 4 μT, an offset temperature coefficient of only 8 nT/K and an offset drift of less than 0.25 μT even after aggressive thermal cycling [38]. In a compass application, this offset drift corresponds to an angular error of less then 0.5°. To date, this represents the best offset performance reported for a CMOS Hall sensor.

1.5.3   Recent Work

Standard (horizontal) Hall sensors are only sensitive to magnetic fields normal to the chip's surface. A 3D magnetic compass then requires three orthogonally-oriented chips. An alternative is to combine horizontal and so-called vertical Hall plates on a single chip [40], but the latter have much higher offset and so are not suited for compass applications. Recently, single-chip

3D sensors based on thin-film integrated magnetic concentrators have been developed. These concentrators bend in-plane magnetic field components towards the perpendicular where they can be sensed by horizontal Hall sensors [41, 42].

Another recent development is the co-integration of auxiliary stress and temperature sensors with Hall sensors. The information provided by these auxiliary sensors can then be used to compensate for the cross-sensitivity of the Hall sensors to temperature changes and packaging stress [4, Chapter 9; 43, 44].

1.6   Conclusions

The designs described above show that, at least for integrated temperature, flow and magnetic sensors, it is possible to design *transparent* interface electronics in standard CMOS. Compared to electronic circuits, most sensors are quite slow, which means that the effects of typical circuit non-idealities such as offset, gain error and $1/f$ noise, can be reduced to negligible levels by dynamic error correction techniques such as auto-zeroing, chopping, DEM, switched-capacitor filtering and sigma-delta modulation.

For example, by using various combinations of auto-zeroing and chopping, amplifiers with input-referred offsets of less than $100\,\text{nV}$ can be realized, which, for input levels of a few volts, corresponds to a 24-bit DC dynamic range. Also, by using DEM, current and voltage ratios, that is, gain factors, can be defined to better than $100\,\text{ppm}$ accuracy. Finally, ADCs based on sigma-delta modulators can be used to flexibly trade-off resolution for bandwidth, and are capable of achieving up to 22-bit resolution in bandwidths of a few tens of Hertz. As an added bonus, the notches in the frequency response of their decimation filters can be used to completely suppress the AC residuals produced by chopping and DEM.

So what can we do with all this precision? It can be used to realize novel sensors based on transduction mechanisms that result in very small, and previously undetectable, signals. One example is the implementation of temperature sensors based on the well-defined thermal diffusivity of bulk silicon, which require the detection of the small temperature variations created by the diffusion of heat pulses through a chip [39]. In existing smart sensors, precision can be traded off against other performance criteria, such as chip area and power dissipation. For example, since DEM mitigates the effects of component mismatch, larger initial mismatch can be tolerated, which means that smaller components can be used. Similarly, since chopping suppresses $1/f$ noise, a given signal-to-noise ratio can be obtained at lower power consumption.

The design of smart sensor systems involves meeting the engineering challenge associated with the design of accurate, reliable systems using inaccurate, low-cost components. Due to the wide variety of sensing principles, packaging methods and circuit techniques that can be used to realize such systems, their design is more of an art than a science. The dynamic techniques described above have been shown to be of great value in meeting this challenge, and will undoubtedly continue to be of use as we further master the art of designing smart sensor systems.

References

[1] S. Middelhoek and S. Audet, *Silicon Sensors*, London: Academic Press, 1989.
[2] Z. Tan, R. Daamen, A. Humbert, Y.V. Ponomarev, Y. Chae, and M.A.P. Pertijs, "A 1.2-V 8.3-nJ CMOS humidity sensor for RFID applications," *Journal of Solid-State Circuits*, vol. 48, no. 10, pp. 2469–2477, October 2013.

[3] M. Graf, U. Frey, S. Taschini, and A. Hierlemann, "Micro hot plate-based sensor array system for the detection of environmentally relevant gases," *Analytical Chemistry*, vol. 78, no. 19, pp. 6801–6808, 2006.

[4] G.C.M. Meijer, Ed. *Smart Sensor Systems*. John Wiley & Sons Ltd, 2008.

[5] N. Yazdi, F. Ayazi, and K. Najafi, "Micromachined inertial sensors," Proc. IEEE, vol. 86, no. 8, pp. 1640–1659, August 1998.

[6] P.G.R. Costa, L.J. Breems, K.A.A. Makinwa, R. Roovers, and J.H. Huijsing "A 118 dB dynamic range continuous-time IF-to-baseband sigma-delta modulator for AM/FM/IBOC radio receivers," *Journal of Solid-State Circuits*, vol. 42, pp. 1076–1089, May 2007.

[7] K.A.A. Makinwa, M.A.P. Pertijs, J.C. van der Meer, and J.H. Huijsing, "Smart sensor design: the art of compensation and cancellation," **Plenary**, *Proc. ESSCIRC*, pp. 76–82, September 2007.

[8] C.C. Enz and G.C. Temes, "Circuit techniques for reducing the effects of op-amp imperfections: autozeroing, correlated double sampling and chopper stabilization," Proc. IEEE, vol. 84, no. 11, 1584–1614, November 1996.

[9] C.G. Yu and Randall L. Geiger. "An automatic offset compensation scheme with ping-pong control for CMOS operational amplifiers." *Journal of Solid-State Circuits*, vol. 29, no. 5, pp. 601–610, May 1994.

[10] Q. Fan, J.H. Huijsing, and K.A.A. Makinwa, "A 21 nV/$\sqrt{Hz}$ chopper-stabilized multipath current-feedback instrumentation amplifier with 2 μV offset," *Journal of Solid-State Circuits*, vol. 47, no. 2, pp. 464–475, February 2012.

[11] R. Schreier and G.C. Temes. *Understanding Delta-sigma Data Converters*. vol. 74. Piscataway, NJ: IEEE Press, 2005.

[12] V. Quiquempoix, P. Deval, A. Barreto, G. Bellini, J. Markus, J. Silva, and G. Temes, "A low-power 22-bit incremental ADC," *Journal of Solid-State Circuits*, vol. 41, pp. 1562–1571, July 2006.

[13] Y.C. Chae, K. Souri, and K.A.A. Makinwa, "A 6.3 μW 20bit incremental zoom-ADC with 6 ppm INL and 1 μV offset," *Journal of Solid-State Circuits*, vol. 48, pp. 3019–3027, December 2013.

[14] M.A.P. Pertijs, K.A.A. Makinwa, and J.H. Huijsing, "A CMOS temperature sensor with a 3σ inaccuracy of ±0.1°C from −55°C to 125°C," *Journal of Solid-State Circuits*, vol. 40, pp. 2805–2815, December 2005.

[15] G.C.M. Meijer, G. Wang, and F. Fruett, "Temperature sensors and voltage references implemented in CMOS technology," *IEEE Sensors Journal*, vol. 1, no. 3, pp. 225–234, October 2001.

[16] M.A.P. Pertijs and J.H. Huijsing, *Precision Temperature Sensors in CMOS Technology*. Dordrecht, The Netherlands: Springer, 2006.

[17] C. Hagleitner, D. Lange, A. Hierlemann, O. Brand, and H. Baltes, "CMOS single-chip gas detection system comprising capacitive, calorimetric and mass-sensitive microsensors," *IEEE Journal of Solid-State Circuits*, vol. 37, pp. 1867–1878, 2002.

[18] M.A.P. Pertijs and J.H. Huijsing, "A sigma-delta modulator with bitstream-controlled dynamic element matching", Proc. ESSCIRC 2004, pp. 187–190, September 21–23, 2004.

[19] K.A.A. Makinwa, "Smart Temperature Sensor Survey", [Online]. Available: http://ei.ewi.tudelft.nl/docs/TSensor_survey.xls.

[20] M.A.P. Pertijs, A.L. Aita, K.A.A. Makinwa, and J.H. Huijsing, "Low-cost calibration techniques for smart temperature sensors," *IEEE Sensors Journal*, vol. 10, no. 6, pp. 1098–1105, June 2010.

[21] K. Souri, Y. Chae, and K.A.A. Makinwa, "A CMOS temperature sensor with a voltage-calibrated inaccuracy of ±0.15°C (3s) From −55 to 125°C," *Journal of Solid-State Circuits*, vol. 47, no. 12, pp. 292–301, January 2013.

[22] K.A.A. Makinwa and J.H. Huijsing, "A smart CMOS wind sensor," Digest of Technical Papers ISSCC 2002, pp. 432–479, February 2002.

[23] B.W. van Oudheusden and J. H. Huijsing, "An electronic wind meter based on a silicon flow sensor," *Sensors and Actuators A*, 21–23 (1990), pp. 420–424.

[24] B.W. van Oudheusden, "Silicon thermal flow sensor with a two-dimensional direction sensitivity," *Measurement Science and Technology*, vol. 1, no. 7, pp. 565–575, July 1990.

[25] Mierij Meteo B.V., Datasheet: Solid state wind sensor MMW 005, http://www.mierijmeteo.nl/.

[26] S.P. Matova, K.A.A. Makinwa, and J.H. Huijsing, "Compensation of packaging asymmetry in a 2-D wind sensor," *IEEE Sensors Journal*, vol. 3, no. 6, pp. 761–765, December 2003.

[27] T.S.J. Lammerink, N.R. Tas, G.J.M. Krijnen, and M. Elwenspoek, "A new class of thermal flow sensors using $\Delta T = 0$ as a control signal," *Proceedings of MEMS 2000*, pp. 525–530, January 2000.

[28] K.A.A. Makinwa and J.H. Huijsing, "A wind sensor interface using thermal sigma-delta modulation," Proceedings of Eurosensors XIV, pp. 24–252, September 2000.

[29] K.A.A. Makinwa and J.H. Huijsing, "A smart wind sensor using thermal sigma-delta modulation techniques," *Sensors and Actuators A*, vol. 97–98, pp. 15–20, April 2002.

[30] J. Wu, Y. Chae, C. Van Vroonhoven, and K.A.A. Makinwa, "A 50 mW CMOS wind sensor with $\pm 4\%$ speed and $\pm 2°$ direction error," *Digest ISSCC*, pp. 106–108, February 2011.

[31] J. Wu, C.P.L. van Vroonhoven, Y. Chae, and K.A.A. Makinwa, "A 25 mW smart CMOS sensor for wind and temperature measurement," *Proceedings of IEEE Sensors Conference*, pp. 1261–1264, October 2011.

[32] J.C. van de Meer, F.R. Riedijk, E. van Kampen, K.A.A. Makinwa, and J.H. Huijsing, "A fully integrated CMOS Hall sensor with a 3.65 µT 3σ offset for compass applications," *Digest of Technical Papers ISSCC* 2005, pp. 246–247, February 2005.

[33] R.S. Popovic. *Hall Effect Devices*. CRC Press, 2010.

[34] P. Munter, *Spinning-Current Method for Offset Reduction in Silicon Hall Plates*, Delft University Press, 1992.

[35] J.C. v.d. Meer, K.A.A. Makinwa, and J.H. Huijsing, "A low-cost epoxy package for compass applications," *Proceedings of IEEE Sensors 2005*, pp. 65–68, October 2005.

[36] A. Bakker, K. Thiele, and J.H. Huijsing, "A CMOS nested-chopper instrumentation amplifier with 100-nV offset," *IEEE Journal of Solid-State Circuits*, vol. 35, pp. 1877–1883, December 2000.

[37] C. Menolfi and Q.Huang, "A 200 nV 6.5 nV/$\sqrt{Hz}$ noise PSD 5.6 kHz chopper instrumentation amplifier," *Digest of Technical Papers ISSCC* 2000, pp. 362–363, February 2000.

[38] Xensor Integration B.V., Datasheet: Compass sensor XEN1200, http://www.xensor.nl/.

[39] C.P.L. van Vroonhoven, and K.A.A. Makinwa, "Thermal diffusivity sensing: a new temperature sensing paradigm," *Proceedings of CICC*, September 2011.

[40] J. Pascal, L. Hebrard, V. Frick, J. P. Blonde, J. Felblinger, and J. Oster, "3D Hall probe in standard CMOS technology for magnetic field monitoring in MRI environment," *Proceedings of European Magnetic Sensors and Actuators Conference (EMSA '08)*, 2008.

[41] R. Popovic, P. Drljaca, and P. Kejik, "CMOS magnetic sensors with integrated ferromagnetic parts," *Sensors and Actuators A*, pp. 94–99, 2006.

[42] C. Schott, R. Racz, A. Manco, and N. Simonne, CMOS single-chip electronic compass with microcontroller. *Journal of Solid-State Circuits*, vol. 42, pp. 2923–2933, 2007.

[43] U. Ausserlechner, M. Motz, and M. Holliber, "Compensation of the piezo-Hall effect in integrated Hall sensors on (100)-Si," *IEEE Sensors Journal*, vol. 7, no. 11, pp. 1475–1482, 2007.

[44] S. Huber, C. Schott, and O. Paul, "Package stress monitor to compensate for the piezo-hall effect in CMOS Hall sensors," *Proceedings of IEEE Sensors*, pp. 1–4, October 2012.

2

Calibration and Self-Calibration of Smart Sensors

Michiel Pertijs
Electronic Instrumentation Laboratory, Delft University of Technology, Delft, The Netherlands

2.1 Introduction

Smart sensors acquire information about a non-electrical quantity of interest (the measurand) and convert this information to a useful electrical output signal. In order to do so, they combine a sensing element and the associated interface electronics on a single chip or in a single package. The sensing element performs the conversion from the non-electrical domain of the measurand to an electrical signal, while the interface electronics further process this signal to produce an output that can readily be used in a measurement or control system. Errors introduced in these steps affect the performance and reliability of the overall system. Therefore, it is very important to determine how large these errors are. The process of doing so is generally referred to as *calibration*, and is the topic of this chapter.

Calibration is of interest both to manufacturers of smart sensors and to their customers. Manufacturers need optimized calibration procedures to guarantee the desired level of accuracy at minimum cost. Users need at least a basic understanding of these procedures to be able to correctly interpret the specifications of a sensor, including their limited validity, and to be able to evaluate when re-calibration is required.

The more smart sensors become plug-and-play devices with standardized interfaces, the more the issues associated with calibration are shielded from the user. Users of conventional (non-smart) sensors typically need to obtain calibration coefficients from the manufacturer to be able to interpret the output signal of the sensor. In modern smart sensors, in contrast, such coefficients are often programmed into the sensor, and a corrected output signal is provided to the user. While this makes the use of such sensors easier, it also reduces the user's awareness of the calibration, and its limitations.

Smart Sensor Systems: Emerging Technologies and Applications, First Edition.
Edited by Gerard Meijer, Michiel Pertijs and Kofi Makinwa.
© 2014 John Wiley & Sons, Ltd. Published 2014 by John Wiley & Sons, Ltd.

In the first part of this chapter, the basics of calibration will be reviewed, and the specifics of smart sensor calibration will be discussed. This discussion will be illustrated using the example of a smart temperature sensor. In the second part of the chapter, the possibility of making self-calibrating smart sensors will be explored. It will be shown that, while complete self-calibration is impossible, a significant reduction in the required calibration efforts can be obtained by employing additional co-integrated sensors and/or actuators. This will be illustrated using two detailed examples: a smart magnetic field sensor, and a smart wind sensor. The chapter concludes with a summary and a discussion of future trends.

2.2   Calibration of Smart Sensors

2.2.1   Calibration Terminology

In many measurement and instrumentation systems, sensors are required that measure a quantity of interest with a known, objective level of accuracy. A fever thermometer, for instance, is expected to measure body temperature to within $\pm0.1°C$. In order to verify how accurate a given thermometer is, one can compare its reading with that of a more accurate thermometer. Alternatively, one can submerge the thermometer in ice water or boiling water, and see whether it correctly indicates 0°C or 100°C. These procedures are essentially *calibration* procedures [1, 2].

The International Organization for Standardization (ISO) defines calibration as follows [3]:

> *Calibration* is the operation that, under specified conditions, in a first step, establishes a relation between the quantity values with measurement uncertainties provided by measurement standards and corresponding indications with associated measurement uncertainties and, in a second step, uses this information to establish a relation for obtaining a measurement result from an indication.

These standards are maintained by national standard laboratories, such as the National Institute of Standards and Technology (NIST) in the USA, the National Physical Laboratory (NPL) in the UK, the Physikalisch-Technische Bundesanstalt (PTB) in Germany and the Dutch Metrology Institute VSL. These laboratories have implemented so-called primary reference standards, which define the highest level of accuracy achievable in the measurement of a given physical quantity. For temperature measurement, for instance, the primary reference standards are specially-designed platinum resistance thermometers that are used to interpolate between the so-called fixed points of the International Temperature Scale (ITS-90) [4]. These fixed points are the professional equivalents of the ice water and boiling water mentioned above. An example is the triple point of water, which defines 0.01°C, or 273.16 K.

Primary reference standards are used to calibrate the secondary reference standards found in standards laboratories. These, in turn, are used to calibrate the working standards with which the accuracy of measuring instruments or sensors is determined. A calibration procedure thus establishes a documented chain of comparisons leading from the sensor or instrument being calibrated, via instruments of increasing accuracy, to an international standard. This makes the readings of a sensor or instrument *traceable* [3]:

Traceability is the property of a measurement result whereby the result can be related to a reference through a documented unbroken chain of calibrations, each contributing to the measurement uncertainty.

For conventional (non-smart) sensors, the result of a calibration is typically used by the user to perform a correction on the measurement results obtained with sensor [3]:

A *correction* is a modification applied to the result of a measurement to compensate for a systematic effect.

Note that according to these definitions, neither calibration nor correction involves *adjustment* of sensors or instruments to correct for errors. This may be somewhat confusing, because the word calibration is often used with this meaning in literature. Instead, here, we will follow the ISO terminology [3]:

Adjustment is the set of operations carried out on a measuring system so that it provides prescribed indications corresponding to given values of a quantity to be measured.

Typical examples of adjustments are offset and gain adjustments. Often, after an adjustment, recalibration is required to ensure that the measuring system, instrument or sensor that has been adjusted now performs within its specifications.

Conventional sensors are often not adjusted after calibration. Instead, the user is responsible for applying the appropriate corrections to the measurement results. As will be explained in more detail later in this chapter, it is a unique property of *smart* sensors that they take away at least part of this burden from the user, in the sense that they store the required correction data internally, or even in the sense that they perform the correction internally, so that the whole process of calibration and correction becomes transparent to the user. This necessarily involves some form of adjustment to the smart sensor after calibration, although this adjustment can be a 'mild' form of adjustment that does not involve making changes to the sensor element, but only to the interface electronics, for example, to store an offset- or gain-correction coefficient.

2.2.2   Limited Validity of a Calibration

It is important to realize that a calibration procedure can never cover all conditions under which a sensor can potentially be used. In fact, the conditions of use are never exactly identical to the calibration conditions. Therefore, the accuracy found under calibration conditions is not necessarily the accuracy that will be obtained during the actual application of the sensor. This is, first of all, due to the fact that a calibration procedure can never cover the whole range of measurand values that a sensor can be exposed to. The calibration is necessarily restricted to a limited number of points. Usually, it is fairly safe to make interpolations between such points. Making extrapolations, however, can be more dangerous.

A second reason why the accuracy of a sensor in an application may differ from that found during calibration is that the operating conditions during use are not necessarily the same as during calibration. Ideally, a sensor should be sensitive only to the measurand, but in practice, it will also exhibit some degree of cross-sensitivity to other quantities, such as operating temperature, humidity, supply voltage, mechanical stress and interference. Such cross-sensitivities affect the accuracy of the sensor's output.

Finally, a sensor's accuracy will also degrade after calibration as a result of ageing. The rate of such degradation is typically related to the frequency of use, mechanical wear, exposure to dirt or dust, temperature changes, or humidity, but degradation can even occur if a sensor is kept on the shelf [1]. A sensor manufacturer will typically perform an accelerated ageing experiment on samples of a sensor to obtain an estimate of how fast the accuracy degrades. Such an experiment may consist of exposing a group of sensors to a large number of temperature cycles, or exposing them to high temperature and humidity for a long period of time. Based on the results, extra margin is included in the accuracy specification, along with an indication of how long this specification is valid, or how much degradation can be expected over time. In any case, re-calibration will be needed after a certain time.

2.2.3   Specifics of Smart Sensor Calibration

A calibration procedure, according to the definition given in Section 2.2.1, provides information about the accuracy of a sensor. This information is typically provided in the form of a calibration report that states the measured output of the sensor under calibration conditions, along with the associated measurement uncertainty. Such a report provides the user with a means of interpreting the readings of the sensor in relation to international standards.

Consider, as an example, the calibration of a thermistor, which is a sensor whose resistance changes with temperature [4]. This typically involves exposing the thermistor to a number of well-defined temperatures distributed across the temperature range of interest, and measuring its resistance. The actual temperature is measured using an accurate, traceable reference thermometer that is carefully kept at the same temperature as the thermistor, for instance by immersing the two in a well-stirred liquid bath. Based on the resulting list of temperatures (with uncertainty) and measured resistances (again with uncertainty), an equation can describe the relationship between the sensor's temperature and its resistance. The parameters of such an equation are sometimes referred to as calibration coefficients. All this information goes into a report, based on which a user can calculate temperature from measured resistance.

A smart sensor is typically used somewhat differently. Rather than relying on the user to interpret its output signal and make corrections based on a calibration report, it ideally provides a readily interpretable digital output signal [5]. In other words, the translation of the output of the sensor (the resistance of the thermistor) to the value of the measurand (temperature) should take place *inside* the smart sensor, and is performed by the sensor's interface electronics. This implies that the results of a calibration procedure have to be treated differently. Rather than being passed on to the user of the sensor, they should be stored, in some form, inside the sensor.

2.2.4   Storing Calibration Data in the Sensor

There are various ways in which calibration data can be stored inside a smart sensor. A first way is to use the electronic equivalent of a calibration report: a so-called transducer electronic datasheet (TEDS). Smart sensors that comply with the IEEE 1451 smart sensor standard store such a datasheet in non-volatile memory inside the sensor (Figure 2.1) [6, 7]. Even if the user is still required to post-process the sensor's readings based on the calibration data contained in the TEDS, he or she does not have to worry anymore about keeping sensors and their calibration data together. The calibration data are obtained from the sensor itself via the same interface that is also used to obtain the sensor readings. This makes it much easier to replace or re-calibrate such a smart sensor, as there is no longer a need to separately update the associated calibration data. Especially in systems with many sensors, this is a substantial improvement, as it reduces the risk of misinterpretation of sensor readings due to the use of the wrong calibration data.

A second way of storing calibration data inside a smart sensor goes a bit further: not only is the data stored in non-volatile memory inside the sensor, it is also used to perform the correction internally [8]. As a result, the sensor becomes truly plug-and-play: no more post-processing is needed. The internal correction can be implemented in various ways, for instance by digitally processing the sensor reading based on the calibration data, or by performing a correction in the analog domain (Figure 2.2). The latter often requires a relatively simple offset or gain adjustment, and is often referred to as 'trimming'.

A traditional way of making an analog adjustment is laser trimming, a technique that is frequently used in precision analog integrated circuits (e.g. low-offset amplifiers or bandgap references), and that can also be used in the analog readout electronics of smart sensors. It consists of adjusting the value of a resistor by making cuts in it using a laser beam. Typically, this resistor defines the offset or gain of the circuit, which can thus be fine-tuned by means of the laser. Tolerances down to 0.01% can be achieved using this trimming technique [9].

Laser-trimming is a procedure that needs to be performed before packaging, which implies that any errors introduced by the packaging procedure cannot be corrected for. A related

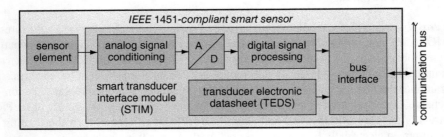

Figure 2.1 Block diagram of a smart sensor with a built-in transducer electronic datasheet (TEDS)

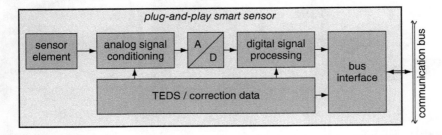

Figure 2.2 Block diagram of a smart sensor that internally corrects readings based on calibration data

technique that can be applied after packaging is the adjustment of polysilicon resistors by passing current pulses through them [10]. The local heating associated with these pulses results in a permanent decrease of the resistor value, and thus an associated permanent adjustment of the transfer function of the circuit in which the resistor is applied.

Sensors realized in CMOS technology, in which high-quality switches and digital circuits can easily be realized, are usually adjusted, or trimmed, by programming a digital non-volatile memory that drives either analog switches or digital correction circuitry. Such digital non-volatile memory can either be erasable or non-erasable. Erasable non-volatile memory is particularly of use when re-calibration is to be performed during the lifetime of a sensor.

The two most common non-erasable digital non-volatile memory techniques are zener zapping and fusible links:

- *Zener zapping* (Figure 2.3) changes a zener diode, which initially acts as an open-circuit, to a short circuit. This is done by bringing it in avalanche mode by means of a programming pulse, which destroys the junction and creates a reliable metallic connection [11]. Relatively high programming voltages ($>6\,\text{V}$) are required to bring the diode into avalanche mode. These voltages have to be processed with special care to avoid breakdown of other junctions.
- *Fusible links* consist of a metal or polysilicon connection that can be physically destroyed by passing a large current through it [12]. An initial short-circuit is thus converted to an open-circuit. An advantage compared to zener zapping is that low-voltage pulses can be used ($<7\,\text{V}$), which are below the junction-breakdown voltage of most CMOS devices. Fusible links may, however, be less reliable than zener zapping, because metal regrowth can (partially) restore the connection. Links can also be broken by cutting them with a laser beam [12]. This simplifies the circuitry, but again has the disadvantage that it cannot be performed after packaging.

Most *erasable* non-volatile memories are based on floating-gate technology [13]: the threshold of a MOS transistor is altered by storing charge on an extra floating polysilicon gate in between the selection gate of the transistor and its channel. In the case of EPROM (Electrically Programmable Read-Only Memory), programming consists of injection of hot-electrons from the drain into the floating gate as a result of a high voltage being applied to the selection gate. This charge can be released by exposing the chip to ultraviolet light. In the case of EEPROM (Electrically Erasable PROM) and so-called flash EPROM, the charge on the floating gate can be removed electrically by means of tunneling through a thin oxide layer.

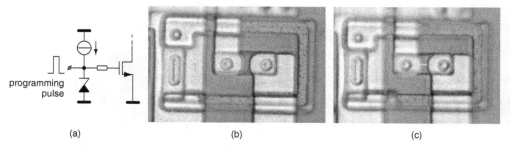

(a)   (b)   (c)

Figure 2.3   Zener zapping: (a) circuit diagram, and micrograph of a zener diode (b) before and (c) after zapping. Photograph: Wil Straver, reproduced by permission of Smartec BV

2.2.5   Calibration in the Production Process

There are several ways to include calibration and adjustment steps in the production process of smart sensors. Most smart sensors are produced in a standard IC production flow augmented with pre- and/or post-processing steps to manufacture the sensor element. This implies that many of the production steps are wafer-level batch-processing steps. In many cases, only the packaging of the individual sensors, after dicing of the wafers, is a non-batch processing step.

To take advantage of the batch processing, calibration and adjustment can be performed at wafer-level, that is, before dicing and packaging. In some cases, all sensors on a wafer can be simultaneously exposed to the same well-defined calibration conditions. An example is the calibration of smart temperature sensors using a temperature-stabilized wafer chuck [14]. An advantage of this parallel approach is that the time and costs associated with creating the calibration conditions are shared by many sensors. An important disadvantage, on the other hand, is that additional errors introduced by dicing and packaging are not taken into account. Many smart temperature sensors, for instance, exhibit a so-called packaging shift when they are encapsulated in a plastic package. When calibration and adjustment have been performed at wafer-level, this packaging shift limits the accuracy of the sensors after packaging [15–17].

In many cases, calibration and adjustment are performed at the end of the production line, that is, after packaging, because the package is required for the sensor to properly operate. While the cost benefit of wafer-level parallel processing is lost, this approach makes it possible to correct for individual errors including packaging shifts. The range of suitable non-volatile memory techniques is then restricted to those techniques that do not require direct access to the die.

A possible consequence of batch processing is that sensors from the same batch will have similar errors. If the mean error of a batch is significant compared to the variation of the error within the batch, as illustrated in Figure 2.4a, batch calibration is an option. This consists of calibrating a limited number of samples from a production batch (either before or after packaging) to estimate the mean error of that batch. Based on this estimate, the same correction is then applied to all sensors from the batch (Figure 2.4b). This technique can result in significant cost savings, since the number of sensors that need to undergo an actual calibration procedure is strongly reduced.

If the costs of adjusting every individual sensor are significant, for instance because of the associated production time, or because of the extensions to the IC process required to realize non-volatile memory, *binning* may be an interesting alternative. Binning implies that the sensors are not adjusted, but sorted into various accuracy bins based on the results of the calibration procedure (Figure 2.5). Sensors with large errors, for instance, end up in a 'low accuracy' bin,

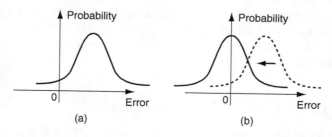

Figure 2.4   Error distribution (a) before and (b) after batch calibration

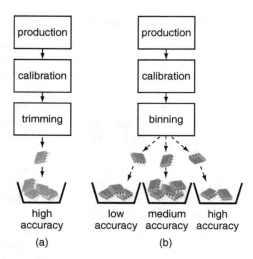

Figure 2.5 Production flow in case of (a) trimming and (b) binning

while sensors with small errors are sorted into a 'high accuracy' bin. Whether this approach is economically attractive depends mainly on whether the error distribution of the production process matches the market demand for the various accuracy grades.

2.2.6 Opportunities for Smart Sensor Calibration

In a smart sensor, the sensor and its readout electronics are combined on a single chip, or in a single package. This combination creates interesting opportunities that go beyond the mere advantage of integration density [5, 18]. Small, sensitive sensor signals can be locally amplified and digitized, making it easier to read out the sensor in the presence of parasitics, and to transport the associated information without degradation in the presence of interference. Moreover, the fact that sensor and electronics are so close to each other may enable new sensor readout approaches.

In the context of calibration, a substantial advantage of the availability of local signal processing is that the correction of sensor readings can be performed locally, as discussed earlier in this chapter. A further opportunity is that some sensors can perform self-testing, so as to warn the user in case the sensor is no longer in a condition under which the calibration data are valid, and its corrected readings as a result are no longer within the accuracy specifications. This can be a cue for the user that the sensor needs to be replaced or re-calibrated. A final opportunity, which will be explored in more detail in Section 2.3, is to make sensors that are (partially) self-calibrating.

2.2.7 Case Study: A Smart Temperature Sensor

To illustrate the principles discussed in the previous sections, we will now take a look at the calibration of a smart temperature sensor. This smart sensor, which has been introduced in

Chapter 1 and is described in detail in [14, 19], is manufactured in standard CMOS technology, and provides a digital output that can be directly interpreted as a temperature reading in degrees Celsius. For completeness, first, its operating principle will be briefly summarized here.

Figure 2.6 shows a simplified block diagram of the smart temperature sensor. It produces a temperature reading by digitizing the ratio of a voltage that is proportional to absolute temperature, V_{PTAT}, and a temperature-independent reference voltage, V_{REF}. Both of these voltages are generated on-chip using bipolar transistors [14]. The voltage V_{PTAT} is obtained by amplifying the difference ΔV_{BE} between the base-emitter voltages of two bipolar transistors $Q_{1,2}$ biased at a $1:p$ current ratio (see Figure 2.7). It can be shown that this voltage hardly spreads due to fabrication tolerances, provided that the amplification and biasing are performed accurately [14]. To achieve this, as discussed in Chapter 1, precision circuit techniques such as dynamic offset cancellation and dynamic element matching are applied in this design. The reference voltage V_{REF} is formed by adding V_{PTAT} to the base-emitter voltage V_{BE} of a third transistor Q_3. The negative temperature dependence of V_{BE} is then compensated for by the positive temperature dependence of V_{PTAT}, resulting in a reference voltage that is nominally temperature independent (Figure 2.7). In contrast with V_{PTAT}, V_{BE}, and hence V_{REF}, suffers from considerable fabrication tolerances. As a result, this temperature sensor needs to be trimmed to obtain accuracies better than $\pm 2°C$. In order to obtain the information required for trimming, the sensor is calibrated during production.

Smart temperature sensors are usually calibrated by comparing them with a traceable reference thermometer of known accuracy [4, 14]. To save production costs, this is typically done at only one temperature. It can be done either at wafer-level, or after packaging.

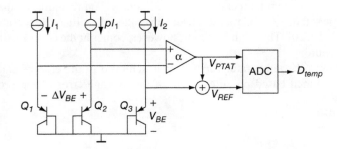

Figure 2.6   Block diagram of the smart temperature sensor [19]

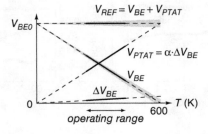

Figure 2.7   Temperature dependence of the voltages shown in Figure 2.6; the shaded regions indicate production tolerances

When calibrating at wafer-level, the temperature of a complete wafer, which may contain thousands of sensors, is stabilized and measured using a number of reference thermometers (e.g. thermistors or platinum resistors) mounted in the wafer chuck. A wafer prober then steps over the wafer, making contact to the bondpads of each of the sensor chips. It usually performs some electrical tests, takes a temperature reading from the chip, and then trims it to adjust its reading. The time required to stabilize the temperature of the whole wafer may be significant, but it is shared by many sensors.

An important limitation of wafer-level calibration lies in the fact that the subsequent dicing and packaging can introduce temperature errors, which are mainly due to mechanical stress [15]. When a chip is packaged in plastic without a stress-relieving cover layer, packaging shifts of up to ±0.5°C can occur. Therefore, calibration and trimming have to take place after packaging if high accuracy is to be combined with low-cost packaging.

Calibration after packaging requires that every individual packaged sensor is brought to the same temperature as a reference thermometer. This typically means that the two are brought in good thermal contact by means of a thermally conducting medium, such as a liquid bath or a metal block. Some stabilization time will be needed, since the sensor will not be at the desired temperature when it enters the calibration setup. For inaccuracies in the order of ±0.1°C, this time will be much longer (minutes) than the time spent on electrical tests (less than a second), due to the thermal time-constants involved. Unlike in the case of wafer-level calibration, this costly stabilization time is now associated with a single sensor, or at best with a handful of sensors that are stabilized together, and, as a result, dominates the total production costs.

Figure 2.8a shows the temperature error of 20 samples of the smart temperature sensor that were individually calibrated against a platinum thermometer and then trimmed. This calibration was performed at room temperature after packaging. The resulting temperature error is state-of-the-art: less than ±0.1°C for all sensors, over the full military temperature range from −55°C to 125°C [19, 20]. This high accuracy, however, comes at the cost of a time-consuming calibration procedure.

Figure 2.8b shows the temperature error that is achieved in the case of batch calibration: all sensors are trimmed based on an estimate of the average error of the production batch

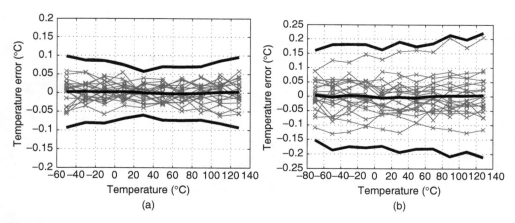

Figure 2.8 Temperature error after (a) individual calibration, and (b) batch calibration (20 samples, bold lines indicate 3σ limits) [20]

(see Figure 2.4). The resulting error after trimming is less than ±0.25°C over the military range. For applications in which this moderate accuracy is acceptable, this and other indirect calibration approaches can significantly reduce the production costs [21].

2.3 Self-Calibration

2.3.1 Limitations of Self-Calibration

Given that smart sensors offer the opportunity to co-integrate intelligence with the sensor, it is interesting to investigate if thus self-calibrating smart sensors can be made. Clearly, referring to the definition of calibration given in Section 2.2.1, complete self-calibration is impossible to achieve: after all, a proper calibration establishes the relation between the readings of a sensor and international standards, and without a comparison to a reference external to the sensor, such a relation cannot be found. However, if we stretch the term self-calibration a bit, some interesting constructions are possible in which a 'partial' calibration is performed by the sensor itself, for instance by calibrating the sensor element against a co-integrated actuator. Such self-calibration techniques are the topic of this section. While they cannot completely take the place of an actual calibration, they may reduce the number of calibration points needed to obtain a given level of accuracy, or extend the time between (re)calibrations.

2.3.2 Self-Calibration by Combining Multiple Sensors

It is sometimes possible to obtain more accurate measurement results by combining the outputs of multiple sensors. In this section, three techniques based on this principle will be discussed: compensation for cross-sensitivity, differential sensing, and background calibration.

Cross-sensitivity compensation – The principle of cross-sensitivity compensation is illustrated in Figure 2.9, which shows a main sensor (sensor 1) that is not only sensitive to a desired measurand X, but also to an interfering quantity C. An additional sensor is used to sense C so as to be able to correct the main sensor's output for its dependency on C.

The effectiveness of this approach depends on how reproducible the first sensor's cross-sensitivity to C is. For instance, if this cross-sensitivity varies substantially with time, the improvement that can be obtained may be limited. In the case of a sensor with a well-defined cross-sensitivity, the addition of an extra sensor may substantially increase the overall performance.

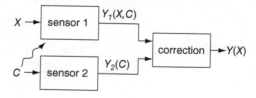

Figure 2.9 Compensation for cross-sensitivity by means of an additional sensor that senses an interfering quantity C

An example of this approach is the correction of a pressure sensor's cross-sensitivity to temperature using a co-integrated temperature sensor. A CMOS smart pressure sensor based on this principle is described in [22]. In this design, a micro-machined piezo-resistive pressure sensor and a temperature sensor based on bipolar transistors are co-integrated with readout electronics on a single CMOS chip. During a calibration procedure, the non-linear temperature-dependent offset and sensitivity of the pressure sensor are stored in an on-chip lookup table. During operation, this information is used to adjust the pressure sensor's readings based on the measured temperature. Thus, the pressure sensor's temperature coefficient is reduced from 1315 ppm/°C to 86 ppm/°C.

Magnetic field sensors based on the Hall effect also suffer from cross-sensitivity to temperature. Moreover, they are also sensitive to the mechanical stress exerted by a plastic package on the sensor die [23]. While errors induced by such stress can be compensated for based on a production calibration, the errors associated with the *drift* of stress cannot. Such drift may be caused, for instance, by the absorption of moisture in the package during the sensor's lifetime. In [24], an integrated magnetic field sensor is described in which a Hall sensor is combined with a temperature sensor *and* a stress sensor to compensate for such drift. The outputs of all three sensors are digitized, after which the drift is determined by comparing the measured stress with the initial stress, which was measured during a production calibration procedure and stored in an on-chip EEPROM. The measured magnetic flux density is then compensated for this drift. Thus, the drift in magnetic sensitivity is reduced from several percent to less than ±0.5%. Note that an alternative self-calibration approach for Hall sensors will be discussed in detail in Section 2.3.4.

A variation of the cross-sensitivity compensation approach is applied in the self-calibrating electrochemical gas sensor described in [25]. Electrochemical gas sensors, which are used to measure the concentration of gases like oxygen and carbon monoxide, are very sensitive to drift in their catalytic surface area. Because of the resulting variation in sensitivity and response time, they typically need to be recalibrated every few weeks. This problem can be mitigated by deriving the surface area from the electrical impedance of the sensing electrode measured during operation. This indirectly measured surface area can be used to correct the sensor's readings. Thus, initial drift-induced errors as large as 50% have been reduced to less than 10%.

Differential sensing – Figure 2.10a shows an alternative way of compensating for cross-sensitivity: two identical sensors are used, which are arranged in such a way that they detect the same measurand, but with opposite sign. Both sensors are cross-sensitive to a

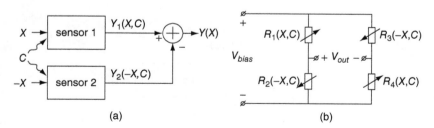

(a)   (b)

Figure 2.10 (a) Compensation for cross-sensitivity by means of two identical sensors, exposed to measurands of opposite sign; (b) implementation of this principle in a Wheatstone bridge

quantity C. Provided that C introduces an additive error in the sensors' outputs (i.e. provided that the cross-sensitivity is independent of the value of the measurand X), this error can be compensated for by taking the difference between the outputs. This compensation is only effective to the extent that the sensors are matched, and to the extent that they are equally exposed to C. While such matching is never perfect, often an order-of-magnitude reduction in overall cross-sensitivity is feasible. An additional advantage of differential sensing is that it also compensates for offset and even-order non-linearity of the sensors, again to the extent that these properties match.

A classic example of the approach shown in Figure 2.10a is the compensation of resistive sensors (for instance strain gauges) for temperature cross-sensitivity by incorporating them in a Wheatstone bridge, as shown in Figure 2.10b. Four resistors R_{1-4} are all sensitive to the measurand X (e.g. strain): resistors R_1 and R_4 increase, while R_2 and R_3 decrease with increasing values of X. Thus, changes in X cause an unbalance in the bridge, which results in a change in the differential output voltage V_{out}. Changes in the interfering quantity C (e.g. temperature), in contrast, will not change the balance in the bridge, and therefore will not affect V_{out}, provided that all resistors are equally cross-sensitive to the interfering quantity C.

This technique is used, for example, to compensate for the temperature cross-sensitivity in integrated pressure sensors in which a micro-machined membrane is exposed to a pressure difference, and the resulting deflection is measured by means of piezoresistors integrated in the membrane [26]. The desired difference in the sign of their response is obtained by using resistors with different orientations (e.g. perpendicular vs. parallel to the edge of the membrane). Another, more recent, example of the use differential sensing to compensate for the temperature cross-sensitivity of a micro-machined accelerometer can be found in [27].

Of course, not all types of sensors can be arranged such that they detect X with opposite signs. In cases where this is not possible, it might be possible to shield one of the sensors from the measurand instead. As illustrated in Figure 2.11, the output of that sensor will then only be a function of the interfering quantity C, while the 'main' sensor is still a function of both X and C. Compensation can again be achieved by taking the difference between the two outputs.

This compensation principle has been used, for instance, to reduce the temperature cross-sensitivity of surface acoustic wave (SAW) sensors [28]. SAW sensors can be used to measure the concentration of gaseous chemical compounds in a carrier gas. These compounds change the acoustic velocity in a chemically sensitive layer deposited on the SAW device. This change is detected as a change in propagation delay between a piezo-electric acoustic actuator and a sensor. If the actuator and sensor are incorporated in a delay-line oscillator, changes in gas concentration can be detected as changes in the oscillation frequency. Unfortunately, the acoustic velocity is also a function of temperature. A first-order compensation of this temperature-dependence can be obtained by using two SAW devices: one chemically-sensitive

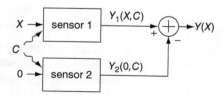

Figure 2.11 Compensation for cross-sensitivity by means of two identical sensors, only one of which is exposed to the measurand

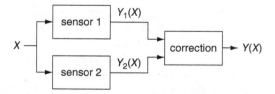

Figure 2.12   Combining two different sensors to improve performance

device, and a reference device without a chemically-sensitive layer. Assuming that both will depend on temperature in the same way, the temperature dependency can be cancelled by measuring the difference in propagation delay, or the difference in oscillation frequency. Note that an alternative self-calibration approach for SAW sensors (temperature stabilization) will be discussed in Section 2.3.7.

Background calibration – A final way in which a combination of multiple sensors can be used to improve performance and reduce calibration needs is shown in Figure 2.12. Here, both sensors are sensitive to the same measurand, but they have substantially different properties. For instance, one may be intrinsically more accurate than the other, but also much slower. In that case, the slow and accurate sensor can be used to *background calibrate* the fast and inaccurate sensor. This entails that the low-frequency content at the output of the fast sensor is compared to that at the output of the slow sensor. From this comparison, the correction required for the fast sensor's response can be derived. Thus, the combination of the two sensors yields a fast *and* accurate measurement system.

An example of this approach is the background calibration of a resistive temperature detector (RTD) by means of a Johnson noise thermometer [29]. RTDs provide a fast way of measuring temperature, but are subject to drift, in particular when they are applied in harsh environments. Johnson noise thermometers, in contrast, are insensitive to drift, because they are based on a first-principles way of measuring temperature: the fact that the noise of a resistor is proportional to absolute temperature [30]. Johnson noise thermometers are relatively slow, because of the long measurement time needed to accurately estimate the noise power. When used for the background calibration of an RTD, however, this slow response is not a problem, because the real-time temperature information is provided by the RTD. The Johnson noise thermometer only calibrates the average temperature measured by the RTD to correct for drift. In practice, a single resistor can be used both as RTD and as a sensitive device of the Johnson noise thermometer [29].

2.3.3   Self-Calibrating Sensactors

A self-calibration approach that comes close to literally integrating a conventional calibration setup inside a smart sensor is shown in Figure 2.13. This figure shows an example of a so-called 'sensactor': the combination of a sensor and an actuator. The actuator generates a calibration signal X_{REF}, which is added to the external measurand X_{EXT} at the input of the sensor. At the output of the sensor, its response Y_{REF} to the reference signal is separated out and compared with the response that is expected based on the signal that was applied to the actuator. Based on

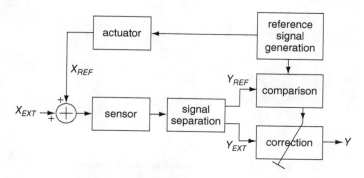

Figure 2.13   Block diagram of a self-calibrating sensactor

the result of this comparison, the sensor's response Y_{EXT} to the external signal is corrected for any errors introduced by the sensor.

This approach only works under two conditions. First, an accurate actuator needs to be available, since the overall accuracy of the system will be at best as good as that of the actuator. This approach therefore makes most sense if it is more feasible to realize an accurate actuator than an accurate sensor for the quantity of interest. Second, it has to be possible to distinguish between the sensor's response to the calibration signal generated by the actuator, and its response to the external measurand. This can be done, for instance, by modulating the calibration signal, so that it occupies a different part of the frequency spectrum than the external measurand, and can be separated out by means of filtering or synchronous detection.

Sometimes, the excitation signal can be generated in an indirect way. In a self-calibrating inertial sensor, for instance, the actuator is not likely to generate acceleration or rotation. Instead, it may exert an (electrostatic) force on the sensor's proof mass similar to the force that an external acceleration or rotation would produce. A recent example of this can be found in [31], which describes a gyroscope in which the effect of the Coriolis force is mimicked by the application of a rotating excitation to the device's drive and sense modes.

2.3.4   Case Study: A Smart Magnetic Field Sensor

This section describes an example of a self-calibrating sensactor for measurement of magnetic fields. The basic principle used is the Hall effect, which is illustrated in Figure 2.14: when a current I_{BIAS} is passed through a plate of conductive or semiconductive material, while the plate is exposed to an external magnetic field, a voltage develops across the plate transverse to the current flow [23]. This so-called Hall voltage V_H is proportional to I_{BIAS} and to the magnetic flux density B_{EXT}, and can thus be used to sense the flux density. The sensitivity of such Hall sensors, however, is typically not very well-defined, and is for instance a strong function of temperature. One way of compensating for this cross-sensitivity is co-integrating a temperature sensor (see Section 2.3.2), which can be used, for instance, to adjust I_{BIAS} in a temperature-dependent fashion so as to obtain a temperature-independent overall sensitivity [23, 24].

An interesting alternative approach, which can in principle eliminate *any* sensitivity errors of the Hall sensor, is shown in Figure 2.15a: a magnetic sensactor is created by co-integrating a coil around the Hall sensor [32–34]. A reference current I_{REF} passed through this coil will

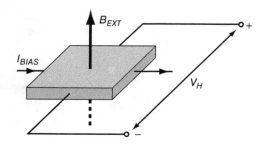

Figure 2.14   Basic principle of a Hall sensor

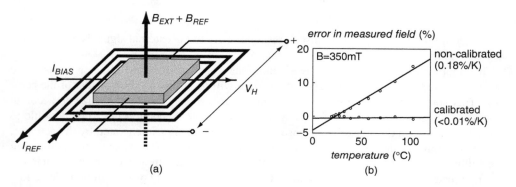

Figure 2.15   (a) Sensactor consisting of a Hall plate and a coil; (b) measured cross-sensitivity to temperature with and without self-calibration [32]

generate a magnetic field B_{REF} that adds to the external magnetic field. The measured Hall voltage, as a result, will be proportional to the sum of these two fields. If we now modulate or pulse the reference current, so as to be able to separate the response to the reference field from the response to the external field, we can construct a self-calibrating magnetic sensactor, in accordance with Figure 2.13. Experimental results (Figure 2.15b) show that this can indeed be done [32]: with self-calibration, the sensitivity is no longer determined by the Hall plate, but by the coil, and, in consequence, the cross-sensitivity to temperature is reduced from 0.18%/K to less than 0.01%/K.

This principle is taken several steps further in the self-calibrating Hall sensor for current measurement described in [33]. A block diagram of this CMOS chip is shown in Figure 2.16. In this design, the Hall plate is biased by a square-wave modulated current. Thus, the direction of the current flow in the Hall plate is periodically reversed. This is an implementation of the so-called spinning-current principle [23], which allows offsets in the Hall plate and the readout amplifier A to be distinguished from the Hall plate's response to a magnetic field. As a result of the bias-current modulation, the latter will give rise to a modulated component in the Hall voltage, while the offsets give rise to a DC component. With an appropriate demodulation scheme, these two components can be separated.

The bias current of the integrated coil is also square-wave modulated, at half of the frequency used for the sensor biasing. Thus, the Hall plate's response to the reference field B_{REF} generated

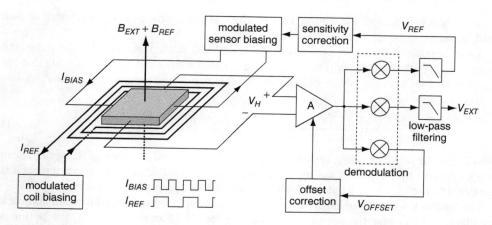

Figure 2.16 Block diagram of a Hall sensactor that self-calibrates offset and sensitivity errors [33]

by the coil can be distinguished from its response to the external field B_{EXT}, and from the offsets in the system. The corresponding three components of the amplified Hall voltage are detected by three demodulators. The offset component V_{OFFSET} is fed back to the amplifier A. This negative feedback loop prevents the output of the amplifier from being overloaded by the offsets, which are typically much larger than the signal of interest. The component V_{REF}, which corresponds to the reference field, is a measure of the Hall plate's sensitivity and is used to adjust the amplitude of the modulated bias current, so as to obtain a well-defined desired sensitivity. The component associated with the external field, finally, is low-pass filtered to eliminate spurious modulated components. The result is a voltage proportional to the external magnetic field with very small offset and gain errors. For instance, the measured gain drift with temperature for this design was only 30 ppm/°C, more than an order-of-magnitude less than the typical uncompensated drift of 500 ppm/°C [33].

2.3.5 Null-Balancing Sensactors

The self-calibrating sensactors described in the previous section perform a feed-forward correction: the error detected by the calibration system is used to correct the reading of the sensor. It is also possible to incorporate a sensor and an actuator in a *feedback* loop. This results in a system as shown in Figure 2.17. In this system, an amplifier with gain A drives the actuator

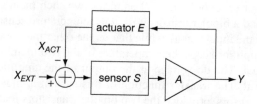

Figure 2.17 Block diagram of a null-balancing sensactor

so as to null the input of the sensor. This is why this configuration is often referred to as a null-balancing system.

Assuming that the amplifier gain A is high enough, and that the feedback loop is stable, the output X_{ACT} of the actuator will be equal to the external measurand X_{EXT}. Assuming that the transfer E of the actuator is well-defined, its input, which is also the overall output Y, will be a good measure of X_{EXT}:

$$Y = \frac{S \cdot A}{1 + S \cdot A \cdot E} X_{EXT} \cong \frac{1}{E} X_{EXT} \quad (S \cdot A \cdot E \gg 1) \tag{2.1}$$

Just like in the self-calibrating system of Figure 2.13, the overall accuracy of the system is no longer defined by the sensor, but by the actuator. Therefore, this null-balancing approach is particularly useful if an accurate actuator can be made for the quantity of interest.

The feedback approach of Figure 2.17 differs considerably from the feed-forward approach of Figure 2.13 in how the sensor is operated. In the feed-forward approach, the sensor is exposed to the superposition of the measurand and a calibration signal. In order to be able to separate these signals at the sensor's output, it generally has to process them in a linear fashion. In the feedback approach, in contrast, the sensor's input is reduced to zero by the feedback loop. This means that gain errors and non-linearity of the sensor no longer have any impact. Moreover, the bandwidth can be extended beyond that of the sensor. This comes at the cost of potential stability issues associated with the feedback loop.

Null balancing is widely used in inertial sensors, such as accelerometers and gyroscopes, where the approach is often referred to as force feedback [35]. Typical accelerometers employ a mechanical proof mass suspended above a substrate by springs. Acceleration of the substrate results in displacement of the proof mass, which is detected, for instance, by means of capacitive or piezo-resistive sensing. In the case of force feedback, in contrast, the proof mass is kept in place by means of a feedback loop that senses the displacement and applies a restoring force to the proof mass, for instance by means of electrostatic actuation. Thus, the accelerometer becomes insensitive to variations in the mechanical spring constants and the sensitivity of the displacement sensor. Moreover, its bandwidth can extend beyond that of the mechanical structure. If the force feedback is implemented by means of a delta-sigma modulator architecture, resulting in an electro-mechanical delta-sigma modulator, a direct digital output can be obtained [35, 36]. Chapter 5 introduces an advanced version of the use of force feedback in a gyroscope and shows how it enables a wider bandwidth than an open-loop architecture.

Another example of the application of null balancing can be found in thermal rms-to-dc converters [37]. To measure the true root-mean-square (rms) value of an electrical input signal, thermal rms-to-dc converters measure the temperature rise associated with the power that this signal dissipates in a resistor. Typically, the resistor is mounted on a thermally-isolated structure, such as a micro-machined membrane on a chip, so as to obtain a substantial temperature rise. The converter reported in [37] employs two such membranes, each containing a polysilicon resistor and a bipolar transistor used as temperature sensor. The resistor on one membrane is driven by the input signal, while that on the other membrane is driven by a feedback amplifier. This amplifier keeps the temperature of the two membranes, as sensed by the bipolar transistors, equal. The dc voltage at the output of the amplifier then equals the rms value of the input signal. The transfer function of this system is independent of that of the transistor temperature sensors, provided the two bipolar transistors match.

2.3.6 Case Study: A Smart Wind Sensor

This section describes an example of a null-balancing sensactor: a smart wind sensor [38, 39]. This sensor has been introduced in Chapter 1. Its operating will be briefly summarized using the schematic diagram shown in Figure 2.18a. It consists of a CMOS chip which is bonded to a thin ceramic disc to protect it from direct contact with the airflow. The chip contains four heaters, four thermopiles, and interface circuitry. By heating the chip, a hot spot is created on the surface of the disc. Airflow will cool the disc asymmetrically, moving the hotspot away from the center and inducing a temperature difference in the chip. This can be measured by means of the thermopiles, and from their outputs, flow speed and direction can be derived.

Rather than allowing the wind to create a temperature difference, however, the heaters and thermopile are incorporated in a feedback loop so as to null the temperature difference. In this so-called temperature-balance mode, the heaters are driven asymmetrically so as to drive the output of the thermopiles to zero. Wind speed and direction can then be derived from the asymmetry in the power dissipated in the heaters.

Figure 2.18b shows how this asymmetry in North-South direction is directly digitized by using a clocked feedback loop called a thermal sigma-delta modulator. An identical modulator (not shown) is used for the East-West direction. In these modulators, the thermopiles are read out using a comparator, rather than using an amplifier. This comparator detects the sign of the thermopile voltage, and determines thus whether the North or the South side of the chip is hotter. The comparator's output is then turned into a sequence of bits (1's and 0's) by means of a clocked flip-flop, and this bitstream and its inverse drive the heaters. The heaters are thus periodically switched fully on or fully off, depending on whether they are on the cold or the hot side of the chip. Because the system is clocked at a much faster rate than that of typical flow-induced temperature variations, the heaters will provide an *average* power difference that

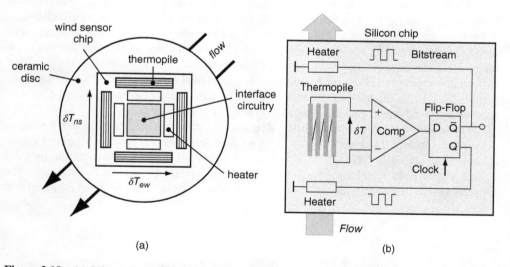

(a)   (b)

Figure 2.18 (a) Schematic diagram of the smart wind sensor; (b) block diagram of one of the thermal sigma-delta modulators [38, 39]

cancels the flow-induced temperature gradient. It is also possible to incorporate an electrical integrator in the feedback loop to obtain more aggressive averaging of the heat pulses (or, in sigma-delta terminology, high-order noise shaping) [40]. Information about the wind speed can then easily be derived by counting the fraction of 1's in the bitstream.

Just like in a continuous null-balancing loop, the sensitivity of this thermal sigma-delta modulator is fully determined by the actuators (the heaters), and not by the sensors (the thermopiles). This makes the smart wind sensor insensitive to production tolerances in the sensitivity of the thermopiles. An even more important advantage of the null-balancing approach is that it automatically cancels temperature gradients resulting from thermal asymmetry in the sensor's package. In a first-generation of the wind sensor, which did not use null-balancing, these gradients had to be removed by manual trimming. In the smart wind sensor, this expensive procedure is no longer needed: the sensor can now be automatically calibrated, and correction can be implemented in the digital post-processing of the bitstreams. After this procedure, errors in wind speed and direction are less than ±5% and ±3°, respectively, for wind speeds between 1 m/s and 25 m/s. Thus, the form of self-calibration applied in this sensor does not eliminate the need for calibration, but it greatly simplifies the calibration and correction procedure, thus substantially reducing the sensor's production costs.

2.3.7   Other Self-Calibration Approaches

Various other approaches of realizing a self-calibrating sensactor can be envisioned, depending of course on the availability of a suitable actuator. Two approaches will be discussed in this section: auto-zeroing and stabilization.

Auto-zeroing – If an actuator is available by means of which the sensor's input can be modulated, a system can be made that can autonomously calibrate out offset errors. This concept is shown in Figure 2.19. The actuator in this case plays the role of a switch or multiplexer in the (non-electrical) domain of the measurand, and prevents the sensor from being exposed to the measurand during auto-zeroing. Thus, the sensor's response $Y(0)$ in the absence of an input signal can be measured and subtracted from subsequent measurements. This is similar to the auto-zeroing and chopping techniques often used to eliminate offset errors in electronic circuits [41]. If the modulation is performed fast enough, it can also be applied to eliminate any low-frequency noise that the sensor may have, such as flicker noise.

The applicability of this technique to a given sensor depends on the availability of a suitable modulating actuator. The classical example of such an actuator is the optical chopper: a slotted rotating disc placed in front of an optical detector that periodically interrupts the incident light. Using MEMS technology, a micro-machined equivalent of this can be realized in the form of an electrostatically-actuated comb drive which displaces a film over an integrated

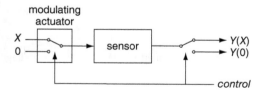

Figure 2.19   Smart sensor that uses auto-zeroing to eliminate offset errors

photodiode to give alternating transmission and blocking of the impinging light [42]. Thus, the signal of interest can be separated from the low-frequency noise of the photodiode and its readout circuitry.

Also for magnetic field sensors a modulating actuator can be realized, in the form of a ring of ferromagnetic material that encloses the sensor [43]. This material normally acts as a magnetic shield and prevents an external field from reaching the sensor, so that the sensor's offset can be measured. When the ferromagnetic material is driven into magnetic saturation, by sending a large enough current through windings around the ring, its permeability drops and it loses its shielding property, allowing the sensor to measure the external field.

Even if an explicit actuator that modulates the input signal is not available, the auto-zeroing technique can sometimes still be applied. In the self-calibrating capacitive fingerprint sensor described in [44], for example, it is essentially the human user of the sensor who acts as the 'actuator' by placing his or her finger on the sensor. This fingerprint sensor captures the topography of a finger by means of an array of pixels, each of which measures the small capacitance between a sensor plate in the detector and the finger placed on top of the sensor. Dirt that accumulates on top of the sensor surface affects this capacitance, and therefore degrades the quality of the captured fingerprint image. To mitigate this problem, a self-calibration scheme is applied when there is no finger on the sensor. Every pixel is equipped with a programmable capacitance that is adjusted during self-calibration so as to make the output signals of all pixels equal. Thus, any capacitance associated with dirt is compensated for by the programmable capacitances, and will not give rise to a visible pattern in a subsequently captured fingerprint image.

Three-signal auto-calibration – For linear systems, the auto-zeroing technique can be extended to correct for both offset and gain errors if an appropriate reference input X_{REF} can be generated. The system then performs three successive measurements: the sensor's response $Y(X)$ to the physical input X, its response $Y(0)$ in the absence of an input, and its response $Y(X_{REF})$ to the reference input. By combining these three measurements, additive and multiplicative errors of the sensor and its readout circuit can be corrected for. This approach is referred to as the three-signal auto-calibration technique [45]. In most implementations of this technique, however, the multiplexing takes place in the electrical domain, which means that errors of the sensor's readout circuit can be corrected, but errors in the transfer function of the sensor itself are not corrected. However, if suitable actuators are available that can implement the multiplexing and generation of a reference input in the non-electrical domain of the measurand, the three-signal approach can, in principle, also correct for offset and gain errors introduced by the sensor.

Stabilization – An actuator can sometimes also be used to create a stable environment for a sensor to operate in, thus improving its accuracy and reducing calibration needs. This principle is illustrated in Figure 2.20. A sensor that is cross-sensitive to a parameter C is placed in an environment in which C is regulated by means of an additional sensor and an actuator in a feedback loop. This feedback loop is similar to that found in a null-balancing sensactor (see Figure 2.17), except that the parameter of interest is not necessarily regulated to zero, but to some suitable value C_{stab}. Variations in the external value C will be suppressed by the loop gain in the feedback loop, so that effectively the cross-sensitivity of the main sensor's output Y_1 to C will be reduced by a factor equal to this same loop gain. The output Y_2 of the feedback amplifier A, incidentally, can be used as a measure of C, should that information be of interest.

An example of this technique is the use of a heater and a temperature sensor to stabilize a sensor's temperature. Typically, a temperature beyond the ambient temperature range

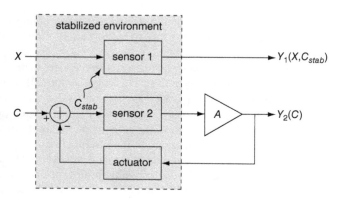

Figure 2.20 Creating a stabilized operating environment for a sensor by means of an additional sensor and an actuator in a feedback loop

of the sensor is chosen, because efficient cooling is very difficult to realize on chip. Thus, a temperature-stabilized micro-oven is realized, inside of which the sensor is operated at a fixed, well-defined temperature. Such an *ovenized* sensor is shielded from ambient temperature variations, reducing the need to take cross-sensitivity for such variations into account in the calibration procedure, or to compensate for such cross-sensitivity. Note that the block diagram of such an oversized sensor is slightly different from that shown in Figure 2.20, because the heater does not produce a temperature that adds to the ambient temperature. Instead, it produces a heat flux that determines the temperature inside the micro-oven via the thermal resistance from the micro-oven to the ambient.

Temperature stabilization can be applied, for example, to reduce the temperature cross-sensitivity of surface acoustic wave (SAW) gas sensors [46, 47]. As discussed in Section 2.3.2, a first-order compensation of this cross-sensitivity can be obtained by using the temperature-dependence of a SAW device without a chemically sensitive layer to cancel that of a chemically sensitive device. However, since these devices will never match perfectly, a residual temperature dependency will remain. This can be eliminated by stabilizing the temperature of the SAW device. In the design described in [46], this is done using an integrated smart temperature sensor and an aluminum heater. These are encapsulated in good thermal contact with the SAW device in a package that provides thermal insulation from the environment. The temperature sensor and heater are incorporated in a control loop which stabilizes the temperature of the SAW sensor to within $\pm 0.01°C$ of a set-point temperature that is programmable between $40°C$ and $120°C$. Thus, the SAW sensor can be operated at a temperature at which its sensitivity or response time is optimal, without being affected by variations in ambient temperature.

2.4 Summary and Future Trends

2.4.1 Summary

Calibration is, in short, the procedure of establishing the accuracy of a sensor. A proper calibration procedure ensures that readings of a sensor can be traced to international standards, and enables correction of these readings when necessary. Smart sensors ideally perform this correction internally, so as to provide a readily interpretable output signal. Calibration and correction data are therefore stored inside the sensor in non-volatile memory, for instance in the form of a transducer electronic datasheet (TEDS). While the costs associated with calibration are often a substantial component in the production costs of smart sensors, they can sometimes be reduced by techniques such as wafer-level calibration, batch calibration, and binning.

Self-calibration is a term that is used for a variety of techniques that improve the accuracy of a sensor by adding intelligence to it. A first form of self-calibration is the co-integration of additional sensors, for instance to compensate for cross-sensitivity. If an accurate actuator can be fabricated, it can be used to generate a reference signal on a chip, for instance to calibrate the sensor's sensitivity. It is then the actuator, rather than the sensor, that determines the overall accuracy. This is also the case in a null-balancing configuration, in which an actuator is used in a feedback loop to null the sensor's input. An important advantage of this approach is that errors in the sensitivity and linearity of the sensor are attenuated by the loop gain. A modulating actuator can be used to periodically shield the sensor from the input signal, so as to be able to measure and subtract its offset error and low-frequency noise. An actuator incorporated in a feedback loop with an additional sensor, finally, can be used to create a stabilized environment for a sensor to operate in. Thus, optimized operating conditions for the sensor can be realized, shielding it from ambient variations. While such self-calibration techniques can never take the place of an actual calibration procedure, which is needed for traceability, they can reduce the number of calibration points needed, and extend the time between (re)calibrations.

2.4.2 Future Trends

Calibration is essential to the field of instrumentation and measurement. While it is a relatively old discipline, it is by no means static. A continuous drive for cost reduction as well as an increasing demand for higher performance leads to the development of more efficient and effective calibration procedures.

A challenge that has received a lot of attention in the past, and will continue to do so in the future, is maintaining traceability and consistency in a sensor market that becomes more and more global. The sensors in a modern measurement system can originate from manufacturers all over the world. In spite of this, their specifications are expected to be consistent, and traceable to common international standards. Moreover, when several smart sensors have to work together in such a system, standardization of interfaces, communication protocols and calibration data formats is essential. Therefore, such standardization, for instance in the context of the IEEE 1451 Standard, can be expected to receive continued attention in the future.

Smart sensors will likely be applied at a much larger scale in the future than today. In applications such as body-area networks, environmental monitoring, structural health monitoring, and automotive sensor systems, large numbers of sensors will be applied, which in some cases will be wireless and autonomous. In such applications, the traditional approach of keeping calibration data separate from the sensor is no longer a reliable solution, nor economically or logistically feasible. Sensors will have to become more modular, so that they can easily be interchanged or replaced when necessary.

Self-calibration techniques can be expected to become more important in the future, given their potential for cost reduction and performance improvements. An important challenge will be to develop better self-calibration schemes. Since the more or less obvious approaches have been explored by now, this will require substantial creativity and out-of-the-box thinking. Inspiration can be found, perhaps, in nature, considering the amazing capabilities of some natural sensor systems, such as our own senses.

References

[1] Morris, A.S. (1991). *Measurement & Calibration for Quality Assurance*, Prentice Hall, New York.

[2] Morris, A.S. (2001). *Measurement and Instrumentation Principles*, Butterworth-Heinemann, Oxford.

[3] ISO (2007). *International vocabulary of metrology – Basic and general concepts and associated terms (VIM)*, ISO/IEC Guide 99, online: http://www.iso.org/sites/JCGM/VIM-JCGM200.htm.

[4] Nicholas, J.V. and D.R. White (1994). *Traceable Temperatures*, Wiley, New York.

[5] Huijsing, J.H. *et al.* (1994). Developments in integrated smart sensors, *Sensors and Actuators A*, **43**, 276–288.

[6] IEEE (2007). *IEEE Standard 1451.0: IEEE Standard for a Smart Transducer Interface for Sensors and Actuators*, IEEE, New Jersey.

[7] Cummins, T. *et al.* (1998). An IEEE 1451 standard transducer interface chip with 12-b ADC, Two 12-b DAC's, 10-kB Flash EEPROM, and 8-b Microcontroller, *IEEE Journal of Solid-State Circuits*, **33**, 2112–2120.

[8] Van der Horn, G. (1997). *Integrated smart sensor calibration*, Ph.D. Thesis, Delft University of Technology.

[9] Elshabini-Riad, A. and I.A. Bhutta (1993). Lightly trimming the hybrids, *IEEE Circuits and Devices Magazine*, **9**, 30–34.

[10] Babcock, J.A., D.W. Feldbaumer, and V.M. Mercier (1993). Polysilicon resistor trimming for packaged integrated circuits. In *Proc. IEDM*, 247–250.

[11] Erdi, G. (1975). A precision trim technique for monolithic analog circuits, *IEEE Journal of Solid-State Circuits*, **SC-10**, 412–416.

[12] Rincón-Mora, G.A. (2002). *Voltage References*, IEEE Press/Wiley, New York.

[13] Murray, A.F. and L.W. Buchan (1998). A user's guide to non-volatile, on-chip analogue memory, *IEE Electronics & Communication Engineering Journal*, **10**, 53–63.

[14] Pertijs, M.A.P. and J.H. Huijsing (2006). *Precision Temperature Sensors in CMOS Technology*, Springer, Dordrecht.

[15] Fruett, F. and G.C.M. Meijer (2002). *The Piezojunction Effect in Silicon Integrated Circuits and Sensors*, Springer, Dordrecht.

[16] Creemer, J.F. *et al.* (2001). The piezojunction effect in silicon sensors and circuits and its relation to piezoresistance, *IEEE Sensors Journal*, **1**, 98–108.

[17] Meijer, G.C.M., G. Wang, and F. Fruett (2001). Temperature sensors and voltage references implemented in CMOS technology, *IEEE Sensors Journal*, **1**, 225–234.

[18] Häberli, A. (1997). *Compensation and Calibration of IC Microsensors*, Ph.D. Thesis, Swiss Federal Institute of Technology.

[19] Pertijs, M.A.P. *et al.* (2005). A CMOS smart temperature sensor with a 3σ inaccuracy of $\pm0.1°C$ from $-55°C$ to $125°C$, *IEEE Journal of Solid-State Circuits*, **40**, 2805–2815.

[20] Aita, A.L. *et al.* (2012). A low-power CMOS smart temperature sensor with a batch-calibrated inaccuracy of $\pm0.25°C$ ($\pm3\sigma$) from $-70°$ to $130°C$, *IEEE Sensors Journal*, **13**, 1840–1848.

[21] Pertijs, M.A.P., *et al.* (2010). Low-cost calibration techniques for smart temperature sensors, *IEEE Sensors Journal*, **10**, 1098–1105.

[22] Machul, O. *et al.* (1997). A smart pressure transducer with on-chip readout, calibration and non-linear temperature compensation based on spline-functions. In *Digest of Technical Papers ISSCC*, 198–199.

[23] Kajik, P. and R. Popovic (2008). Integrated Hall magnetic sensors. In *Smart Sensor Systems*, Wiley, New York.

[24] Ausserlechner, U., M. Motz, and M. Holliber (2007). Compensation of the Piezo-Hall effect in integrated Hall sensors on (100)-Si, *IEEE Sensors Journal*, **7**, 1475–1482.

[25] Makadmini, L. and M. Horn (1997). Self-calibrating electrochemical gassensor. In *Digest of Transducers*, 299–302.

[26] Ishihara, T. *et al.* (1987). CMOS Integrated silicon pressure sensor, *IEEE Journal of Solid-State Circuits*, **SC-22**, 151–156.

[27] Trusov, A. *et al.* (2013). Silicon accelerometer with differential frequency modulation and continuous self-calibration. In *Proc. Int. Conf. on Micro Electro Mechanical Systems (MEMS)*, 29–32.

[28] Drafts, B. (2000). Acoustic wave technology sensors, *Sensors*, online: www.sensorsmag.com /sensors.

[29] Eryurek, E. and G. Lenz (1998). Temperature transmitter with on-line calibration using Johnson noise, US Patent 5,746,511.

[30] Michalski, L. *et al.* (2001). *Temperature Measurement*, Wiley, New York.

[31] Casinovi, G. *et al.* (2012). Electrostatic self-calibration of vibratory gyroscopes. In *Proc. Int. Conf. on Micro Electro Mechanical Systems (MEMS)*, 559–562.

[32] Simon, P.L.C., P.H.S. de Vries, and S. Middelhoek (1996). Autocalibration of silicon Hall devices, *Sensors and Actuators A*, **52**, 203–207.

[33] Pastre, M. *et al.* (2007). A Hall sensor analog front end for current measurement with continuous gain calibration, *IEEE Sensors Journal*, **7**, 860–867.

[34] Badaroglu, M. *et al.* (2008). Calibration of integrated CMOS Hall sensors using coil-on-chip in ATE environment. In *Proc. Design, Automation and Test in Europe*, 873–878.

[35] Lemkin, M. and B.E. Boser (1999). A three-axis micromachined accelerometer with a CMOS position-sense interface and digital offset-trim electronics, *IEEE Journal of Solid-State Circuits*, **34**, 456–468.

[36] Petkov, V. and B.E. Boser (2005). A fourth-order $\Sigma\Delta$ interface for micromachined intertial sensors, *IEEE Journal of Solid-State Circuits*, **40**, 1602–1609.

[37] Klaassen, E.H, R.J. Reay, and G.T.A. Kovacs (1996). Diode-based thermal r.m.s. converter with on-chip circuitry fabricated using CMOS technology, *Sensors and Actuators A*, **52**, 33–40.

[38] Makinwa, K.A.A. and J.H. Huijsing (2002). A smart wind sensor using thermal sigma-delta modulation techniques, *Sensors and Actuators A*, **97–98**, 15–20.

[39] Makinwa, K.A.A. and J.H. Huijsing (2002). A smart CMOS wind sensor. In *Digest of Technical Papers ISSCC*, 432–433.

[40] Wu, J. *et al.* (2011). A 50 mW CMOS wind sensor with ±4% speed and ±2° direction error. In *Digest of Technical Papers ISSCC*, 106–107.

[41] Enz, C. C. and G. C. Temes (1996). Circuit techniques for reducing the effects of op-amp imperfections: autozeroing, correlated double sampling, and chopper stabilization, *Proceedings of the IEEE*, **84**, 1584–1614.

[42] Wolffenbuttel, R.F. and G. de Graaf (1995). Noise performance and chopper frequency in integrated micromachined chopper-detectors in silicon, *IEEE Transactions on Instrumentation and Measurement*, **44**, 451–453.

[43] Popovic, R.S. and J.A. Flanagan (1997). Sensor microsystems, *Microelectronics and Reliability*, **37**, 1401–1409.

[44] Morimura, H. *et al.* (2002). A pixel-level automatic calibration circuit scheme for capacitive fingerprint sensor LSIs, *IEEE Journal of Solid-State Circuits*, **37**, 1300–1306.

[45] Meijer, G.C.M. (2008). Interface electronics and measurement techniques for smart sensor systems, in *Smart Sensor Systems*, Wiley, New York, 23–54.

[46] Meer, P.R. van der, *et al.* (1998). A temperature-controlled smart surface-acoustic-wave gas sensor, *Sensors and Actuators A*, **71**, 27–34.

[47] Meijer, G.C.M. (2008). Smart temperature sensors and temperature-sensor systems. In *Smart Sensor Systems*, Wiley, New York.

3

Precision Instrumentation Amplifiers

Johan Huijsing
Electronic Instrumentation Laboratory, Delft University of Technology, Delft, The Netherlands

3.1 Introduction

This chapter gives an overview of techniques that achieve low offset, low noise, and high accuracy in CMOS operational amplifiers and instrumentation amplifiers. These techniques are essential for the accurate amplification of small sensor output voltages. It will be shown how auto-zero and chopper techniques can be used, both separately and in combination with each other, to achieve offset voltages lower than 1 µV. Frequency-compensation techniques will be described that result in straight first-order roll-off frequency characteristics in the multi-path architectures of chopper-stabilized amplifiers. Thus, these amplifiers can be used in combination with standard feedback networks.

The combination of an accurate voltage gain A_v, a low input offset voltage V_{os} and a high common-mode rejection ratio (CMRR) is not easily implemented [1]. The closest type of amplifier that can achieve low offset and high CMRR is the operational amplifier (OpAmp). But the gain of an OpAmp is not well defined. it is normally so high that feedback around the OpAmp is needed to produce an accurate result [1]. This configuration is depicted in Figure 3.1.

The feedback network connects the input common-mode (CM) voltage to the output CM voltage, and thus potentially destroys the CMRR, as will be explained in more detail in Section 3.3. Therefore, other ways have to be found to combine an accurate voltage gain, a low offset, and a high CMRR.

Ideally, the combination of accurate gain, low offset voltage V_{os}, and high common-mode rejection ratio is found in instrumentation amplifiers (InstAmps). But they are more difficult to implement than operational amplifiers. A general symbol for an instrumentation amplifier is given in Figure 3.2.

Smart Sensor Systems: Emerging Technologies and Applications, First Edition.
Edited by Gerard Meijer, Michiel Pertijs and Kofi Makinwa.
© 2014 John Wiley & Sons, Ltd. Published 2014 by John Wiley & Sons, Ltd.

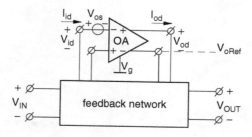

Figure 3.1 Operational amplifier in feedback network ($V_{id} = 0$, $I_{id} = 0$, CMRR = High)

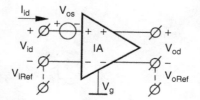

Figure 3.2 Instrumentation Amplifier ($V_{id} \neq 0$, $I_{id} = 0$, $V_{od} = A_v V_{id}$, CMRR = High)

The sections of this chapter discuss the following aspects of the design of InstAmps:

- Applications of InstAmps (Section 3.2).
- Three-OpAmp InstAmps (Section 3.3).
- Current-Feedback InstAmps (Section 3.4).
- Auto-zeroed OpAmps and InstAmps (Section 3.5).
- Chopper OpAmps and InstAmps (Section 3.6).
- Chopper-Stabilized OpAmps and InstAmps (Section 3.7).
- OpAmps and InstAmps that combine chopper stabilization and auto-zeroing (Section 3.8).

3.2 Applications of Instrumentation Amplifiers

All applications of InstAmps require the combination of accurate gain and high CMRR. A first application example is a general one: to overcome a ground loop. This occurs when we want to transfer a voltage signal referred to a different ground potential V_{sRef} than that of the destination potential V_{oRef}. This situation is depicted in Figure 3.3.

This is the case, for instance, when an instrument has to interface a sensor, like a thermocouple, that is connected to a remote ground [2]. The small output voltage of the thermocouple requires a low offset voltage of the amplifier, while the remote ground can have a large potential difference with respect to the ground of the sensing instrument. This requires a high CMRR.

A second common application is the interfacing of the small differential output voltage V_{Bd} of a sensor bridge superimposed on a large common-mode voltage V_{BCM}, as shown in Figure 3.4 [3]. Accuracy and low offset of the measurement is of high priority in this application.

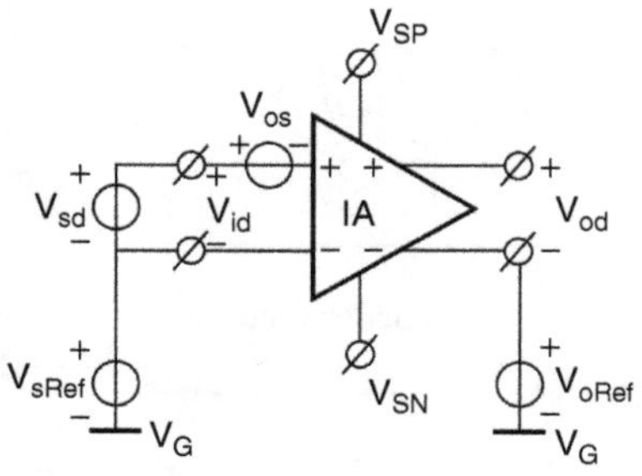

Figure 3.3 Instrumentation amplifier bridging the common-mode voltage between V_{sRef} and V_{oRef}

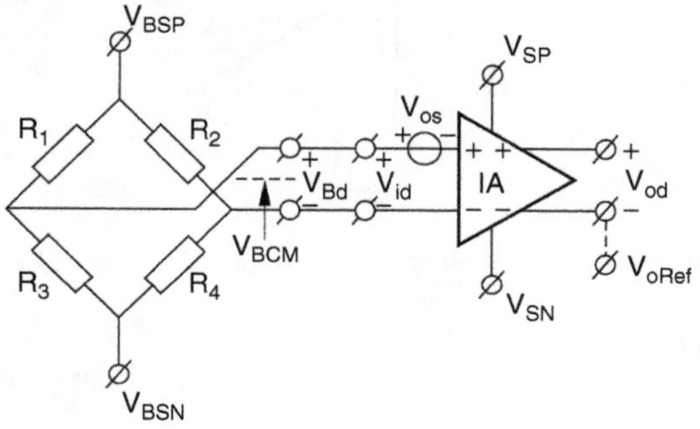

Figure 3.4 Instrumentation amplifier for the readout of a sensor bridge

A third application example is monitoring of the voltage V_{Rsd} across a current-sense resistor R_s in the supply lines of battery-powered systems, like cell phones and laptops (see Figure 3.5) [4]. Power management and battery life makes this application rapidly more important.

A high dynamic range is required for the current-sense application, given the desire to measure both high and low supply currents reasonably accurately without dissipating significant power across the sense resistor at high currents. This implies that small sense resistors have to be used, with an associated small voltage drop. As a result, the InstAmp or "current-sense" amplifier needs to have a low offset voltage under high CM input voltages. The CM input voltage range may even extend far above the supply voltage, or in other cases, needs to include both supply rails. This thoroughly complicates the design of the InstAmp.

A final application example is sensing of differences in voltages of skin electrodes for measuring an ECG, EEG, or EMG of a person (see Figure 3.6) [5, 6]. These differential voltages are in the order of 100 μV to 1 mV in the presence of mains-operated equipment inducing CM voltages on the order of 10 V to 100 V. A high CMRR and patient safety (e.g., low leakage into the InstAmp) are main requirements here.

3.3 Three-OpAmp Instrumentation Amplifiers

The most common approach to an InstAmp is the three-OpAmp topology as shown in Figure 3.7 [1].

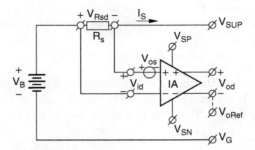

Figure 3.5 Instrumentation amplifier for interfacing a current-sense resistor

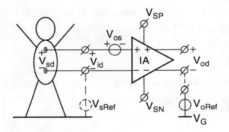

Figure 3.6 Instrumentation amplifier for interfacing medical electrodes

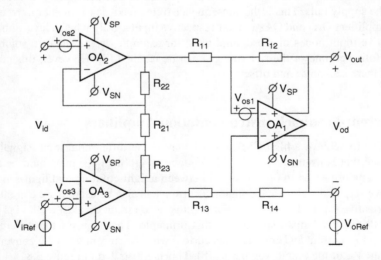

Figure 3.7 Three-OpAmp instrumentation amplifier with resistor-bridge feedback and input buffer amplifier

The actual InstAmp consists of an OpAmp combined with a feedback network consisting of a resistor bridge R_{11}, R_{12}, R_{13} and R_{14}. If the bridge is in balance, the gain for differential signals is:

$$A_d = -R_{12}/R_{11} \approx -R_{14}/R_{13} \tag{3.1}$$

To achieve a high input impedance, buffer amplifiers OA_2 and OA_3 have been placed in front of the bridge resistors. These amplifiers are connected in a non-inverting gain configuration with resistors R_{21}, R_{22}, and R_{23}. They thus provide an additional gain of:

$$A_{d2} = (R_{22} + R_{23})/R_{21} \tag{3.2}$$

The total voltage gain is:

$$A_V = -(R_{22} + R_{23})R_{12}/(R_{11}R_{21}) \tag{3.3}$$

The main problem of the three-OpAmp approach is its limited CMRR. In this topology, the CMRR depends on the matching of the feedback bridge resistors (see [1]):

$$CMRR = (R/\Delta R)A_V \tag{3.4}$$

in which $\Delta R/R$ is the relative error in one of the bridge resistors with respect to its ideal value if the bridge were balanced. For instance, for resistor R_{11}, this relative error equals:

$$\Delta R_{11}/R_{11} = R_{11} - R_{14}/(R_{12}R_{13}) \tag{3.5}$$

With a relative error of 0.1%, a typical value for well-matched on-chip resistors, the CMRR is therefore limited to 1000 A_V.

Another shortcoming of the three-OpAmp approach is that the input CM range does not extend to the supply rails. This is the consequence of the feedback connection from the output of buffer amplifiers OA_2 and OA_3 to their respective inputs. Only when a level shift is built-in at the positive input nodes of these amplifiers, for example, by means of a source follower or emitter follower, one of the rail voltages can be reached [7]. However, this comes at the expense of increased noise and offset.

3.4 Current-Feedback Instrumentation Amplifiers

The best way to achieve a high CMRR is to convert the differential input signal V_{id} into a type of signal that is insensitive to the CM voltage V_{iCM}. Such a signal could be a magnetic signal in a transformer, or an optical signal between a light-emitting and light-sensing diode. But when we stay in the electrical domain, a current can also be used, if we can make it sufficiently insensitive for the CM voltage. For an integrated circuit this last method is preferable. Current-feedback InstAmps make use of this principle. They convert the differential input voltage V_{id} into a current and compare this current with the current from the conversion of the feedback part V_{fb} of the output voltage V_o [8]. This is illustrated in Figure 3.8.

The first voltage-to-current converter G_{m21} converts the differential input voltage V_{id} into a first current. The second converter G_{m22} converts the feedback output signal V_{fb} into a second current. Both currents are subtracted and compared by a control amplifier G_{m1} that drives the

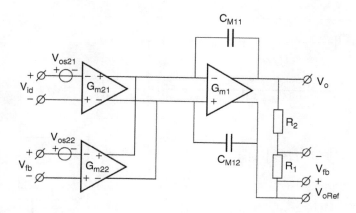

Figure 3.8 Current-feedback instrumentation amplifier

output voltage. A resistive divider R_2, R_1 determines the fraction V_{fb} of the output voltage V_o that is fed back. In consequence, the gain of the whole amplifier will be:

$$A_V = (G_{m21}/G_{m22})(R_2 + R_1)/R_1 \tag{3.6}$$

Often it is difficult to obtain an accurately-defined ratio of transconductances G_{m21} and G_{m22}, unless we make them equal. In that case the gain of the amplifier simplifies to:

$$A_V = (R_2 + R_1)/R_1, \text{while: } G_{m21} = G_{m22} \tag{3.7}$$

The CMRR is now not determined by matching of feedback elements, but exclusively by the ratio of the G_m and small parasitic conductances. As a result, much larger CMRRs can be obtained than those of three-OpAmp instrumentation amplifiers.

To ensure stability of the feedback loop, the InstAmp is Miller compensated by means of capacitors C_{M11} and C_{M12}.

A simple example of a transistor-level implementation of a current-feedback InstAmp is given in Figure 3.9.

The input and feedback VI converters are as simple as possible. They can be degenerated to increase the differential input voltage range if needed. Their intrinsic linearity is rather poor and their transconductance not very well-defined. Nevertheless, good overall linearity and gain accuracy can be obtained, as only mismatch between the transfer functions of the *VI* converters affects the output voltage. The input CM voltage range may include the negative supply-rail voltage V_{SN}. This allows the output voltage V_o being referenced to V_{SN}. The input stages are followed by folded cascodes loaded by a current mirror. The push-pull output transistors are biased in class-AB by a class-AB mesh composed from M_{39} and M_{40} and proper bias voltages V_{B5} and V_{B6}. See Figure 9.4.6 in [1].

A general symbol for a current-feedback InstAmp is given in Figure 3.10. It shows that inside the InstAmp there are two G_m stages: one for the input G_{mi} and one for the feedback G_{mfb}.

It is interesting that the output as well as the input has a high CMRR. This means that we can connect the output reference voltage V_{oRef} terminal to any voltage. An example of this is shown in Figure 3.11. The voltage across the measuring resistor R_M and the current through R_M are

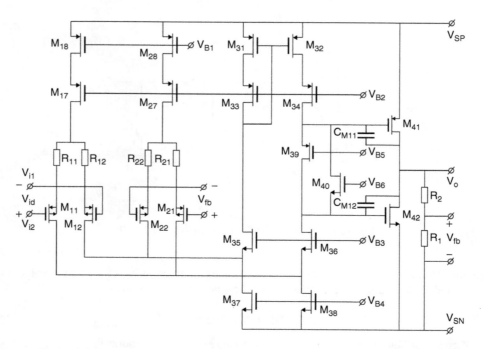

Figure 3.9   Simple circuit-diagram of a current-feedback instrumentation amplifier

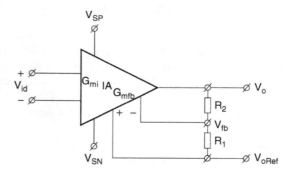

Figure 3.10   Symbol for a current-feedback instrumentation amplifier

not influenced by the voltage on V_{oRef}. Hence, we obtain a voltage controlled current source at the V_{oRef} terminal. The whole topology of Figure 3.11 acts as an accurate general-purpose $V - I$ converter with a transconductance of $1/R_M$. Hence $I_o = V_{id}/R_M$.

3.5   Auto-Zero OpAmps and InstAmps

In Section 3.2 we have seen several applications that need low offset. Auto-zeroing and chopping are the main tools to obtain low offset [9]. In this section we start with auto-zeroing.

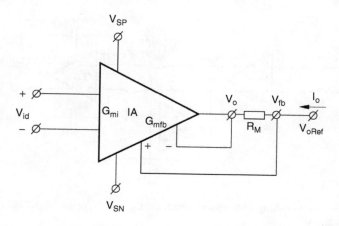

Figure 3.11 Universal voltage-to-current converter with a current-feedback instrumentation amplifier

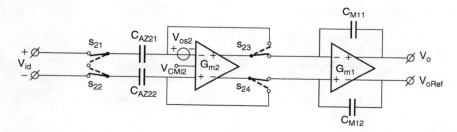

Figure 3.12 Switched-cap auto-zero OpAmp. $V_{os} = 100\,\mu V$

Firstly, we will apply auto-zeroing to an OpAmp in order to reduce its offset. Of the many ways to implement auto-zeroing, we first examine the simple method with switched capacitors at the input, as shown in Figure 3.12.

The auto-zero OpAmp consists of an auto-zeroing input stage G_{m2} with input CM control and a Miller-compensated output stage G_{m1}.

Auto-zeroing has two phases. In phase 1, the forward path is broken, and the output of G_{m2} is fed back, so that its offset appears at its input. The auto-zero capacitors C_{AZ21} and C_{AZ22} store this offset voltage as their inputs are short-circuited together. In phase 2, G_{m2} is inserted back into the signal path, and the auto-zero capacitors are connected to the input. The offset voltage stored on these capacitors now compensates for the offset of G_{m2}. Thus, the input-referred offset of the OpAmp is strongly reduced in phase 2.

An improved auto-zero topology with capacitors at output is shown in Figure 3.13.

When the switches S_{21} and S_{22} short-circuit the input, and the auto-zero switches S_{23} and S_{24} are in auto-zero position, the output current of G_{m2} charges the capacitors C_{31} and C_{32} at its output until the correction amplifier G_{m3} compensates this current. The output common-mode voltage of G_{m2} is controlled at its output.

The advantage of this topology is that the capacitors can store a larger voltage than in the preceding case. This is achieved by making G_{m3} smaller than G_{m2}, so that an amplified version

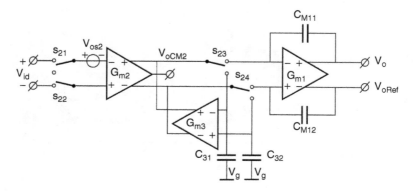

Figure 3.13 Auto-zero OpAmp with storage capacitors C_{31} and C_{32} at the output and correction amplifier G_{m3}. $V_{os} = {\sim}20\,\mu V$

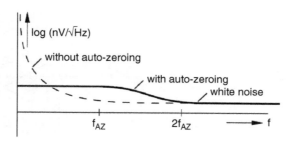

Figure 3.14 Noise densities with and without auto-zeroing

of the input-referred offset voltage appears across the auto-zeroing capacitors. As a result, smaller capacitors can be used for the same kT/C noise and charge injection errors. The offset of G_{m3} is not of interest because it is automatically taken into account in the voltage stored on the auto-zeroing capacitors.

Very important is that the auto-zero action removes not only offset but also $1/f$ noise. This comes at the expense of extra noise V_{naz} generated in the frequency range below $2f_{AZ}$ due to noise folding back from the bandwidth f_{BW} of the local auto-zero feedback loop [9]. This is depicted in Figure 3.14. This increased noise can be mitigated by combining auto-zeroing with chopping, so as to modulate the increased noise away from DC [10, 11].

$$V_{naz} = V_n(\text{white}) \, (f_{BW}/f_{az})^{1/2} \tag{3.8}$$

A problem is that the auto-zero OpAmp has no continuous-time transfer: the signal path is periodically interrupted to perform the auto-zeroing. This means that when the output has to follow a ramp, a staircase with steps at the clock frequency is the result. Moreover, the noise is multiplied by a factor $\sqrt{2}$, as the amplifier is only used half of the time effectively. To overcome these problems the Ping-Pong auto-zero concept of Figure 3.15 has been invented [12].

In Figure 3.15 two auto-zero input stages G_{m21} and G_{m22} alternately are connected between the input and the output stage in order to obtain a continuous-time solution. The stage that is

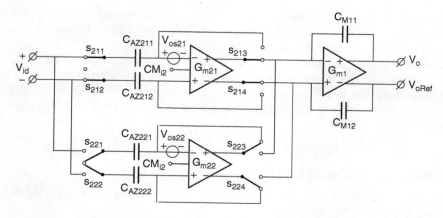

Figure 3.15   Ping-pong auto-zero OpAmp. $V_{os} = \sim 100\,\mu V$

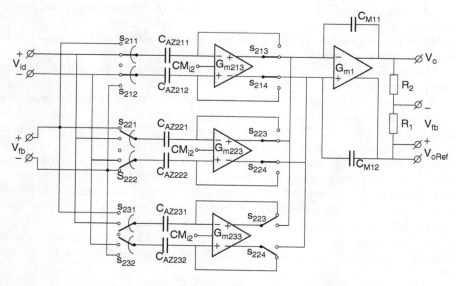

Figure 3.16   Ping-pong-pang auto-zero InstAmp. $V_{os} = 100\,\mu V$

not connected gets time to auto-zero itself. This allows the OpAmp to be generally used in continuous-time feedback configurations.

We can extend the principle of ping-pong to ping-pong-pang in order to obtain a suitable current-feedback InstAmp topology, as shown in Figure 3.16 [13].

In Figure 3.16 three auto-zero input stages G_{213}, G_{223} and G_{233} are used. Sequentially, two stages are connected to the output stage G_{m1}, while one stage is in auto-zero mode. In this way a continuous-time InstAmp is created while its offset and $1/f$ noise is strongly reduced by auto-zeroing.

Offset reduction is limited by parasitic capacitors of capacitors and switches. When the input switches change from auto-zero mode to transfer mode and vice versa, parasitic capacitors to ground are charged and discharged. Any unbalance in this process will change the offset voltage stored on the AZ capacitors. Storing an amplified offset on an intermediate node, as shown in Figure 3.13, would therefore be preferable.

In practice the offset can maximally be reduced by a factor on the order of 100 to 500 with auto-zeroing, reducing a 10 mV offset to 100 µV to 20 µV.

It is very interesting to see that the use of AZ also drastically increases the CMRR, since finite CMRR is equivalent to a common-mode dependent offset voltage, which is reduced by the AZ function.

3.6 Chopper OpAmps and InstAmps

Before we discuss the chopper InstAmp we will look at a chopper OpAmp [14]. Such an OpAmp is depicted in Figure 3.17.

The choppers Ch_2 and Ch_1 alternately reverse the polarity of the signals through the input stage G_{m2}. This means that the input voltage V_{id} will appear as a continuous-time current at the output. But the input offset voltage V_{os2} appears as a square wave current, superimposed on the output, as shown in Figure 3.18.

If the OpAmp is placed in a feedback application, the input voltage will consist of a residual offset voltage with a low-pass filtered square wave ripple on top of it.

In the input-referred noise spectrum, the offset and $1/f$ noise are now shifted to the clock frequency f_{cl}, as shown in Figure 3.19.

The residual offset has mainly two origins. Firstly, non-50% duty cycle in the chopper clocks. If we suppose a 6σ input offset of 10 mV and an asymmetry in the duty cycle of 10^{-4}, the resulting offset is 1 µV. Secondly, imbalance of parasitic capacitors in the choppers also gives rise to residual offset. These parasitic capacitors are shown in Figure 3.20.

Suppose that chopper Ch_1 (in between the input- and output stage) has only the capacitors C_{p11} and C_{p12} around transistor M_1. The capacitor C_{p12} produces alternating positive and negative current spikes at the output of the chopper Ch_1. This does not contribute to the offset. However, capacitor C_{p11} produces similar alternating spike currents at the input of chopper Ch_1. When going to the output, these spike currents are rectified by the chopper Ch_1.

The rectified spikes represent an average DC current, which, when referred to the input, translates into residual offset voltage. Fortunately, the chopper is balanced. Hence, charge

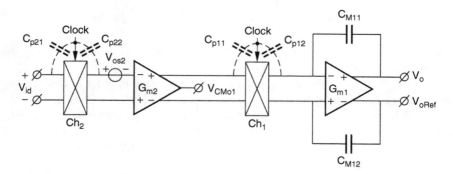

Figure 3.17 Chopper OpAmp with continuous-time transfer. $V_{os} = \sim 10\,\mu V$, $V_{rip} = \sim 10\,mV$

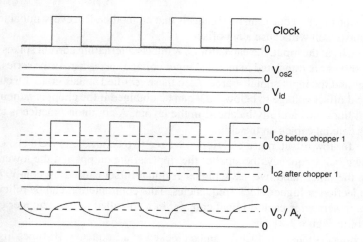

Figure 3.18 Voltage and current signals as function of time in a chopper amplifier

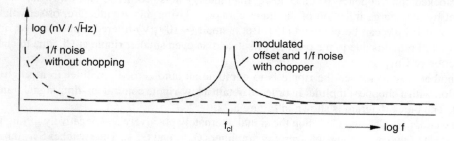

Figure 3.19 Noise densities in an amplifier with and without chopping

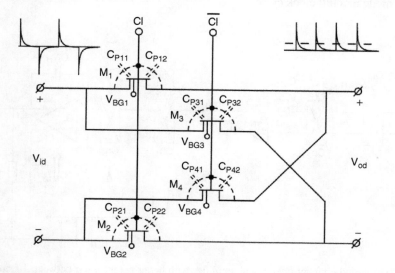

Figure 3.20 Charge injection current in C_{p11} of chopper Ch_1 gets rectified at output

injection from the clock in one transistor cancels that of another. But every imbalance in layout
or transistor mismatch will cause a net offset.

For chopper Ch_2 at the input, the capacitor C_{p22} injects alternating current spikes on the clock
edges. These spikes are translated in rectified input voltage spikes across the series impedances
of the chopper and the input signal source. Also these rectified spikes go to the output as a net
offset. Practical offset voltage to below 1 µV can be obtained if the choppers, their clock lines,
and the signal lines are carefully balanced in the layout. A common practice is to lay out the
clock lines as coaxial cables on the chip.

In our quest for low offset, noise and ripple, we see two contradictory effects. On one hand,
the higher the clock frequency, the smaller the ripple at the output and the lower the residual
1/f noise. On the other hand, we see a higher residual offset caused by non-50% duty cycle
and charge injection at higher clock frequencies. This contradiction can be relieved by using
two choppers in series for each original chopper in a so-called nested-chopper configuration
[15], as shown in Figure 3.21.

The inner choppers Ch_{211} and Ch_{11} can be clocked at a frequency 10-times higher than the
1/f noise corner to overcome $1/f$ noise and ripple, while the outer choppers C_{221} and C_{12}
are clocked at a frequency 10 times lower than the $1/f$ noise corner to take away the residual
offset by the charge injection of the inner choppers. Using this architecture, offset voltages
as low as 0.1 µV can be obtained [15]. But a small (~ 100 µV) filtered input-referred ripple
at Cl_H still remains due to the original offset, and an even smaller ripple at Cl_L due to charge
injection of Ch_{11}.

Another way to reduce the ripple is to combine an auto-zeroed amplifier in a ping-pong
fashion with a chopper amplifier in order to obtain a low-ripple continuous-time signal transfer
[10]. The block diagram is shown in Figure 3.22.

The choppers Ch_1 and Ch_2 chop the signal alternately positively and negatively through the
whole set of two ping-pong auto-zeroing amplifiers G_{m21} and G_{m22}. The switches S_{211} through
S_{222} and S_{213} through S_{224} sequentially switch the amplifiers G_{m21} and G_{m22} in a transfer or
auto-zero mode in a full clock cycle.

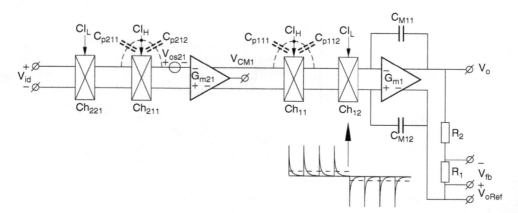

Figure 3.21 Nested-chopper operational amplifier with better compromise between 1/f noise, ripple,
and offset. $V_{os} = \sim 0.1$ µV, $V_{rip} = \sim 100$ µV

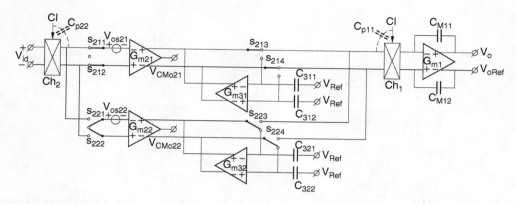

Figure 3.22 Operational chopper amplifier with ping-pong auto-zero input stages. $V_{os} = \sim 3\,\mu V$, $V_{rip} = \sim 10\,\mu V$.

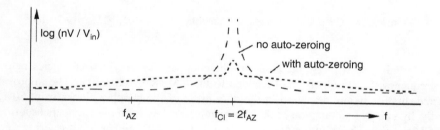

Figure 3.23 Noise in an operational chopper amplifier with ping-pong auto-zero input stages

The capacitors C_{311} through C_{322} differentially store the auto-zero correction voltages. The transconductances G_{m31} and G_{m32} correct the amplifiers G_{m21} and G_{m22} for their offsets, respectively. The auto-zero switches S_{213} through S_{224} switch the outputs of G_{m21} and G_{m22} between the stored voltages on the auto-zero capacitors and the input offset voltage of the output stage. This causes some extra charge injection. An offset of $3\,\mu V$ and an input referred ripple on the order of $10\,\mu V$ can be obtained [10]. The noise of the auto-zero amplifier is now up-converted by the choppers to the clock frequency, which keeps the low frequencies cleaner, as shown in Figure 3.23.

An advantage of the ping-pong continuous-time topology is the simplicity of the frequency compensation. It is restricted to one set of Miller-compensation capacitors.

A chopper instrumentation amplifier can be constructed if we use two input stages G_{m21} and G_{m22}, each preceeded by a chopper, Ch_{21} and Ch_{21}, respectively. This is illustrated in Figure 3.24.

The gain A_v is:

$$A_v = ((R_1 + R_2)/R_2)(G_{m21}/G_{m22}) \tag{3.9}$$

The accuracy of the instrumentation amplifier fully depends on the matching of G_{21} and G_{22}. Even with an ordinary differential pair in weak inversion, and well-matched tail currents, an accuracy better than 1% can easily be achieved without trimming.

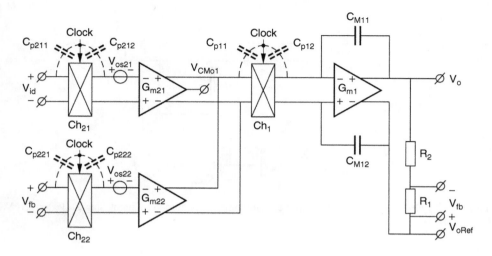

Figure 3.24   Chopper instrumentation amplifier. $V_{os} = \sim 20\,\mu V$, $V_{rip} = \sim 20\,mV$

The CMRR is also strongly increased by the chopper function for frequencies below the clock frequency. Easily 60 dB can be added to the CMRR by chopping. The improvement is limited, firstly, by any deviation in the duty cycle of the chopper clocks from 50%, and secondly, by unequal modulation of the charge injection spikes in the choppers as a function of the CM voltage. The resulting offset can be as low as 20 µV and an input-referred ripple of 20 mV, These numbers are roughly twice as high as those of a chopper OpAmp, which results from the fact that there are two parallel input stages.

To improve the offset and ripple, we may also apply the nested-chopper [15] principle to the chopper instrumentation amplifier, as shown in Figure 3.25. Thus, a better compromise of chopper ripple and $1/f$ noise on one hand and residual offset on the other hand can be achieved (as explained with Figure 3.23). An offset on the order of 0.2 µV can be achieved and a residual ripple on the order of 200 µV.

3.7   Chopper-Stabilized OpAmps and InstAmps

The output ripple from a chopper amplifier invites us to search for ways to reduce it. The chopper-stabilized amplifier is one of the best approaches [9]. A basic chopper-stabilized OpAmp topology is shown in Figure 3.26.

The basic OpAmp is composed of two stages G_{m1} and G_{m2}. The output stage G_{m1} is Miller compensated by C_{M11} and C_{M12}. The input stage G_{m2} forms the 'high-frequency' path. The CM level at the output of G_{m2} is controlled at V_{CMo2}.

The input stage G_{m2} offset V_{os2} is taken into account. When the OpAmp is placed in a feedback loop, this offset appears at the input. This input error voltage V_{id} is now measured and corrected by the chopper amplifier's low-frequency high-gain path. This path starts with an input chopper Ch_2 that translates the input error voltage V_{id} into a square wave. The sense amplifier G_{m5} produces a square-wave output current proportional to V_{id} together with a DC

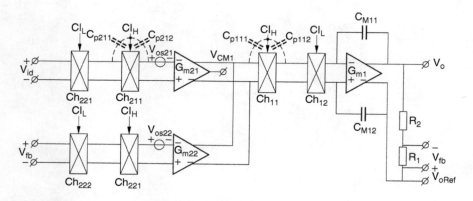

Figure 3.25 Nested chopper instrumentation amplifier with better compromise between 1/f noise, ripple, and offset. $V_{os} = \sim 0.2\,\mu V$, $V_{rip} = \sim 200\,\mu V$

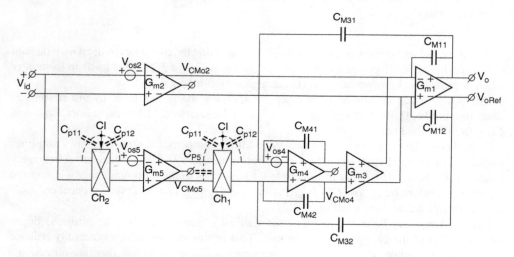

Figure 3.26 Chopper-stabilized operational amplifier with multipath hybrid-nested Miller compensation. $V_{os} = \sim 10\,\mu V$, $V_{rip} = \sim 100\,\mu V$

output current due to its own DC offset V_{os5}. The chopper Ch_1 chops the square-wave current back to a DC error current, while the DC offset current is changed into a square-wave current. The square-wave current due to offset G_{m5} is filtered out by integrator G_{m4}, while the DC current as a function of the input error voltage V_{id} is integrated and strongly amplified by the DC gain of the integrator G_{m4}. Finally the integrated error voltage is added through G_{m3} to the output current of the input amplifier G_{m2}. It should be noted that the output CM levels of G_{m5} and G_{m4} have to be controlled to their CM levels V_{CMo5} and V_{CMo4}, respectively.

We have now obtained a two-path amplifier: a high-frequency low-gain path through G_{m2}, and a low-frequency high-gain path through G_{m5}, G_{m4}, and G_{m3}. The offset can only be reduced to the extent that the high-gain path has a higher gain than the low-gain path.

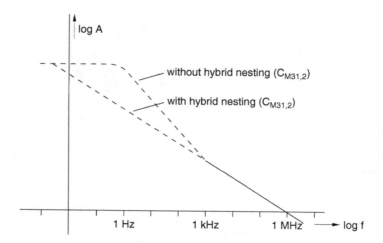

Figure 3.27 Amplitude characteristic of a chopper-stabilized amplifier with and without hybrid-nested Miller capacitors C_{M31} and C_{M32}

One of the old struggles with chopper-stabilization is that the two poles in the gain path lead to a second-order 6 dB-per-octave role-off, as shown in Figure 3.27. This leads to instability when large feedback factors are used (i.e. at high closed-loop gains).

This problem can be solved in practice by applying the principle of hybrid nesting as described in [16, 1]. To that end we connect two hybrid-nested Miller capacitors C_{M31} and C_{M32} from the final output to the input of the integrator G_{M4}.

If we choose the bandwidth of the two-stage Miller-compensated high-frequency amplifier path equal to the bandwidth of the four-stage hybrid-nested Miller loop, the overall frequency characteristic becomes straight from very low frequencies to the unity-gain bandwidth of the OpAmp. Therefore we choose $G_{m2}/C_{M11,12} = G_{m5}/C_{M31,32}$. The result is a classical one-pole frequency roll-off, as shown in Figure 3.27.

The low-frequency behavior, and thus the residual offset of the whole amplifier, is determined by that of the chopped high-gain path. That means that we have to carefully balance the parasitic capacitors C_{p11} and C_{p22} of the choppers Ch_1 and Ch_2 and their lay-out. Also a non-50% duty cycle of the chopper clocks contributes to residual offset. If the asymmetry in the duty cycle is 10^{-4}, and the 6σ offset of the chopper amplifier is 10 mV, an offset of 1 µV remains.

There is one more source of offset we have to watch for. That is caused by a combination of the parasitic capacitor C_{p5} between the outputs of G_{m5} and the offset V_{os4} of the integrator built using G_{m4}. The chopper Ch_3 chops this offset voltage back and forth on C_{p5}, while it rectifies the associated current spikes into a DC current I_{p5} at the input of the integrator equal to:

$$I_{p5} = 4V_{os4}C_{p5}f_{cl} \tag{3.10}$$

This current cannot be distinguished from the DC output current of the chopper sense amplifier, which is also present at the input of the integrator. As a result, it causes an equivalent input offset V_{osi} of:

$$V_{osi} = I_{p5}/G_{m5} = 4\,V_{os4}C_{p5}f_{cl}/G_{m5} \tag{3.11}$$

This offset is smaller than $1\,\mu V$ only if we take measures to make C_{p5} small, that is, in the order of 0.1 pF. We can always chopper-stabilize the integrator amplifier to further reduce this offset component.

The input-referred ripple has now been reduced by a factor of 100 from a square wave of about $10\,mV$ in the chopper amplifier into a triangular wave of about $50\,\mu V$ in the chopper-stabilized amplifier. If we want to decrease the ripple further, we can auto-zero the chopper amplifier [17], as shown in Figure 3.28.

We now have a combination of a chopper-stabilized amplifier in which the chopper amplifier is auto-zeroed. In this way the ripple can further be reduced to the $1\,\mu V$ level. The noise spectrum of such an amplifier is shown in Figure 3.29.

An interesting alternative way to reduce the ripple is using a sample-and-hold after the integration [18], as shown in Figure 3.30.

In this design two passive integrators have been connected as a ping-pong sample and hold with C_{41}, C_{42}, and C_H. The design is simple and elegant and has an offset of $3\,\mu V$, while the ripple is on the order of $20\,\mu V$.

Next, these blocks are integrated into an instrumentation amplifier. To this end, the chopper-stabilized OpAmp must be transformed into a current-feedback InstAmp architecture [4]. The circuit is shown in Figure 3.31.

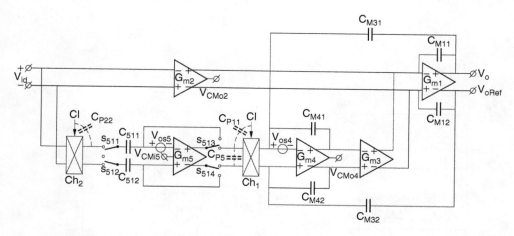

Figure 3.28   Chopper-stabilized OpAmp with auto-zero G_{m5}. $V_{os} = \sim 1\,\mu V$, $V_{rip} = \sim 1\,\mu V$

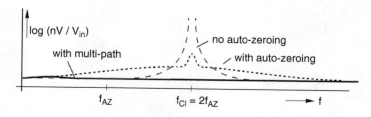

Figure 3.29   Noise densities of a chopper-stabilized multi-path instrumentation amplifier with and without auto-zeroing

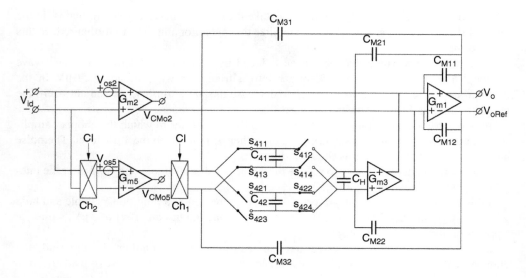

Figure 3.30 Chopper-stabilized OpAmp with passive integrator and sample and hold. $V_{os} = \sim 3\,\mu V$, $V_{rip} = \sim 20\,\mu V$

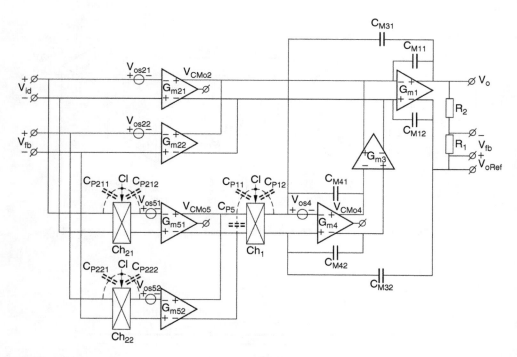

Figure 3.31 Chopper-stabilized InstAmp with multipath hybrid-nested Miller comp. $V_{os} = \sim 20\,\mu V$, $V_{rip} = \sim 200\,\mu V$

The InstAmp has a high-frequency path through G_{m21} and G_{m22} and a low-frequency high-gain path through G_{m51} and G_{m52}. The latter not only determines the offset and CMRR, but also sets the gain accuracy at low frequencies.

The gain A_{VL} at low frequencies is:

$$A_{VL} = (G_{m51}/G_{m52})(R_2/(R_1 + R_2)),\qquad(3.12)$$

and that at high frequencies is:

$$A_{VH} = (G_{m21}/G_{m22})(R_1/(R_1 + R_2)).\qquad(3.13)$$

An offset in the order of $20\,\mu V$ and a ripple of $200\,\mu V$ can be obtained. The offset and ripple are roughly a factor 2 larger than in the OpAmp case because we have two input stages in parallel in both the high- and low-frequency path. The noise is $\sqrt{2}$ times larger than in the OpAmp case.

If we want to further reduce offset and ripple the chopper amplifiers can be auto-zeroed as in the OpAmp case [4]. The resulting block diagram is shown in Figure 3.32.

This topology may result in an input-referred offset voltage lower than $2\,\mu V$ and a ripple lower than $2\,\mu V$.

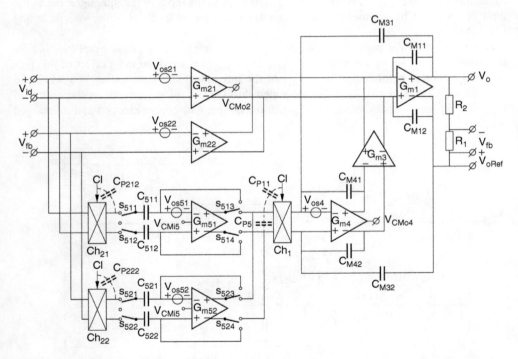

Figure 3.32 Chopper-stabilized InstAmp with auto-zero sense amplifiers. $V_{os} = \sim 2\,\mu V$, $V_{rip} = \sim 2\,\mu V$

3.8   Chopper-Stabilized and AZ Chopper OpAmps and InstAmps

The smooth continuous-time chopper amplifier is the best approach to low offset. However, a 0.01% asymmetry in the duty cycle of the chopper clock multiplied by an initial 6σ offset voltage of 10 mV of the first stage of a CMOS amplifier presents a lower limit to the residual offset on the order of 1 µV. Moreover, the main disadvantage of the chopper amplifier is the chopper-induced square wave ripple, which referred to the input is equal to the initial offset on the order of 10 mV at 6σ. Hence, the ripple and offset of the input amplifier must be further reduced.

The next step of improvement is the chopper-stabilized chopper amplifier [19], which combines the low offset of a chopper-stabilized amplifier with full chopping of the whole amplifier. The topology is shown in Figure 3.33.

If an amplifier has a high loop gain, the differential input voltage becomes zero, except for the input offset voltage. This means in the case of the chopper-stabilized chopper amplifier of Figure 3.33 that the right-hand side of chopper Ch_2 sees V_{os2}. Hence, the left-hand input side carries a square wave voltage equal to V_{os2}. This allows us to directly connect the correction amplifier G_{m5} to the input without extra chopper. The chopper-stabilizer loop has already been discussed in the previous section (see Figure 3.26). However, there are major differences.

The first stage of the main amplifier now determines the noise at low frequencies, while the correction loop determines the ripple at the clock frequency. Further, the hybrid nested capacitors C_{M31} and C_{M32} are not connected anymore to the input of the integrator, but to the input of chopper Ch_3, in order to maintain continuous negative feedback in the loop including Ch_1 [16].

This means that the parasitic capacitor C_{p5}, at the output of the sense amplifier, is now increased with the series connection of C_{M31} and C_{M32}. To avoid the extra offset of this parasitic capacitor in combination with V_{os4} of the integrator, either the offset V_{os4} has to be reduced, or C_{M31} and C_{M32} can be connected through the folded cascode at the output of G_{m5}. Thirdly, the parasitic capacitor C_{p2} before chopper Ch_1 is now charged and discharged

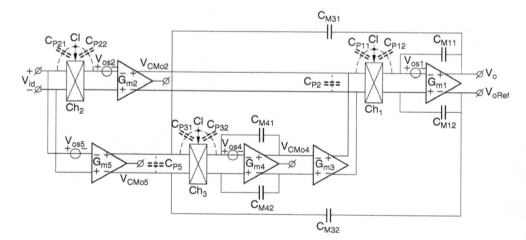

Figure 3.33   Chopper-stabilized chopper OpAmp with multipath hybrid-nested Miller compensation. $V_{os} = \sim 1\,\mu V$, $V_{rip} = \sim 50\,\mu V$

to the offset voltage V_{os1} of the output stage G_{m1}. This causes spikes at the output through the first set of Miller capacitors C_{M11} and C_{M12} at the size of $V_{os1}C_{p2}/C_{M1}$, while $C_{M1} = C_{M11}C_{M12}/(C_{M11} + C_{M12})$. Therefore, the parasitic capacitor C_{p2} at the output of G_{m2} and G_{m3} needs to be small.

The offset of G_{m5} causes a triangular ripple at the output of the integrator and a saw-tooth like ripple through Ch_1 at the output. This can be eliminated if the offset of the sense amplifier G_{m5} is auto-zeroed, similar to the chopper-stabilized amplifier of Figure 3.28. To further reduce the offset caused by the parasitic capacitor C_{p5} in combination with the offset of the integrator amplifier G_{m4} this amplifier can also be auto-zeroed by an extra loop around it [19]. These features are shown in Figure 3.34. In this way an offset of $0.1\,\mu V$ can be achieved with a ripple lower than $2\,\mu V$. Nanosecond chopper spikes of several mV can be observed at the input and output.

A chopper-stabilized chopper instrumentation amplifier appears when the high- and low-frequency amplifier paths are doubled [20] according to Figure 3.35. In contrast to the chopper-stabilized InstAmp of Section 3.7, the gain in a chopper InstAmp is not set by the ratio of G_{m51} and G_{m52} of the correction loop, but by the ratio of G_{m21} and G_{m22} of the main amplifier in cooperation with the feedback network.

$$A_v = G_{m21}(R_1 + R_2)/G_{m22}R_1 \qquad (3.14)$$

The gain is not determined by the ratio of the transconductances G_{m51} and G_{m52} of the sense amplifiers because their influence is shifted by the choppers around the main amplifier to the

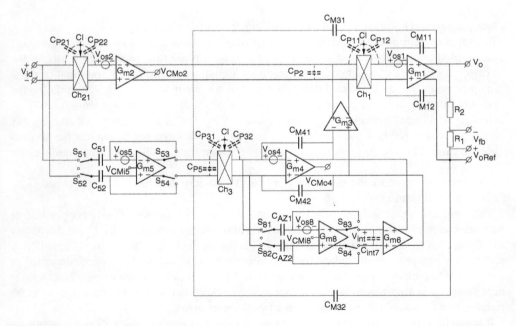

Figure 3.34 Chopper-stabilized chopper OpAmp with multipath hybrid-nested Miller compensation, auto-zero G_{m5} and G_{m4}. $V_{os} = \sim 0.1\,\mu V$, $V_{rip} = \sim 1\,\mu V$

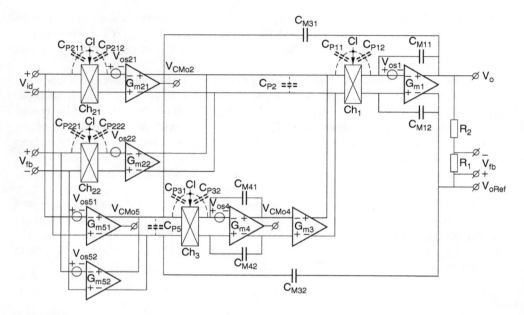

Figure 3.35 Chopper-stabilized chopper InstAmp with multipath hybrid-nested Miller compensation. $V_{os} = 2\,\mu V$, $V_{rip} = \sim 200\,\mu V$

clock frequency. G_{m52} is sensing the feedback ripple as a result of the offset of G_{m21} and G_{m22}. The output current of G_{m52} is rectified by chopper Ch_3 and amplified by the integrator G_{m4} and coupled by G_{m3} to the output of G_{m21} and G_{m22} in order to compensate the ripple due to offset in the main chopper path. The feedback signal-dependant part at the input of G_{m52} is compensated for by the signal-dependant part at the input of G_{m51}. Therefore the signal does not interfere with the offset cancellation. The offset of the correction amplifiers G_{m51} and G_{m52} is chopped into a square wave by chopper Ch_3. The integrator does not amplify this square wave, but reduces it to a small triangular wave. Referred to the input it has to pass chopper Ch_{21}. This means that the shape now becomes a small saw-tooth at the double clock frequency.

The next step to reduce the saw-tooth ripple is to auto-zero the sense stages G_{m51} and G_{m52} [20]. This is shown in Figure 3.36.

The most important residual offset contribution in the chopper-stabilized chopper instrumentation amplifier comes from the combination of the parasitic capacitance C_{p5} at the output of G_{m5} in combination of the offset voltage V_{os4} at the input of G_{m4}, (see Figure 3.26). This is particularly important as the hybrid nested Miller capacitors C_{M31} and C_{M32} are connected in parallel to the parasitic capacitor C_{p5} at the output of G_5. To further reduce this offset component also G_{m4} is auto-zeroed too, as shown in Figure 3.36. In this way the final offset can be reduced to values well below $0.2\,\mu V$ with a ripple lower than $2\,\mu V$.

It has to be kept in mind that the voltage gain of the correction loop G_{m5}, G_{m4}, G_{m3} must be taken 10^4 times larger than the voltage gain of G_{m2} in order to reduce its offset to $0.4\,\mu V$ and ripple from $10\,mV$ to a level of $10\,\mu V$.

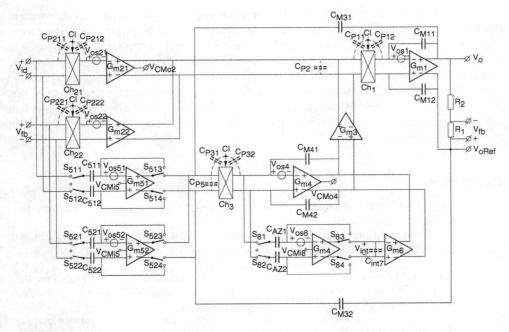

Figure 3.36 Chopper-stabilized chopper InstAmp with multipath hybrid-nested Miller comp. and auto-zero G_{m5} and G_{m4}. $V_{os} = 0.2\,\mu V$, $V_{rip} = \sim 2\,\mu V$

Table 3.1 Summary of offset voltage V_{os} and ripple voltage V_{rip} that can be obtained. AZ = Auto-Zeroing, N = Nested, ChSt = Chopper-Stabilized, Ch = Chopping

OpAmps	V_{os}	V_{rip}	InstAmps	V_{os}	V_{rip}
AZ	$20-100\,\mu V$		AZ	$20-100\,\mu V$	
Chopper	$10\,\mu V$	$10\,mV$	Chopper	$20\,\mu V$	$20\,mV$
N Chopper	$0.1\,\mu V$	$100\,\mu V$	N Chopper	$0.2\,\mu V$	$200\,\mu V$
ChSt	$10\,\mu V$	$100\,\mu V$	ChSt	$20\,\mu V$	$200\,\mu V$
ChSt + AZ	$1\,\mu V$	$1\,\mu V$	ChSt + AZ	$2\,\mu V$	$2\,\mu V$
Ch + ChSt	$1\,\mu V$	$100\,\mu V$	Ch + ChSt	$2\,\mu V$	$200\,\mu V$
Ch + ChSt + AZ	$0.1\,\mu V$	$1\,\mu V$	Ch + ChSt + AZ	$0.2\,\mu V$	$2\,\mu V$

3.9 Summary and Future Directions

Table 3.1 gives an overview of the offset and noise of the OpAmps and InstAmps described in Sections 3.5 to 3.8.

Chopping generally can reduce offset by a factor of 10,000. But the amplitude of the ripple stays at the same order of magnitude as the initial offset without other measures. Auto-zeroing reduces the offset by a factor of 100 to 500, depending whether the AZ store capacitors are placed at the input or at the output. Similar improvements in CMRR can be obtained. Further improvement can be obtained when we combine chopping and auto-zeroing.

Future work in the field of precision instrumentation amplifiers will include the improvement of gain accuracy, which in many cases limits the precision once the offset-cancellation techniques discussed in this chapter are applied. Extending the input common-mode voltage range is also of interest, first to the supply rails and then also beyond. Finally, since in many cases the output of an instrumentation amplifier is input to an ADC, it is also interesting to explore the possibilities of embedding instrumentation-amplifier functionality into the front-end of ADCs. This will lead to precision instrumentation ADCs that can directly interface with sensors that are currently read out using instrumentation amplifiers. Many of the techniques discussed in this chapter will be applicable in this area.

References

[1] J.H. Huijsing, *Operational Amplifiers, Theory and Design*, Kluwer Academic Publishers, Dordrecht, The Netherlands, 2001.

[2] C. Menolfi and Q. Huang, "A fully integrated, untrimmed CMOS instrumentation amplifier with submicrovolt offset," *IEEE Journal of Solid-State Circuits*, vol. 34, no. 3, pp. 415–420, March 1999.

[3] R. Wu, K.A.A. Makinwa, and J.H. Huijsing, "A chopper current-feedback instrumentation amplifier with 1 mHz 1/f noise corner and an AC-coupled ripple reduction loop," *IEEE Journal of Solid-State Circuits*, vol. 44, no. 12, pp. 3232–3243, December 2009.

[4] J.F. Witte, J.H. Huijsing, and K.A.A. Makinwa, "A current-feedback instrumentation amplifier with 5 µV offset for bidirectional high-side current-sensing," *IEEE Journal of Solid-State Circuits*, vol. 43, no. 12, pp. 2769–2775, December 2008.

[5] T. Denison *et al.*, "A 2 µW 100 nV/$\sqrt{Hz}$ chopper stabilized instrumentation amplifier for chronic measurement of neural field potentials," *IEEE Journal of Solid-State Circuits*, vol. 42, no. 12, pp. 2934–2945, Dec. 2007.

[6] Q. Fan *et al.*, "A 1.8 µW 60 nV/$\sqrt{Hz}$ Capacitively-Coupled Chopper Instrumentation Amplifier in 65 nm CMOS for Wireless Sensor Nodes," *IEEE Journal of Solid-State Circuits*, vol. 46, no. 7, pp. 1534–1543, July 2011.

[7] G. Brisebois, "New instrumentation amplifiers maximize output swing on low voltage supplies", Linear Technology Design Note 323, online: www.linear.com.

[8] B. van den Dool and J. Huijsing, "Indirect current-feedback instrumentation amplifier with a common-mode input range that includes the negative rail", *IEEE Journal of Solid-State Circuits*, vol. 28, no.7, July 1993, pp.743–749.

[9] C. Enz and G. Temes, "Circuit techniques for reducing the effect of OpAmp imperfections: Autozeroing, correlated double sampling and Chopper Stabilization", *Proceedings of the IEEE*, vol. 84, no. 11, November 1996.

[10] A.T.K. Tang, "A 3 µV-offset operational amplifier with 20 nV/$\sqrt{Hz}$ input noise PSD at DC employing both chopping and autozeroing," in *Dig. Technical Papers ISSCC*, February 2002, pp. 386–387.

[11] M.A.P. Pertijs and W.J. Kindt, "A 140 dB-CMRR current-feedback instrumentation amplifier employing ping-pong auto-zeroing and chopping," *IEEE Journal of Solid-State Circuits*, vol. 45, no. 10, pp. 2044–2056, October 2010.

[12] I.E. Opris and G.T.A. Kovacs, "A rail-to-rail ping-pong OpAmp", *IEEE Journal of Solid-State Circuits*, vol. 31, no. 9, pp.1320–1324, September 1996.

[13] S. Sakunia, F. Witte, M. Pertijs and K. Makinwa, "A ping-pong-pang current-feedback instrumentation amplifier with 0.04% gain error," in *Dig. Symposium on VLSI Circuits*, pp. 60–61, June 2011.

[14] C. Enz, E. Vittoz, and F. Krummenacher, "A CMOS chopper amplifier", *IEEE Journal of Solid-State Circuits*, vol. 22, no.3, pp. 708–715, June 1987.

[15] A. Bakker, K. Thiele, and J. Huijsing, "A CMOS nested chopper instrumentation amplifier with 100 nV offset", *IEEE Solid-State Circuits*, vol. 35, no.12, December 2000.

[16] J. Huijsing, J. Fonderie, and B. Shahi, "Frequency stabilization of chopper-stabilized amplifiers", US patent Nr. 7,209,000, April 24, 2007.

[17] J. F. Witte, K. Makinwa, and J. Huijsing, "A CMOS chopper offset-stabilized OpAmp", *2006 European Solid–State Circuits Conference, Proceedings*, pp. 360–363.

[18] R. Burt and J. Zhang, "A micropower chopper-stabilized operational amplifier using a SC notch filter with synchronous integration inside the continuous-time signal path", *IEEE Journal of Solid-State Circuits*, vol. 41, no.12, pp. 2729–2736, December 2006.

[19] J. Huijsing and J. Fonderie, "Chopper chopper-stabilized operational amplifiers and methods", US patent Nr. 6,734,723, May 11, 2004.

[20] J. Huijsing and B. Shahi, "Chopper chopper-stabilized instrumentation and operational amplifiers", US patent Nr. 7,132,883, November 7, 2006.

4

Dedicated Impedance-Sensor Systems

Gerard Meijer[1,4], Xiujun Li[2], Blagoy Iliev[3], Gheorghe Pop[3], Zu-Yao Chang[1], Stoyan Nihtianov[1,4], Zhichao Tan[1], Ali Heidari[5] and Michiel Pertijs[1]

[1]*Electronic Instrumentation Laboratory, Delft University of Technology, Delft, The Netherlands*
[2]*Electronic Instrumentation Laboratory, Exalon, Delft, The Netherlands and Sensytech, Delft, The Netherlands*
[3]*Martil Instruments, Heiloo, The Netherlands*
[4]*SensArt, Delft, The Netherlands*
[5]*Guilan University, Rasht, Iran*

4.1 Introduction

Impedance sensors can be defined as being a set of electrodes which can be used to measure electrical properties of materials or structures. Once these properties are known, it appears that the features of measurements performed with such sensors depend for a large part on the properties of the material or structure to be characterized and only partly on the characteristics of the electrodes. The electrical properties of the sensor in its application can be modeled with passive elements in equivalent electrical circuits. The challenging task for the designer is to make such a sensor system sensitive for the measurands and to obtain immunity for other parameters. In this chapter, we consider impedance sensors to be sensors in a certain measurement environment, and that in the electric model presentation of this setup there is at least one resistive or one reactive component of interest which has to be measured. Impedance sensors can be found in a very wide range of applications. For instance, they are used to measure the following quantities:

- Mechanical quantities, such as displacement, speed and acceleration, the quantities of which can be measured with capacitive [1–3] inductive or resistive sensors (this book Chapter 5). In this chapter such sensors are considered sub-sets of impedance sensors.

Smart Sensor Systems: Emerging Technologies and Applications, First Edition.
Edited by Gerard Meijer, Michiel Pertijs and Kofi Makinwa.
© 2014 John Wiley & Sons, Ltd. Published 2014 by John Wiley & Sons, Ltd.

- Physical quantities, such as relative humidity, which can be measured with capacitive sensors.
- Chemical activity of micro-organisms in food products, which can be monitored by selectively measuring certain resistive and capacitive components of impedances [4, 5].
- Water content of soil, which can be measured by determining the capacitive component of impedance/admittance [6, 7].
- Blood viscosity, which can be monitored *in-vivo* by measuring electrical impedance parameters with excitation signals at specific frequencies [8, 9].

In all of these sensors, certain electrical properties of the sensing elements are modulated by the physical or chemical signals to be measured. Note that in some applications, the electrical properties of the material to be characterized are modulated by the measurand, while in other sensors, such as the first example, the geometry of the electrode structure is modulated. The electrode structures are optimized to achieve the best performance for the desired signals and immunity to possible cross effects and interference.

To measure impedance components, at least two electrodes are required. An AC or DC excitation signal (voltage or current) is supplied over or through a pair of electrodes and the resulting current through, or voltage at, the sensor terminals is measured. Often, *sinusoidal* excitation signals are chosen. An attractive property of using a *sinusoidal* excitation signal is that in linear time-invariant (LTI) systems, all resulting currents and voltages in the circuit are also sinusoidal, and hence Fourier transforms can be applied.[1] This makes it easy to characterize each signal with just two parameters: amplitude and phase. In this case, impedance is a complex ratio $Z(j\omega)$ between the complex voltage and the complex current, with frequency-dependent magnitude and phase. The real part of an impedance is referred to as its resistive part, while the imaginary part is called the reactive part.

In its simplest form, an impedance sensor is implemented with just two electrodes (Figures 4.1(a) and 4.1(b)). With a single measurement *two* parameters are found: magnitude and phase. Consequently, with a single measurement, it is possible to find both the reactive part and the resistive part of an impedance. To improve the accuracy, vector-impedance analyzers measure these parts across a frequency range of interest. To reduce the influence of parasitic impedances associated with connecting cables and multiplexers, often four electrodes (two ports) are used (Figures 4.1(c) and (d)). Such measurements are often referred to as *two-port methods* [1].

Oftentimes, the impedance of practical sensors is more complex than can be represented by a two-parameter model. The sensor can then be modeled using an equivalent lumped-element model containing a larger number of elements (Figure 4.2). The values of these elements can be determined experimentally by applying measurements at different well-chosen frequencies.

As an alternative to using sinusoidal excitation signals, other waveforms can be chosen, such as square-wave signals. In this chapter the capabilities and limitations of these non-sinusoidal signals will be addressed. When using such excitation signals, Laplace transforms can be applied to simplify the analysis. The impedance is then defined as the ratio $Z(s)$ between Laplace-transformed voltages and currents, respectively. The design of a good sensor system should start with an evaluation and characterization of the physical behavior of the material or structure under investigation. In an initial design stage, often a vector-impedance analyzer is

---

[1] This is the case in, for instance, electrical circuits which only contain passive voltage-independent components, such as resistors, capacitors, inductors and transformers.

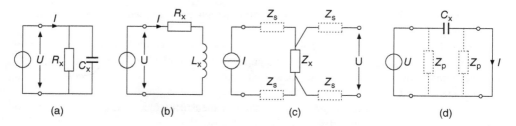

| (a) | (b) | (c) | (d) |

Figure 4.1   (a) and (b) With a single port with two terminals, *two* parameters can be found in a single impedance measurement. (c) and (d) With two-port measurements the effects of parasitic impedances Z_s and Z_p of the connecting wires can be eliminated. Method (c) is applied for a *low-ohmic* sensor impedance Z_x; method (d) is applied for a *high-ohmic* sensor impedance, such as that of a capacitive sensor C_x

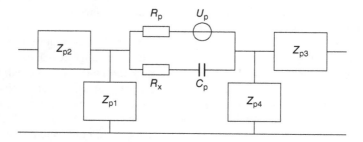

Figure 4.2   An impedance network with more than two elements

used to characterize the impedance sensor in its natural environment. In this way, it is possible to measure the impedance over a wide range of frequencies [10]. By matching the experimental results with the characteristics of a suitable lumped-element model, the specific parameters can be found. It is easy to assume that such an *instrument* with a set of electrodes is also suitable for use as an off-the-shelf *sensor system*. Unfortunately, this is often not true, because such a system would be too expensive, too heavy and not suited for, for instance, industrial on-line use.

To develop a simpler, more efficient instrument, a dedicated measurement technique should be found or developed. In general, when designing a sensor system, the main problems are not due to electronic issues but to physical ones. For instance, without much preliminary study, one might assume that measuring the electrical impedance over a very wide frequency range using an expensive instrument will provide sufficient information for the best design. However, soon it will be obvious that it is not easy or even possible to interpret the huge amount of information generated, and that it is difficult to find the relevant details. Moreover, it may be possible that the set of electrodes and its connection are not suited for the actual sensor task, thereby degrading the value of the measurements. This turns the sensor design into a search for a needle in a hay stack, without the assurance that there is a needle at all.

In this chapter, we will introduce a number of examples of impedance sensors designed for special applications. Using a series of case studies, we will evaluate the physical problems and the corresponding electrical ones together and provide practical solutions. The proposed solutions have been evaluated experimentally at the physical and electronic level.

The examples concern the following applications:

- interface circuits for capacitive sensors employing square-wave excitation signals;
- detection of micro-organisms in milk;
- detection of water content in soil;
- a blood analysis system in which viscosity information is derived from the results of impedance measurements.

4.2 Capacitive-Sensor Interfaces Employing Square-Wave Excitation Signals

4.2.1 Measurement of Single Elements

An impedance measurement with sinusoidal signals enables the measurement of a real and a reactive component in a single measurement. When measuring impedances that consist of just one type of element – for instance, resistance, capacitance or inductance – it should be possible to simplify the measurement. Such a simplification can be achieved by using square-wave excitation signals instead of sinusoidal signals.

Over the last decades, it has been shown that with square-wave excitation signals, even with simple electronics, single sensor elements can be measured with high accuracy [11–15]. For instance, in [15] an integrated circuit, which is called a universal transducer interface (UTI), is presented that is capable of measuring capacitors, resistors and resistive bridges at low and moderate speeds with high accuracy. The voltage-to-period converter (VPC) in this chip consists of a relaxation oscillator (Figure 4.3(a)) the basic principle of which shows some similarity with the simple circuit to be discussed in Section 4.2.2. The front-end is a sensor-specific circuit which for capacitive sensors consists of a low-noise charge amplifier called a capacitance-to-voltage converter (CVC). Furthermore, the front-end contains a multiplexer (MUX) which is used to select one of a number of capacitors to be measured. In the UTI chip the capacitor C_f, which is connected in the feedback loop of the amplifier (CVC), is integrated on-chip. However, a modified circuit will be discussed in Section 4.2.3 which uses an external feedback capacitor.

In this interface a number of advanced measurement techniques have been implemented, such as auto-calibration, advanced chopping, two-port measurement and synchronous detection [1]. In its fastest mode, the acquisition time amounts to about 10 ms [16]. For more details about circuit diagrams, properties, features and applications of this interface, the reader is referred to [1, 17].

In the next sections, it will be shown how with both an application-specific dedicated design, and the use of the same basic concepts applied in the UTI, a very high system performance can be obtained with respect to energy efficiency, speed and resolution, and immunity to parasitic capacitances.

4.2.2 Energy-Efficient Interfaces Based on Period Modulation

The energy-efficient interfaces discussed in this section have been designed for wireless sensors in which, for instance, on-chip capacitive sensing elements are used.

In wireless sensors, given the limited amount of energy available, it is important to minimize energy consumption (see also this book, Chapter 9). As a first step, energy can be saved by putting the sensor system into sleep mode during those periods of time when no measurements are required. As a next step, the energy consumption per measurement can be minimized. To compare the energy efficiency of different sensor interfaces, Pertijs and Tan [18, 19] proposed a Figure of Merit (FOM) that is equal to the one used for ADCs, normalizing the power consumption P of an interface with respect to its effective number of bits (ENOB) and the time T_{meas} needed for a measurement, according to the equation:

$$\text{FOM} = \frac{PT_{\text{meas}}}{2^{\text{ENOB}}} = \frac{E_{\text{meas}}}{2^{\text{ENOB}}}, \tag{4.1}$$

where E_{meas} is the energy required to perform a measurement. This FOM thus expresses the energy efficiency of a sensor interface in terms of the energy required per conversion step.

For the circuit in Figure 4.3(c), a first step to reduce the power consumption can be achieved by omitting the front-end amplifier. In the particular application of Relative Humidity (RH) sensors [18, 20], this was possible because the values of the parasitic capacitances of the on-chip sensing elements were low. Further power reduction could be obtained by redesigning

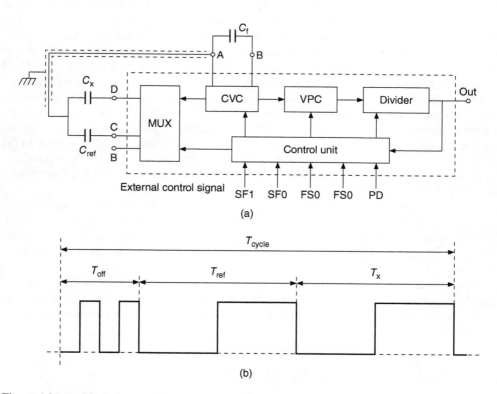

(a)

(b)

Figure 4.3 (a) Block diagram of the UTI system. (b) Period-modulated output signal of the UTI chip. Each time interval contains information about the sensing element selected by the MUX and connected to the input [17]

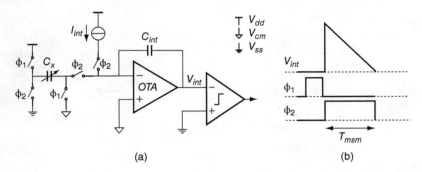

(a)   (b)

Figure 4.4 (a) Operating principle of the period-modulator-based capacitive-sensor interface. (b) Time-modulated output signal V_{int} of the integrator and some control signals

the modulator (the charge-to-period converter). To explain this, we will briefly discuss the basic principles of the modulator (Figure 4.4).[2]

During phase ϕ_1 of a pair of two-phase, non-overlapping clock signals, the sensor capacitor C_x is connected between the supply voltage V_{dd} and a mid-supply common-mode reference V_{cm}. During the subsequent phase ϕ_2, it is connected between V_{ss} and the virtual ground of an active integrator which is also biased at V_{cm}. As a result, a charge $V_{dd}C_x$ is transferred to the integration capacitor C_{int}, causing the output voltage V_{int} of the integrator to step up. A constant integration current I_{int} then removes the charge from C_{int}, bringing V_{int} back to its original level. A comparator at the output of the integrator detects the moment when this happens. The time interval T_{msm} between the start of phase ϕ_2 and this moment is then proportional to C_x:

$$T_{msm} = \frac{V_{dd}}{I_{int}} C_x. \tag{4.2}$$

This time interval can thus be used as a measure of C_x and can be digitized by counting its duration in terms of clock cycles of a faster reference clock. This digitization can be easily performed, for instance, by a counter in a microcontroller, leading to a digital output proportional to C_x [17]. Note that in this type of modulator we are not concerned with the exact waveforms of the voltages, since the signals are in the charge and in the time domain. Due to various circuit non-idealities, such as comparator delay, supply-voltage variations, switch-charge injection, component tolerances and their temperature dependencies, the relation between T_{msm} and C_x is affected by offset and gain errors. While some of these non-idealities, such as comparator delay, could be mitigated at the expense of increased energy consumption, the present design employs simple, energy-efficient building blocks and removes the errors by means of auto-calibration [1]. A further step in reducing energy consumption is to employ a simple cascoded telescopic OTA in the integrator, as opposed to the more current-hungry opamp used in previous designs [1]. This can only be done if the output swing is kept within the limited range of this telescopic topology. As illustrated in Figure 4.4, the output swing of the OTA is directly proportional to the ratio of C_x and C_{int}, which implies that the swing could be reduced by increasing C_{int}. This, however, would come at the cost of an increased die size. Instead,

---

[2] A detailed description of the circuit principles can be found in [17] G. C. M. Meijer and X. Li, "Universal Asynchronous Sensor Interfaces," in *Smart Sensor Systems*: John Wiley & Sons Ltd, 2008, pp. 279–311.

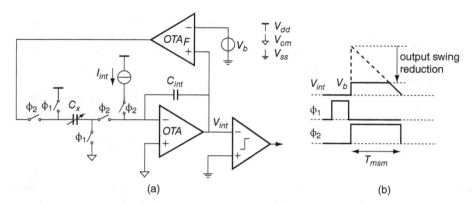

(a)   (b)

Figure 4.5 (a) Use of a negative feedback loop to limit the OTA's output swing. (b) Time-modulated output signal V_{int} of the integrator and some control signals

we employ a negative feedback loop, shown in Figure 4.5(a), that controls the charge transfer between C_x and C_{int} in such a way as to limit the swing at the OTA's output (see also [17], Section 10.3.2).

This negative feedback loop works as follows. Rather than hard-switching C_x to V_{ss} in phase ϕ_2, it is driven by the output of an amplifier OTA_F operating at a current well below that of the main OTA. As soon as OTA_F detects that V_{int} exceeds a maximum level V_b, it limits the current to I_{int}, keeping V_{int} constant. This continues until the drive-side of C_x has almost reached V_{ss}, at which point V_{int} ramps down to the comparator's threshold level. Note again, that in this type of modulator we are concerned not with the wave shapes, but with the charge loss. Since no charge is lost during this entire operation, the total amount of charge transferred from C_x to the integrator is not affected by the feedback loop. Thus, the capacitance-to-time conversion remains the same, that is, Equation (4.2) still holds.

This approach enables the use of a telescopic OTA without the need for a large C_{int}. It comes at the cost, however, of a reduced ability to handle parasitic capacitors: in the presence of large parasitic capacitors around the sensor capacitor, the period modulator may fail to oscillate, or the negative feedback loop can become unstable. The associated trade-offs are discussed in detail in [21, 22]. The prototype discussed here was designed to handle parasitic capacitances of at most five times C_x. Figure 4.6 show a complete transistor-level circuit diagram of the interface.

Experimental results show that the interface achieves 15-bit resolution and 12-bit linearity within a measurement time of 7.6 ms for sensor capacitances up to 6.8 pF, while consuming 64 μA from a 3.3 V power supply. This corresponds to an energy per measurement of 1.6 μJ and a FOM of only 49 pJ/step. The energy efficiency of this interface is much better than that of previous designs based on period modulation [18].

4.2.3 Measurement of Capacitive Sensors with High Speed and High Resolution

In mechanical control systems with mechanical actuators in the forward part and sensors in the feedback part, capacitive sensors are often used to measure position, speed or acceleration.

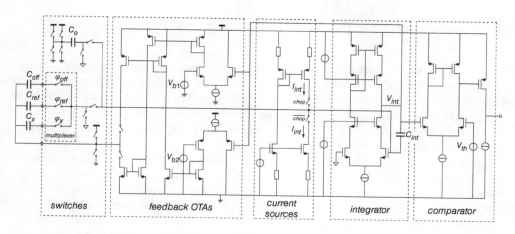

Figure 4.6 Complete transistor-level circuit diagram of the interface

For stability reasons, the acquisition time of these sensors should be much shorter than the time constants of the mechanical actuator. Even with acquisition times shorter than 1 ms, a high resolution is still required. When power dissipation is not a main issue, it is possible to realize very fast sensor systems with a high resolution. In an experimental setup similar to the one of Figure 4.3 [23], a front-end employing high-performance off-the-shelf components was used to measure a capacitor of 2.2 pF within a time frame of 200 μs with an inaccuracy less than 10^{-4}.

Based on the same principles, Heidary developed an integrated-circuit prototype for a flexible interface designed to measure capacitors with a wide choice of capacitance and speed ranges [24]. The interface circuit shows a lot of similarity with that of Figure 4.3. However, the circuit has been improved with respect to its noise performance by optimizing the noise performance of the front-end circuit. Furthermore, in this circuit the feedback capacitor in the capacitance-to-voltage converter (CVC) is accessible for the end user, so that with an off-chip capacitor the dynamic range of the circuit can be optimized, which improves the resolution by a factor one to ten. This interface is suited for capacitive sensors with values up to 220 pF. The measurement time can be set over a wide range from about 50 ms to 100 μs, corresponding to data-acquisition rates from 20 samples/s up to even 10,000 samples/s. For slow measurements the accuracy can be very high. For instance, when averaging the measurement results over a time frame of 10 s, a capacitor in the 1 pF range can be measured with an absolute resolution as good as 0.5 aF, while in the 10 pF range and for a parasitic capacitance up to 680 pF, the measured non-linearity error is less than 5×10^{-5}. On the other hand, for very fast measurements with a data-acquisition rate of 10,000 samples/s, a resolution of 13 bits per measurement is still achieved. These features have been achieved with a chip using 3 mm^2 of silicon area and 5 mW of power dissipation.

Last but not least, in [25] Xia et al. presented a capacitance-to-digital converter in which the so-called zoom-in technique is used. This technique involves removing the offset of the capacitive sensors so that the full dynamic range can be used to change the capacitance in the range of interest. The reported power consumption is less than 15 mW. With a conversion time of 20 μs, a resolution of 17 bits and a FOM of 2.2 pJ/step, this design shows huge potential for capacitive sensors used in high-speed, high-resolution applications.

4.2.4 Measurement of Grounded Capacitors: Feed-Forward Active Guarding

In many sensor systems, the sensing elements are external elements to be connected by the end user, using shielded wires. Sensor interfaces such as those described in the previous sections have been designed for floating capacitive sensor elements, that is, elements with terminals not connected to ground. Using the technique called two-port measurement [1], the basic principle of which is shown in Figure 4.1(d), such elements can be read by interface circuits with a high immunity to stray capacitances to ground.

However, safety reasons and/or operating limitations might require one of the terminals of a sensing element to be grounded. This is the case, for instance, when measuring the level of a conductive liquid in a grounded metallic container with a capacitive sensor [26]. When this happens with high-impedance sensing elements, such as capacitive sensors, the parasitic cable capacitances are in parallel with the sensing element, which can cause huge errors and reduced reliability. For such sensors a common way to reduce the effects of shunting parasitic capacitances is to apply active shielding (Figure 4.7) [27]. In Figure 4.7, C_{p1} and C_{p2} represent the capacitance between the core conductors of a coaxial cable with its shield and the capacitance of the shield to ground, respectively. The active shielding amplifier senses the potential of the shielded wire and replicates this voltage at the shield and guard electrodes. Active guarding and active shielding are well-known techniques which are also applied in many other types of impedance sensors (see, for instance, Section 4.5.3).

A problem with this technique concerns the use of the buffer amplifier because of the applied positive feedback [26], which can yield instability problems. We will show that when using capacitive sensors with square-wave excitation signals it is often not necessary to measure the voltage of the sensing element, because ultimately this voltage will be equal to the excitation signal. Knowing this in advance, instead of using feedback, it is possible to use feed-forward active guarding. This technique prevents instability problems and introduces more design freedom to increase the accuracy. However, one should realize that the successful application of feed-forward active guarding is limited to cases in which precise knowledge about the signal on the guarded wires is available beforehand.

Figure 4.8(a) shows a front-end circuit that employs feed-forward active guarding to convert a sensor capacitance C_x accurately into a voltage V_{out} [28]. This front-end circuit is used as a capacitance-to-voltage converter (CVC) in a period-modulation-based measurement system such as that shown in Figure 4.3. To understand how this circuit works, we first assume that cable capacitances C_{p1} and C_{p2} are zero and that the operational amplifier A_1 and the switches are ideal. During time interval T_1 (Figure 4.8(b)), S_1 is ON, which sets V_{out} to $V_{dd}/2$. At the

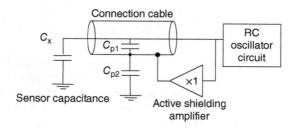

Figure 4.7 Active shielding to reduce the effect of cable capacitance. Reproduced by permission of the Institute of Physics

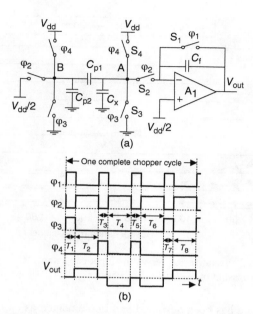

(a)

(b)

Figure 4.8 (a) The capacitance-to-voltage converter (CVC) with feed-forward active guarding and chopping. (b) Some related signals. © IOP Publishing. Reproduced by permission of IOP Publishing.

same time, via S_3, the top electrode (node A) of the sensor capacitance C_x is connected to ground. During time interval T_2, C_x is connected to the negative input of the amplifier. As a consequence, a charge $C_x V_{dd}/2$ will be pumped into C_f, which results in a jump $C_x V_{dd}/(2C_f)$ in the output voltage V_{out}, giving an output voltage proportional to the sensor capacitance. In the voltage-to-period converter (VPC) (Figure 4.3), this output voltage is sampled with a sampling capacitor and converted into the time domain in a process similar to that described for the circuit in Figure 4.4. Similarly, for the time intervals T_3 and T_4, the inverted output voltage V_{out} is generated (Figure 4.8(b)). This is achieved by using different values for the control signals φ_3 and φ_4 for the switches S_3 and $S4$. The use of the signal inversion yields chopping, which helps to reduce the effects of amplifier offset, low-frequency noise and interference. During time intervals $T_5 - T_8$ this process is repeated in the inverted order. This helps to make the chopping even more effective [1].

So far this circuit is identical to the oscillator in the UTI circuit in Figure 4.3. To understand how the principle of active feed-forward is realized we just have to consider a few small details: in the setup in Figure 4.8(a), the voltage at node A has one of three well-known values: 0 V, V_{dd}, and $V_{dd}/2$. Knowing this in advance, without using feedback, we can apply the same voltage to the shielding conductor, which in Figure 4.8(a) is indicated as point B. This reduces the voltage over C_{p1} and thus its overall effect. Furthermore, C_{p2} has no effect at all. Finally, it should also be noted that transient effects, which will occur due to small time differences in the operation of the switches, will not have much effect. This is because the output voltage V_{out} is sampled at the end of a time interval that is so long that transients have already died out. Because the feed-forward principle is used, the effect of cable parasitic capacitances can be eliminated without any instability problems.

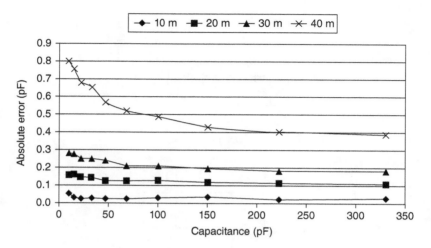

Figure 4.9   The measured absolute error versus the input capacitance C_x for four different lengths of cable. Reproduced by permission of the Institute of Physics

The complete interface has been designed and implemented as an integrated circuit, using standard 0.7 μm CMOS technology. Experimental results show that, for a sensor capacitance down to 10 pF, shielded connection cables up to 30 meters with a parasitic capacitance of 3 nF can be handled with an absolute error of less than 0.3 pF (Figure 4.9). The measured non-linearity of the interface amounts to only 3×10^{-4} even for 30 meters of cable. In this case, with a measurement time of 40 ms, a resolution of 16 bits is obtained.

4.3   Dedicated Measurement Systems: Detection of Micro-Organisms

4.3.1   Characterization of Conductance Changes Due to Metabolism

In the previous section the focus was mainly on sensors with a *capacitance* as the measurand, with capacitive and resistive parasitic components. In other types of sensors, such as conductance sensors, *conductance or resistance* is the measurand, in the presence of capacitive or resistive parasitic components. These parasitic effects can be caused by the need to insulate electrodes. For instance, in some conductance sensors, insulation of electrodes is needed to prevent corrosion, while in others this is done to avoid safety risks. In the case discussed in this section, the electrodes are insulated to measure the conductance of packed food products from the outside, thereby checking the sterility without opening the food containers.

Commercial sterile, aseptically-filled UHT[3] milk can be spoiled by the germination and subsequent growth of bacterial spores that have survived the heat treatment. Additionally, post-processing contamination with spores or vegetative bacterial cells can spoil the product. When using classical sterility-testing techniques, the low probability and low concentration of (post-processing) contamination requires large quantities of the packaged product to be tested. In [29, 30], sterility testing with a thermal method is described. The attractive benefit of thermal

---

[3] UHT means: processed at Ultra-High Temperatures. This is done in a very short time, to kill not only the microorganisms but also the spores.

detection is that it is possible to find a wide range of micro-organisms. On the other hand, a main drawback of thermal detection is that the temperature of the food has to be continuously monitored over a long period of up to four days. In contrast to this, non-invasive detection by means of conductance measurement can be performed in less than a second.[4]

In food products, as bacteria grow and multiply, they break down nutrients and excrete metabolic products. This often causes a change in the electrical conductivity of the product. The use of external electrodes for measuring the conductance of food in packaging enables the detection of bacterial growth without damaging the product or its packaging [4]. As an example, Figure 4.10(a) depicts a plastic bottle containing milk that is blow-molded from high-density polyethylene, and a pair of external electrodes. The electrodes are insulated from the milk by the polyethylene packaging. Figure 4.10(b) shows the equivalent electric circuit of the measured impedance. Resistor R_{food} represents the reciprocal value of the food conductance, which is the measurand. For the microbiological experiment, an automated sterility-testing system was built (Figure 4.11). At an early stage of system development, conductivity was measured with an HP 4192A Impedance Analyzer. Furthermore, the system contained a metal box with controlled temperature, holding up to six plastic bottles and six pairs of external copper electrodes used to measure the conductivity changes of the milk. An HP 3488A Switch/Control Unit selected the pairs of electrodes one-by-one to be connected to the Impedance Analyzer. The equipment was controlled by a PC with an HPIB card in a LabVIEW programming environment. All experiments were carried out in the frequency range from 1 MHz to 5 MHz, which is a high enough frequency to eliminate the influence of the capacitance $C_{1,2}$ of the bottles and low enough to disregard the influence of the milk capacitance C_{milk}. One of the bottles was designated as the reference bottle. At the start of each experiment, the conductivity of the other bottles was zeroed with respect to the reference bottle.

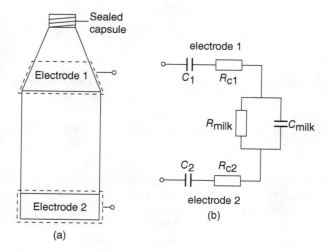

(a)

(b)

Figure 4.10 (a) Plastic bottle with two external electrodes for non-destructive sterility testing. (b) Equivalent circuit diagram of the measured impedance

---

[4] It should be noted that this is only possible after a certain incubation time during which the spores and bacteria grow and multiply. Thus, shortly after packaging, detection is not yet possible.

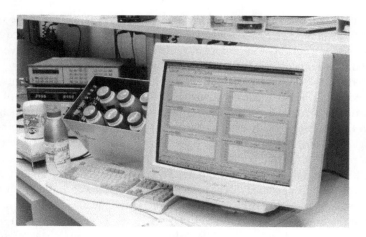

Figure 4.11 Testing system to perform microbiological experiments for non-invasive conductivity measurements of UHF milk in plastic bottles. Each bottle is equipped with a pair of external electrodes

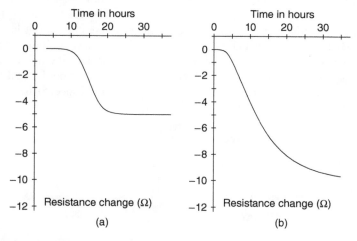

Figure 4.12 Resistance changes in the UHF milk after inoculation with (a) Salmonella ATCC 13311 and (b) Escherichia coli ATCC11775

Figure 4.12 shows the measured change ΔR_{food} after the milk was inoculated with the bacteria Salmonella ATCC 13311 and Escherichia coli ATCC11775 [5]. The horizontal scale represents the time in hours after inoculation. Not all bacteria caused a change in resistance. For instance, the growth of *Serratia marcescens* ATCC 13880 did not influence the resistance at all.

Resistance changes caused by germination and subsequent growth of spores of *B. cereus* (strains ATCC 11778 and MN0089) and *B. subtilis* (MN0226) were minor, though detectable. These measurements demonstrate that an effective sensor system for the detection of micro-organisms can be made using a vector impedance analyzer. However, such an approach will yield a setup which is too complex and too expensive for general use as sensor systems. In the next section, it will be shown how food conductance can be measured with simple circuits.

4.3.2 Impedance Measurements with a Relaxation Oscillator

For the frequency range around 5 MHz, the equivalent circuit diagram of the measured impedance (Figure 4.10(b)) can be simplified to that of just a series connection consisting of a resistor R_x and a capacitor C_x. Simple relaxation oscillators such as those discussed in

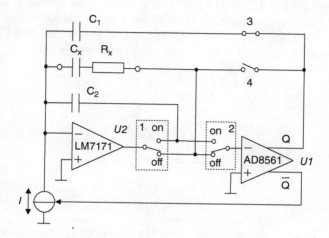

Figure 4.13 Relaxation oscillator for measuring R_x and C_x with autocalibration for offset and scale parameters

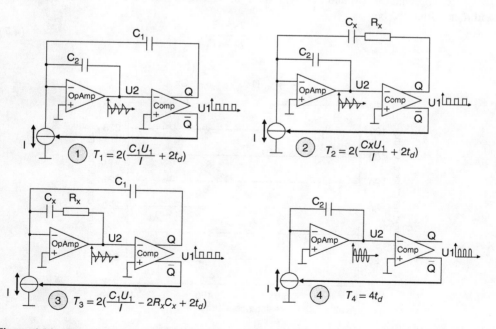

$$① \quad T_1 = 2(\frac{C_1 U_1}{I} + 2t_d)$$

$$② \quad T_2 = 2(\frac{C_x U_1}{I} + 2t_d)$$

$$③ \quad T_3 = 2(\frac{C_1 U_1}{I} - 2R_x C_x + 2t_d)$$

$$④ \quad T_4 = 4t_d$$

Figure 4.14 The four configurations Conf1-Conf4 that are made with the circuit in Figure 4.13 with the switches in various positions

Section 4.2 can be only used to measure a single capacitor. However, with the modified circuit in Figure 4.13, the circuit configuration can be changed to one of four versions [31]. As will be explained, with this circuit both components R_x and C_x can be measured, while autocalibration eliminates the effects of offset and scale parameters of the circuit. Operational amplifier U2 (Figure 4.13) is operated in its linear region as a linear amplifier. The output signal of comparator U1 is HIGH or LOW. The total circuit works as an relaxation oscillator with a period-modulated square-wave signal [17]. With the four switches 1-4, four configurations Conf1-Conf4 for the relaxation oscillator are obtained, which are shown in Figure 4.14. The corresponding period lengths $T_i(i = 1, .., 4)$ amount to, respectively [31]:

$$T_1 = 2\left(\frac{C_1 U_1}{I} - 2t_d\right), \tag{4.3}$$

$$T_2 = 2\left(\frac{C_x U_1}{I} + 2t_d\right), \tag{4.4}$$

$$T_3 = 2\left(\frac{C_1 U_1}{I} - 2R_x C_x + 2t_d\right), \tag{4.5}$$

and

$$T_4 = 4t_d. \tag{4.6}$$

Configurations 4 and 1 are used to determine the time delay t_d of the comparator, and the multiplicative transfer parameter for the case that R_x and C_x are not connected. In configurations 2 and 3, the components R_x and C_x are connected, but in different ways, which results in two different equations (4) and (5). With the four measured periods $T_1 \div T_4$ the values of C_x and R_x are found to be:

$$C_x = C_1 \frac{T_2 - T_4}{T_1 - T_4}, \tag{4.7}$$

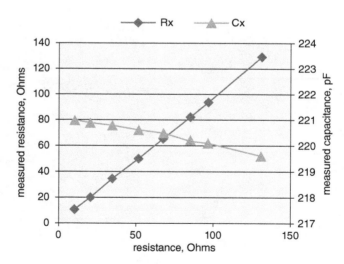

Figure 4.15 The measured resistive and capacitive component of the impedance versus the resistive component for $C_x = 220$ pF

and

$$R_x = \frac{(T_1 - T_3)}{4} \frac{1}{C_x} = \frac{1}{4C_1} \frac{(T_1 - T_3)(T_1 - T_4)}{(T_2 - T_4)}. \tag{4.8}$$

To verify the performance of the interface circuit, measurements were performed with discrete components with values in the range of the values found for the bottles of milk. For configurations 2 and 3, the frequencies reached 5.5 MHz, while for configuration 4, 34 MHz was obtained. Figure 4.15 shows some of the measured values of C_x and R_x versus the nominal resistor value for an unchanged capacitor C_x of 220 pF. More details can be found in [31].

4.4 Dedicated Measurement Systems: Water-Content Measurements

4.4.1 Background

Another application of impedance-sensor systems can be found in water-content measurements in agricultural and horticultural settings. In these sectors, artificial soil is widely used to grow crops instead of using natural soil. In order to optimize a crop-growing process, the amount of water and nutrients must be precisely controlled. To avoid environmental problems and to lower production costs, over-fertilization and excessive use of water must be prevented. Some well-known techniques used to determine water content and nutrient concentration are based on admittance-measurement methods [6]. Nutrient concentration and water content are calculated from the measured values of electrical conductivity and capacitance, respectively. To perform such measurements, long pairs of rod-shaped electrodes (Figure 4.16) are placed in the artificial soil. Depending on the amount of water, nutrients, and water temperature, the electrical conductivity in artificial soil can increase up to 2 S/m. In order to extract the capacitive component, sinusoidal signals with frequencies above 10 MHz are often applied. However, when the electrical conductivity σ is high ($\sigma \geq 0.3$ S/m), the use of long electrode pairs in combination with the use of high signal frequencies gives rise to physical problems that make it difficult to extract the capacitive component accurately. In [7], it is shown that these problems mainly concern the occurrence of the skin effect, the proximity effect, and the presence of parasitic inductances. To reduce the effect of these non-idealities, a probe with a special electrode structure was developed. As will be discussed, with such electrodes the capacitive admittance component can be accurately measured for water conductivity σ up to 2 S/m.

4.4.2 Capacitance Versus Water Content

The complex value of the permittivity ε of a material can be written as:

$$\varepsilon = \varepsilon' - j\varepsilon''. \tag{4.9}$$

The real part of the permittivity ε' is related to the ability to polarize in an electric field and the imaginary part ε'' is related to the energy loss. From an electrical point of view, the admittance between the two electrodes in a salty solution can be modeled as a conductor G in parallel with a capacitor C (Figure 4.17(a)) [6, 32], where it holds that:

$$G = \omega \varepsilon'' \frac{1}{K}, \tag{4.10}$$

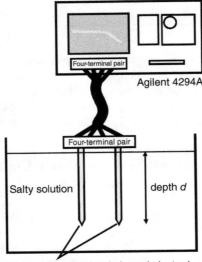

Figure 4.16 Test with a conventional measurement setup: for easy and reliable testing; the artificial soil has been replaced with salty water

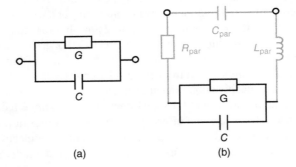

Figure 4.17 Electrical models: (a) for a salty solution and (b) for rod-shaped electrodes in a salty solution, when polarization effects are neglected

and

$$C = \varepsilon' \frac{1}{K}. \tag{4.11}$$

In these equations, ω is the angular frequency and K is the so-called cell-constant, which is a measure for the distance between the electrodes and the electrode areas. For air, $\varepsilon' \approx \varepsilon_0 = 8.854 \times 10^{-12}$ F/m, where ε_0 is the permittivity of free space. For different types of soil, ε' can have values even up to $14\varepsilon_0$. For water, the permittivity ε'_{water} is very high, and amounts to $\varepsilon'_{water} = 80\varepsilon_0$. This high permittivity enables easy detection of the water content, which is quantified as the relative amount of water θ. In our measurements we derived the water content of artificial soil from the measured capacitance C, which is a component of the measured admittance. This was done with linear interpolation between the values of the capacitance for 0% water (C_{air}) and the capacitance for 100% water (C_{water}) using the equation:

$$\theta = \frac{C - C_{air}}{C_{water} - C_{air}},$$ (4.12)

where

$$C_{water} = \varepsilon'_{water}\frac{1}{K}.$$ (4.13)

4.4.3 Skin and Proximity Effects

To measure the capacitive component in the presence of a high shunting conductance, the signal frequency should be as high as possible. However, at high signal frequencies, a number of undesired physical effects pop up, such as the appearance of a series inductance L_{par} that depends on the area of the current loop, parasitic capacitances C_{par}, and the occurrence of skin and proximity effects. The skin effect refers to the phenomenon where, at high frequencies, the main part of a current flows along the surface of a conductor [33, 34]. The depth δ of the skin, where the amplitude is attenuated to e^{-1} (37%), equals:

$$\delta = \sqrt{\frac{2}{\omega\mu\sigma}},$$ (4.14)

where $\omega = 2\pi f$, σ is the conductivity of the conductor and μ is the permeability of the conductor. The skin effect also occurs in salty solutions. With setups like the one shown in Figure 4.15, this poses the problem that the current between the two electrodes will only flow along the surface of the liquid, preventing the water content over a representative volume from being measured. Table 4.1 shows the calculated skin depth δ, according to Equation 4.14, for a signal frequency $f = 20\,\text{MHz}$ and for different conductivities of the salty solutions. From Table 4.1 it can be concluded that with rods of 6 cm in length, at $\sigma \geq 0.94\,\text{S/m}$, the attenuation at the end of the rods is too large. The so-called proximity effect [34] has the same origin as the skin effect, but it occurs on the plane perpendicular to the electrode axis. Both the skin and the proximity effects can be understood as phenomena that tend to cause current loops to minimize their area at higher frequencies.

The effects of parasitic inductance L_{par} and capacitances C_{par} (Figure 4.17(b)) can be reduced by proper design and calibration. To reduce the magnitudes of the skin effect and the proximity effect, as well as series inductance, a special probe with small-sized electrodes has been designed (Figure 4.18) [6, 7]. The reduction of the electrode length to a size far

Table 4.1 The skin depth δ, for a signal frequency $f = 20\,\text{MHz}$, for different liquid conductivities

σ [S/m]	δ [cm]
0.05	50
0.26	22
1.05	11
2.0	8

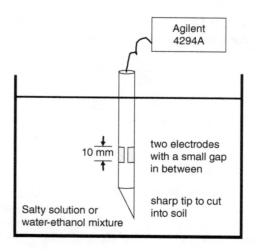

Figure 4.18 Prototype probe with a small-sized electrodes pair

below the skin depth δ resulted in a much more uniform current distribution along the length of the electrodes. Moreover, a significant reduction in the value of the parasitic inductance was obtained. Electrodes (Figure 4.18) with a length of 10 mm were fixed at the surface of an insulating rod with a diameter of 7 mm. The connections of the electrodes to the Agilent 4294A precision impedance analyzer were made with a thin 1 mm thick coaxial cable. The value of the cable capacitance was subtracted from the measured capacitance. To measure over a wider region than the local environment of an electrode set, the probe can be equipped with a number of electrode pairs distributed across its length.

Experimental investigations were performed with both the setup in Figure 4.16 and with the small probe shown in Figure 4.18 [6, 7]. To simplify those measurements, with the setup in Figure 4.16, the effects of water-content changes on impedance were imitated by varying the insertion depths in water for different salinity levels. The frequency of the sinusoidal excitation signal was 20 MHz.

It appeared that with an imitated water content of 60%, increasing the water conductivity from 0.05 S/m to 1 S/m resulted in an error in the measurement of the capacitive component of more than 10%. This error sharply increased for higher conductivities. For salinity greater than 1 S/m, with the long rod-shape electrodes no reliable capacitance measurements were possible.

For the measurements with the setup in Figure 4.18, the effect of water-content changes on impedance was imitated using water-ethanol mixtures. These experiments showed a linear decrease in the capacitance with decreasing water content. In addition to this, experiments were performed with the probe in water with different salt concentrations. The results of these experiments, which are depicted in Table 4.2, show that increasing the water salinity to conductivity levels even up to 2 S/m hardly influences the value of the measured capacitance.

In horticulture, often water with a conductivity > 0.5 S/m is used, while water content is often larger than 50%. It was concluded that for such applications, the use of long rod-shaped electrodes is no longer feasible, and that the use of a series of small electrodes which can measure in a small local environment should yield better results.

4.4.4 Dedicated Interface System for Water-Content Measurements

The measurements reported earlier were performed with an Agilent 4294A impedance ana-lyzer, which is a precision laboratory instrument. For future use in an industrial environment, a dedicated interface circuit for measuring water content with the probe shown in Figure 4.18, has been developed. For this probe, with water content varying from 10% to 90%, and use of the conductivity range in Table 4.2, the target range for C_x ran from 1 pF up to 30 pF, and for the shunting resistance R_x from 22 Ω up to 1 kΩ.

Figure 4.19 shows a block diagram of the developed interface. The applied concept is sim-ilar to what is commonly used in RLC measurement instruments [32]. The principle applied is to measure the voltage U_{Zx} across the unknown impedance Z_x, and the current I_z passing

Table 4.2 The measured capacitance C using a small-sized electrode pair at different specific conductivities of the solution at $T = 23.6°C$

σ [S/m]	C [pF]
0.06	13.1
0.27	13.0
0.95	13.1
2.06	13.2

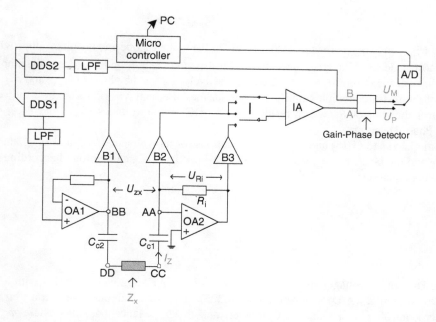

Figure 4.19 Block diagram of the measurement interface

through it. In order not to disturb the DC biasing of the amplifiers, impedance Z_x is separated from the interface circuit by coupling capacitors C_{c1} and C_{c2}. For our application, Z_x can be modeled with an equivalent capacitance C_x and a shunting conductor G_x (Figure 4.17(a)). The current I_z is measured with an *I-U* converter which consists of opamp OA2 and feedback resistor R_i. If the amplifier's input current is negligible, then $I_z = U_{Ri}/R_i$. The unity-gain buffer amplifiers B1 – B3 are there as a preventative measure to make the measurement less sensitive to the parasitic capacitances of the multiplexer and wiring. The multiplexer, sequentially selects the voltages U_{Zx} and U_{Ri} which are amplified with instrumentation amplifier IA. The output voltage of this amplifier is connected to input terminal A of a Gain-Phase Detector (type AD8302).

Two **D**irect-**D**igital-**S**ynthesizer (DDS) chips (type AD9951), designated as DDS1 and DDS2, are controlled via a microcontroller to generate 20 MHz sinusoidal signals. The output signal of DDS1 is "cleaned" by a **L**ow **P**ast **F**ilter and provides the excitation voltage across the unknown measurand Z_x, whereas the filtered output signal of DDS2 provides the reference signal (input B) for the Gain-Phase Detector. Using two DDS chips instead of one is important for applying the proposed calibration procedure, which will be explained below in this section. The Gain-Phase Detector AD8302 provides two DC output voltages which represent the magnitude and the phase of the input signal at node A with respect to that at input B, respectively.

The measurement of voltage U_{Zx} resulted in the two DC output voltages $U_{M,Zx}$ and $U_{P,Zx}$, for which it holds that:

$$U_{M,Zx} = U_{slpM} \log\left(\frac{\hat{U}_{DDS2}}{A_{IA}\hat{U}_{Zx}}\right),\qquad(4.15)$$

and

$$U_{P,Zx} = U_{slpP}(\varphi_{DDS2} - \varphi_{Zx}),\qquad(4.16)$$

where U_{slpM} and $U_{P,Zx}$, are what [35] refers to as the voltage slope and the phase slope with the units V/decade and V/degree, respectively. Furthermore, $\hat{U}_{DDS2}$ and φ_{DDS2} are the amplitude and phase of the output signal of DDS2, respectively; φ_{Zx} is the phase of U_{Zx}; and A_{IA} is the gain of the instrumentation amplifier. A similar measurement was performed for the voltage U_{Ri} over the resistor R_i, which resulted in two DC output voltages $U_{M,Ri}$ and $U_{P,Ri}$ of the Gain-Phase Detector.

From (4.15) and (4.16) and the similar equations for the voltages $U_{M,Ri}$ and $U_{P,Ri}$, the magnitude and the phase of the measured unknown impedance Z_m can be found according to the following equations:

$$|Z_m| = R_i 10^{\frac{1}{U_{slpM}}(U_{M,Ri}-U_{M,Zx})},\qquad(4.17)$$

and

$$\varphi_m = \frac{1}{U_{slpP}}(U_{P,Ri} - U_{P,Zx}).\qquad(4.18)$$

The DC output voltages of the Gain-Phase Detector AD8302 were sampled with a 16-bit Analog-to-Digital (A/D) Converter (ADS8325). The sampled data were transferred via a microcontroller to a Personal Computer, and the impedance $|Z_m|$ and phase φ_m were calculated using (4.17) and (4.18), respectively.

The Gain-Phase Detector AD8302 has gain and phase errors [35]. Since the effects of these errors cannot be eliminated by the ratiometric measurement of $Z = U/I$, they have to be compensated for by other means. To compensate for the gain error, the two DDS chips are programmed to generate signals with different amplitudes. During calibration, the output signal of DDS1 (U_{DDS1}) is directly connected to input A of the Gain-Phase Detector AD8302. This is repeated for different amplitude ratios $U_{\mathrm{DDS2}}/U_{\mathrm{DDS1}}$ and the results are linear-least-square fitted to find an accurate value for the gain slope U_{slpM} of the Gain-Phase detector. A similar calibration method is applied to calibrate the phase output.

In addition to calibrating the Gain-Phase Detector AD8302, open/short compensation is applied, which compensates for the effects of parasitic shunt and series impedances in the measurement setup. Depending on the impedance values, the implemented calibration and compensation techniques reduce the error in the measurement significantly. Experiments with the presented interface system have been performed on calibrated discrete components. Results show that for $C_x = 33$ pF a shunting resistor R_x as low as 22 Ω causes an error in the capacitance C_x measurement amounting to less than 1 pF (3%). When measuring the parallel resistance R_x at a full-scale range of 1 kΩ, the error is less than 1.1%. Finally, experiments with water-ethanol mixtures show that the interface system is suitable for measuring water-volume fractions with a relative error of less than 1.5%.

This case study on a dedicated sensor system for water-content measurements has shown why physical reasons may make it necessary to perform measurements at higher frequencies, and how these higher frequencies can create new physical problems related to skin and proximity effects, and parasitic inductance. Moreover, it has shown the challenges for the electronic-circuit designer, and how he or she can improve the accuracy of the electronics by calibration and compensation techniques.

4.5 Dedicated Measurement Systems: A Characterization System for Blood Impedance

4.5.1 Characteristics of Blood and Electrical Models

This section concerns impedance measurements in blood and describes the development of an *in-vivo*, real-time diagnostic system for hematology and cardiology. The main achievement of this system is the ability of the system to monitor *in-vivo* blood rheological (flow-dynamical) properties and especially whole-blood viscosity, which is derived from measured impedances. It will be shown that in this system, the measurement frequencies are much lower than those found in water-content measurements, which simplifies the design problem. However, blood appears to be a much more complex medium than salty water, which gives rise to new challenges, which are even greater for sensors intended for *in-vivo* applications.

To understand the main requirements of the sensor system, we will briefly describe the main functions of blood flow and the main electrical properties of blood. Blood is an emulsion that is mainly composed of plasma and of red blood cells (Figure 4.20). While circulating through the body, blood performs three basic functions – transportation, regulation and protection:

- Transport of oxygen from the lungs to body tissues (arterial circuit); transport of nutrients, hormones, and enzymes through the body (arterial circuit); and transport of the waste products for eventual removal from the body (venous circuit).

- Regulation of body temperature by adjusting the flow to the skin for optimal heat exchange; regulation of pH through the distribution of buffers; regulation of the renal control of the amount of water and electrolytes.
- Protection against harmful substances by circulating white blood cells, proteins and anti-bodies as an immune-system response to inflammation areas. Additionally, blood contains a subtle balance of clotting substances to prevent bleeding, and lytic substances to prevent clotting.

Whole-blood (natural-blood) viscosity has a significant clinical relevance. The viscosity of blood is intrinsically resistant to flow, which arises from frictional interactions between plasma proteins and blood cells as blood flows through vessels. Red blood cells (RBCs), being the main constituent of the cellular phase of blood, greatly influence whole blood viscosity. At low shear, blood cells aggregate, which induces a sharp increase in viscosity (Figure 4.21), while at high shear, blood cells disaggregate, deform and align in the direction of the flow. In addition to the

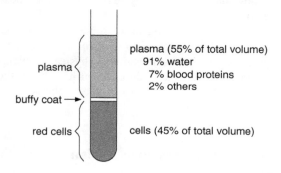

Figure 4.20   Composition of blood

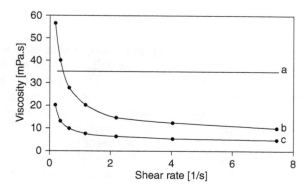

Figure 4.21   Viscosity versus shear rate. Curve "a" is theoretical and represents Newtonian liquid. Curves "b" and "c" show measured values for blood with hematocrits of 46% and 31%, respectively

presence of RBCs, the viscosity is affected by temperature, and by the deformability of the cells [36, 37]. Liquids with a flow-dependent viscosity, such as blood, are called "non-Newtonian".

An increase in blood viscosity at a low shear rate, which indicates increased RBC aggregation, is associated with a thrombotic risk. Furthermore, increased RBC aggregation is also a good marker for inflammatory activity, because due to the presence of inflammatory proteins, blood becomes "stickier" [8, 38]. Therefore, the acquisition of viscosity data can be helpful to detect the likelihood of thrombosis and level of inflammatory activity. Increased RBC aggregation is also a good marker for inflammatory activity.

It appears that there is a strong correlation between electrical characteristics of blood and its viscosity [39]. For instance, both viscosity and electrical impedance increase with a decreasing shear rate and increasing hematocrit. *Hematocrit* is the relative volume concentration of the red blood cells with respect to the total volume. The relation between the flow (rheological) parameters of blood and the electrical parameters is complex. Therefore, using electrical parameters to measure rheological ones requires a good understanding of the nature of their mutual correlation. A detailed study in this field is presented in the PhD thesis of B. Iliev [9], and is briefly summarized below.

RBCs resemble biconcave disks, 7 μm in diameter and about 2 μm thick. Almost the entire weight of an RBC consists of hemoglobin which is enveloped by a thin cell membrane (plasma membrane) [40, 41]. Hemoglobin itself is a globular protein. Its ability to combine with oxygen defines the main role of RBCs, that is, to transport oxygen from the lungs to body tissue. The cell membrane is electrically inert. It is very thin, which results in high specific electrical membrane capacitance, in the order of 0.8 μF.cm^{-2} to 1 μF.cm^{-2} [42].

At low frequencies, the blood impedance can be characterized by the electrical resistance of the plasma around the RBCs. The RBC content has a low resistance. However, this content does not contribute to current conductance at low frequencies, because of the insulating cell membrane. At higher frequencies, the reactive impedance of the cell wall will decrease, which reduces the electrical impedance of blood. With further increasing frequencies, the blood impedance will further drop to a value that is a mixture of the resistances of plasma and RBC intracellular fluid.

This electrical behavior can be modeled with the three-element "macro-model" shown in Figure 4.22(a). In this model, R_p represents the macro effect of plasma resistance, C_m is cell-membrane capacitance and R_i is the hemoglobin resistance. In [43], it is shown that for very high frequencies other effects such as water capacitance will also influence blood impedance. At low frequencies ($f < 20$ kHz), when using electrodes to measure blood impedance, the so-called polarization impedances Z_e have to be taken into account (Figure 4.22(b)) [43]. These polarization impedances are rather complex, which can complicate impedance measurements significantly. Fortunately, polarization effects are only relevant for the low-frequency range. For the intermediate frequency range, used for indirect viscosity measurements, these can be ignored, so that the simple three-element model in Figure 4.22 is sufficient. Because of the non-Newtonian behavior of blood, the values of the three elements are shear-rate dependent. Therefore, impedance measurements should be synchronized with the T-wave of the heartbeat. Moreover, in *in-vivo* measurement systems, the measurements should be fast enough to yield results that are valid for the instantaneous shear rate.

In the next section we will firstly present the technical details of the *in-vivo* blood-analysis system. Then, in Section 4.5.3, the correlation between the electrical parameters and the

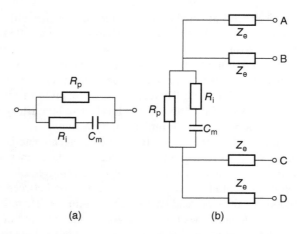

(a)   (b)

Figure 4.22 (a) Three-element model of blood and (b) electrical model of blood and together with the polarization impedances of four electrodes

rheological (flow) parameters will be discussed, while the relation between shear rate, flow and viscosity will be discussed in more detail.

4.5.2   In-vivo *Blood Analysis System*

The blood-analysis system (Figure 4.23), called HemoCard Vision® [9], consists of a central venous catheter, interface electronics, processing software and a computer. Figure 4.24 shows a screenshot of the computer monitor during a measurement.

As a result of the direct measurement and/or additional data post-processing, the system derives a set of results that consists of:

- hematocrit (the relative volume concentration of red blood cells), derived from the plasma resistance R_p;
- blood viscosity, derived from the cell membrane capacitance C_m;
- intracardiac ECG (electrocardiograph) in the heart;
- core body temperature.

In [43], it is shown that for impedance measurements at intermediate frequencies (20 kHz $< f <$ 2 MHz) the simple two-electrode method will perform as well as a more complex four-electrode measurement.

This is valid for experimental *in-vitro* work in a laboratory, where it is possible to check the condition of the electrodes and to clean them when necessary. Unfortunately, with *in-vivo* settings it is not so easy to observe the electrode conditions. When experiments run over a longer time, growth of biolayers (Figure 4.25) give rise to a sharp increase in the contact impedance. To overcome this, the tip of the catheter was coated with heparin, which is a material with anticoagulant properties which were indeed effective in preventing the growth of

Figure 4.23 HemoCard Vision® blood-analysis system: electronic part of the system and central venous catheter. Reproduced by permission of Martil Instruments B.V

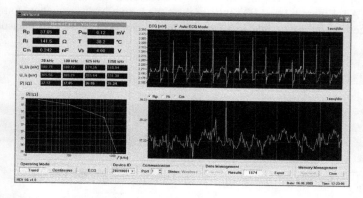

Figure 4.24 HemoCard Vision blood analysis-system: data plotted on the computer screen. Reproduced by permission of Martil Instruments B.V

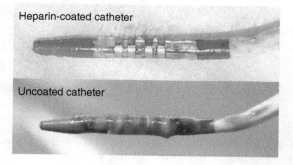

Figure 4.25 A biolayer-free, heparin-coated catheter, and an uncoated catheter encapsulated by a biolayer after being implanted for five days. Reproduced by permission of Martil Instruments B.V

biolayers (Figure 4.25) [9]. To reduce sensitivity to residual effects of biolayers and other contamination thus improving the reliability, four-electrode measurements are preferred to the simpler two-electrode measurements.[5]

Figure 4.26 shows a block diagram of the designed electronic interface [9]. The sensors indicated in the top-right part of the diagram represent the four ring-shaped electrodes at the tip of the catheter (Figure 4.25). The impedance Z_b is the one to be measured. To find Z_b, two quantities are measured simultaneously: the voltage across the inner terminals, and the current through Z_b, which equals the current through coax 1.

Triaxes 1, 2 and 3 (Figure 4.26) are very thin coaxial cables which connect sensor rings and electronics. These cables consist of one inner conductor and two concentric conductive shields. The inner conductors are connected to the electrode rings. The middle conductors (the first shields) are used for active guarding and are connected to the outputs of unity-gain amplifiers A1, A2 and A3, which are buffer amplifiers with a high input impedance. In this way, the effect of parasitic capacitances between the inner conductors and the first shields is significantly reduced. The outer shields are connected to ground and are needed to prevent signals from being emitted into the environment. The resistor R_t is a thermistor in the catheter tip which is used to measure temperature. A sinusoidal excitation voltage is generated with a digital signal generator DDS and filter F1. The buffered differential voltage of A2, A3 is amplified by A4. Filters F2 and F3 separate the signals in a high-frequency component used for impedance measurement, and in a low-frequency component used to measure the ECG.

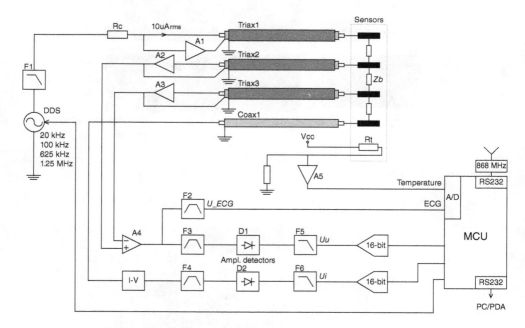

Figure 4.26 Block diagram of the interface electronics of the HemoCard Vision® blood-analysis system. Reproduced by permission of Martil Instruments B.V

---

[5] The better results achieved by the four-electrode measurements can be understood as the fundamental advantage of using two-port (four-wire) measurements [1] instead of one-port (two-wire) measurements.

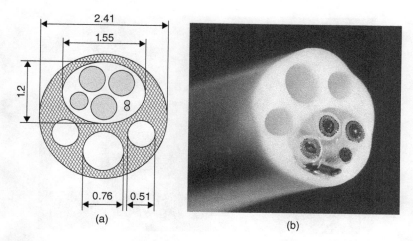

Figure 4.27 (a) Catheter extrusion profile (all sizes are in millimeters), (b) photograph. Reproduced by permission of Martil Instruments B.V

The output signal of filter F3 (for impedance measurement) is processed in amplitude detector F5, which, after low-pass filtering, results in a DC voltage that is converted into a digital signal with a 16-bit AD converter.

The excitation current is converted into a voltage by the I-V converter, which has a low input impedance. This voltage is also filtered, and its detected amplitude is converted into a digital signal by a second 16-bit AD converter. These measurements are performed at three frequencies: 100 kHz, 625 kHz and 1.25 MHz, respectively. Only the amplitude, and not the phase information, is used[6] to calculate the parameters R_p, C_m and R_i. The microcontroller takes care of digital processing and provides an RS232 output interface (Figure 4.27). For further details the reader is referred to [9].

4.5.3   Experimental Results

For the initial tests, an *in-vitro* measurement setup was developed. The *in-vitro* setup consisted of a main tube (Figure 4.28) into which the catheter was inserted, a centrifugal pump with a built-in ultrasonic flow meter, and a heat exchanger for temperature control. The main tube and the setup is optimized to work with flow rates between 0.1 l/min and 1 l/min, so that low shear conditions can be achieved with a Reynolds number Re < 100. A cone at the entrance of the main tube (left side) prevents a jet flow from forming. The shear rate varies proportionally with the flow. Simulations with the Fluid Dynamics Analysis Package (FIDAP) were performed to ensure that the right shear rates are obtained at the location of the catheter. The shear rate varies from 0.2 s^{-1} to 2 s^{-1}. It was found that with these flow conditions, the occurrence of blood-cell aggregation corresponds to that during the *in-vivo* measurements performed at the lowest shear conditions in the right atrium (see the remark at the end of this section).

---

[6] The phase errors are significantly greater than the gain errors, and are due to the presence of a thin biolayer that forms on the electrodes and the polarization impedance. Both behave like a phase element (imperfect capacitor). Therefore, only the modulus of the impedance is used.

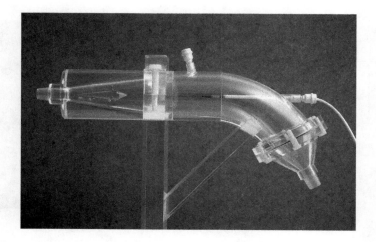

Figure 4.28   Main tube of the *in-vitro* setup. Reproduced by permission of Martil Instruments B.V

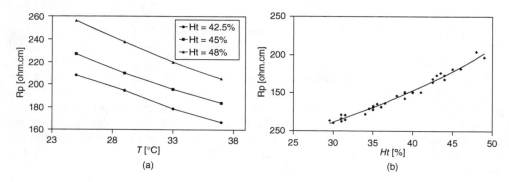

Figure 4.29   The resistance R_p measured with the *in-vitro* setup: (a) R_p versus T measured at shear rate of 1 s^{-1}; (b) Rp versus Ht and the least-squares-fitting curve, as measured at 37 °C. Reproduced by permission of Martil Instruments B.V

Figure 4.29 shows the values of the plasma resistance R_p measured with the *in-vitro* setup. It was found that R_p is a very accurate measure for the hematocrit, provided that the temperature coefficient of the specific blood-plasma resistance is taken into account.

Figure 4.30 shows the measured value of C_m versus blood viscosity. In this case, the viscosity η was measured with a Contraves LS30 viscometer using whole-blood samples and testing them at appropriate shear rates [9]. The least-square fitting curve can be expressed as:

$$C_m = \alpha \ln \frac{\eta}{\eta_0} + \beta, \tag{4.19}$$

or

$$\eta = \eta_0 e^{\frac{1}{\alpha}(C_m - \beta)}, \tag{4.20}$$

where $\alpha = 0.1544$ nF/cm, $\beta = -0.1025$ nF/cm, and $\eta_0 = 1$ mPa.s. For further experimental results, the reader is referred to [9]. The HemoCard Vision® system was successfully used in

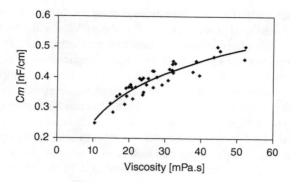

Figure 4.30 The capacitance C_m versus whole-blood viscosity and the least-squares fitting curve. Reproduced by permission of Martil Instruments B.V

a limited *in-vivo* human pilot study to monitor recipients of stem-cell transplants (HSCT) after certain therapies. In this case, the system continuously registered core-body temperature, heart rate, hematocrit and whole-blood viscosity, while using Equation (4.20) to calculate viscosity [9]. Unfortunately, because of the limitations and variations in the specific circumstances of this study, the results were too complex to discuss in this book. For further reading, the interested reader is referred to [9].

4.6   Conclusions

The physical and chemical properties of materials and structures can be derived from electrical characteristics that can be measured with impedance sensors. Because these characteristics tend to be cross-sensitive to many other parameters beyond the properties of interest, the design of such sensor systems requires careful characterization and modeling of the measurement conditions and the actual environment. In this chapter, various case studies have shown how the main problems can be understood and can often be solved using dedicated measurement techniques and design approaches. These case studies are as follows:

Case study 1 concerns sensor elements the electrical behavior of which can be characterized by a single electrical measurand, such as capacitance or resistance. Such sensor elements are applied in, for instance, mechanical sensors to measure position, displacement, speed and acceleration, or other physical sensors to measure relative humidity. The resistance or capacitance of single elements can be measured with simple circuits using square-wave excitation signals. In addition to their simplicity, such interfaces can offer the attractive features of having a low power dissipation and high accuracy. It has been shown that parasitic physical effects can cause accuracy and reliability problems. These problems can often be mitigated or eliminated by using appropriate measurement techniques, such as two-port methods, the fast-discharge technique, or active guarding.

Case study 2 concerns the detection of bacterial growth in packed food products, and specifically testing the sterility of UHF milk. The electrical properties of milk can simply be characterized by a conductance and a capacitance. Bacterial growth affects the conductance parameters. For a properly chosen frequency range, the capacitance parameter can be neglected, which simplifies the measurement problem. In the case of the non-invasive measurement of packed food products, the wall of the package will introduce a series capacitance. By using sinusoidal excitation signals and measuring the phase and magnitude, a reactive and a resistive parameter – for instance, a capacitance or a resistance – can be found in a single measurement. It has been shown that, even with simple relaxation oscillators, capacitance and conductance can be measured, while eliminating the effects of the offset and gain parameters of the interface circuit. This can be implemented by doing four measurements in four different relaxation-oscillator configurations.

Case study 3 concerns the detection of the water content in soil. In this application the liquid to be characterized has both a strong conductance component and a capacitive one, both of which depend on various physical and chemical effects. While the admittance component depends on both water content and mineral concentration, the capacitance component primarily depends on water content, while being less sensitive for mineral concentration. Therefore, the capacitance is a good indicator of water content. To measure this capacitance accurately enough in the presence of a rather low shunting resistance, sinusoidal signals with a relatively high frequency should be used, that is, beyond 20 MHz. To reduce the influence of the skin effect and the proximity effect, a small measurement probe should be used. Consequently, when the mineral concentrations are high, the detection volume is small. To cover a larger volume, a cluster of sensors can be used. A prototype interface circuit has been discussed. With this circuit, a capacitance of 33 pF can be measured with an error of only 1 pF (3%) in the presence of a shunting resistance of as low as 22 Ω.

Case study 4 concerns impedance measurement in human blood. Blood is a complex material because it is an emulsion of conductive material (plasma) and blood cells with a dielectric wall and conductive contents. Due to the aggregation (clustering) of red blood cells, blood is a non-Newtonian liquid, which means that its viscosity depends on the shear rate. With a set of measurements at a number of well-chosen frequencies, it is possible to measure capacitive and resistive components of blood from which important parameters such as blood viscosity and hematocrit (volume concentration of red blood cells) can be derived. In the case of *in-vivo* measurements, these measurements are performed in a fraction of a cardiac cycle and synchronized with the T-wave in the cardiac cycle. Experimental results of *in-vitro* measurements have been reported. A limited pilot study with *in-vivo* testing has been successfully performed.

References

[1] G.C.M. Meijer, "Interface electronics and measurement techniques for smart sensor systems," in *Smart Sensor Systems*: John Wiley & Sons, Ltd, 2008, pp. 23–54.

[2] X. Li and G.C.M. Meijer, "Capacitive sensors," in *Smart Sensor Systems*: John Wiley & Sons, Ltd, 2008, pp. 225–248.

[3] L.K. Baxter, *Capacitive Sensors*. New York: IEEE Press, 1997.

[4] S.N. Nihtianov and G.C.M. Meijer, "Non-destructive on-line sterility testing of long-shelf-life aseptically packaged food products by impedance measurements," in *IEEE AUTOTESTCON'99*, 1999, pp. 243–249.

[5] M. de Nijs, S.N. Nihtianov, M. van der Most, M.D. Northolt, and G.C.M. Meijer, "Indirect conductivity measurement for non-destructive sterility testing of UHT milk," Private Communication, 1998.

[6] M.A. Hilhorst, *Dielectric Characterisation of Soil*. PhD Wageningen, The Netherlands: IMAG-DGO Wageningen, "*Dielectric Characterisation of Soil*". PhD thesis, University Wageningen, The Netherlands, 1998.

[7] Z.-y. Chang, B.P. Iliev, J.F. de Groot, and G.C.M. Meijer, "Extending the limits of a capacitive soil-water-content measurement," *Instrumentation and Measurement, IEEE Transactions on*, vol. 56, pp. 2240–2244, 2007.

[8] G. Pop, D. Duncker, M. Gardien, P. Vranckx, S. Versluis, D. Hassan, and C.J. Slager, "The clinical significance of whole blood viscosity in (cardio)vascular medicine," *Netherlands Heart Journal*, vol. 10, p. 5, 2002.

[9] B. P. Iliev, "*In vivo* blood analysis system" PhD thesis, Delft University of Technology, Delft, The Netherlands, 2012.

[10] Agilent Technologies, *Impedance Measurement Handbook*, 2009.

[11] P. Bruschi, N. Nizza, and M. Piotto, "A current-mode, dual slope, integrated capacitance-to-pulse duration converter," *Solid-State Circuits, IEEE Journal of*, vol. 42, pp. 1884–1891, 2007.

[12] J.H.L. Lu, M. Inerowicz, J. Sanghoon, K. Jong-Kee, and J. Byunghoo, "A low-power, wide-dynamic-range semi-digital universal sensor readout circuit using pulsewidth modulation," *Sensors Journal, IEEE*, vol. 11, pp. 1134–1144, 2011.

[13] A. Flammini, D. Marioli, and A. Taroni, "A low-cost interface to high-value resistive sensors varying over a wide range," *Instrumentation and Measurement, IEEE Transactions on*, vol. 53, pp. 1052–1056, 2004.

[14] A. Depari, A.D. Marcellis, G. Ferri, and A. Flammini, "A complementary metal oxide semiconductor – integrable conditioning circuit for resistive chemical sensor management," *Measurement Science and Technology*, vol. 22, p. 124001, 2011.

[15] F.M.L. van der Goes, "*Low-cost smart sensor interfacing*," PhD thesis, Delft University of Technology, Delft, The Netherlands, 1996, p. 177.

[16] Smartec, "*Universal Transducer Interface UTI*," 2006 [On-line]. Available: http://www.smartec.nl/pdf/DSUTI.pdf.

[17] G.C.M. Meijer and X. Li, "Universal asynchronous sensor interfaces," in *Smart Sensor Systems*: John Wiley & Sons, Ltd, 2008, pp. 279–311.

[18] M.A.P. Pertijs and Z. Tan, "Energy-efficient capacitive sensor interfaces" in: A. H. M. van Roermund, A. Baschirotto, and M. Steyaert, Eds., *Nyquist AD Converters, Sensor Interfaces, and Robustness: Advances in Analog Circuit Design, 2012*, Springer, Dordrecht, The Netherlands, 2012.

[19] T. Zhichao, R. Daamen, A. Humbert, K. Souri, C. Youngcheol, Y.V. Ponomarev, and M.A.P. Pertijs, "A 1.8 V 11 μW CMOS smart humidity sensor for RFID sensing applications," in *Solid State Circuits Conference (A-SSCC), 2011 IEEE Asian*, 2011, pp. 105–108.

[20] Z. Tan, Y. Chae, R. Daamen, A. Humbert, Y.V. Ponomarev, and M.A.P. Pertijs, "A 1.2 V 8.3 nJ energy-efficient CMOS humidity sensor for RFID applications," in *Symposium on VLSI Circuits*, Honolulu, U.S., 2012, pp. 24–25.

[21] A. Heidary and G.C.M. Meijer, "Features and design constraints for an optimized SC front-end circuit for capacitive sensors with a wide dynamic range," *Solid-State Circuits, IEEE Journal of*, vol. 43, pp. 1609–1616, 2008.

[22] Z. Tan, S. Heidary, G.C.M. Meijer, and M.A.P. Pertijs, "An energy-efficient 15-bit capacitive-sensor interface based on period modulation," *Solid-State Circuits, IEEE Journal of*, vol. 47, pp. 1703–1711, 2012.

[23] M. Gasulla, L. Xiujun, and G.C.M. Meijer, "The noise performance of a high-speed capacitive-sensor interface based on a relaxation oscillator and a fast counter," *Instrumentation and Measurement, IEEE Transactions on*, vol. 54, pp. 1934–1940, 2005.

[24] A. Heidary, "*A low-cost universal integrated interface for capacitive sensors*." PhD thesis, Delft University of Technology, Delft, The Netherlands, 2010, p. 149.

[25] S. Xia, K. Makinwa, and S. Nihtianov, "A capacitance-to-digital converter for displacement sensing with 17b resolution and 20 μs conversion time," in *Solid-State Circuits Conference Digest of Technical Papers (ISSCC), 2012 IEEE International*, 2012, pp. 198–200.

[26] F. Reverter, X. Li, and G.C.M. Meijer, "Stability and accuracy of active shielding for grounded capacitive sensors," *IOP Measurement Science and Technology*, vol. 17, p. 5, 2006.

[27] S.M. Huang, A.L. Stott, R.G. Green, and M.S. Beck, "Electronic transducers for industrial measurement of low value capacitances," *Journal of Physics E: Scientific Instruments*, vol. 21, 1988.

[28] F. Reverter, X. Li, and G.C.M. Meijer, "A novel interface circuit for grounded capacitive sensors with feedforward- based active shielding," *Measurement. Science and Technology*, vol. 19, p. 025202, 2008.

[29] G.C.M. Meijer, "Smart temperature sensors and temperature-sensor systems," in *Smart Sensor Systems*: John Wiley & Sons, Ltd, 2008, pp. 185–223.

[30] G.C.M. Meijer, "Thermal detection of micro-organisms with smart temperature sensors," in *Thermal Sensors*: IOP, 1994, pp. 236–242.

[31] S.N. Nihtianov, G.P. Shterev, B. Iliev, and G.C.M. Meijer, "An interface circuit for R-C impedance sensors with a relaxation oscillator," *Instrumentation and Measurement, IEEE Transactions on*, vol. 50, p. 1563, 2001.

[32] T. Flaschke and H.-R. Tränkler, "Dielectric soil water content measurements independent of soil" in *Instrumentation and Measurement Technology Conference, 1999. IMTC/99. Conference Proceedings. IEEE*, 1999.

[33] C.R. Paul, *Introduction to Electromagnetic Compatibility*. Hoboken NJ: Wiley, 1992.

[34] B. Danker, *Fundamentals of Electromagnetic Compatibility*. Hoboken NJ: BICON Laboratories, 2004.

[35] Analog Devices, *"RF/IF Gain and Phase Detector AD 8302,"* 2002. Available: http://www.analog.com/en/rfif-components/detectors/ad8302/products/product.html.

[36] S. Chien, J. Dormandy, E. Ernst, and A. Matrai, *Clinical Hemorheology*. Dordrecht, The Netherlands: Martinus Nijhoff Publishers, 1987.

[37] G.B. Thurston, "Viscoelasticity of human blood," *Biophysical Journal*, vol. 12, p. 13, 1972.

[38] C. Gorman and A. Park, "The secret killer," *TIME*, vol. 24, February 2004.

[39] G.A. Pop, Z.-y. Chang, C.J. Slager, B.-J. Kooij, E.D. van Deel, L. Moraru, J. Quak, G.C. Meijer, and D.J. Duncker, "Catheter-based impedance measurements in the right atrium for continuously monitoring hematocrit and estimating blood viscosity changes; an in vivo feasibility study in swine," *Biosensors and Bioelectronics*, vol. 19, pp. 1685–1693, 2004.

[40] L.A. Geddes and L.E. Baker, *Principles of Applied Biomedical Instrumentation*: Wiley, 1989.

[41] J.D. Bronzino, *The Biomedical Engineering Handbook*: CRC Press, 1995.

[42] L.C. Stoner and F.M. Kregenow, "A single-cell technique for the measurement of membrane potential, membrane conductance, and the efflux of rapidly penetrating solutes in amphiuma erythrocytes," *Journal of General Physiology*, vol. 76, p. 23, 1980.

[43] Z.-y. Chang, G.A.M. Pop, and G.C.M. Meijer, "A comparison of two- and four-electrode techniques to characterize blood impedance for the frequency range of 100 Hz to 100 MHz," *Biomedical Engineering, IEEE Transactions on*, vol. 55, pp. 1247–1249, 2008.

5

Low-Power Vibratory Gyroscope Readout

Chinwuba Ezekwe[1] and Bernhard Boser[2]

[1]Robert Bosch, Palo Alto, California, USA

[2]Berkeley Sensor & Actuator Center, University of California, Berkeley, USA

5.1 Introduction

Inexpensive MEMS gyroscopes are enabling a wide range of automotive and consumer applications. Examples include image stabilization in cameras, game consoles, and improving vehicle handling on challenging terrain. Many of these applications impose very stringent requirements on power dissipation. For continued expansion into new applications it is imperative to reduce power consumption of present devices by an order-of-magnitude.

Gyroscopes infer angular rate from measuring the Coriolis force exerted on a vibrating or rotating mass. For typical designs and inputs, this signal is extremely small, requiring ultralow noise pickup electronic circuits. This low noise requirement directly translates into excessive power dissipation.

This chapter describes a solution that exploits mechanical signal amplification using a technique called mode-matching. The electronic circuit continuously senses the resonance frequency of the mechanical sense element and electrically tunes it to maximize the output signal. A new and robust feedback controller is used to accurately control the scaling factor and bandwidth of the gyroscope while at the same time guaranteeing stability in the presence of undesired parasitic resonances.

5.2 Power-Efficient Coriolis Sensing

After a brief review of the basic operating principle of vibratory gyroscopes, this section explores opportunities for improving the power efficiency of the readout interface, identifying

Smart Sensor Systems: Emerging Technologies and Applications, First Edition.
Edited by Gerard Meijer, Michiel Pertijs and Kofi Makinwa.
© 2014 John Wiley & Sons, Ltd. Published 2014 by John Wiley & Sons, Ltd.

mode-matching as a potential means to reduce power dissipation by orders of magnitude from the levels set by traditional vibratory gyroscopes.

5.2.1 Review of Vibratory Gyroscopes

Figure 5.1 illustrates the basic operating principles of vibratory gyroscopes. A proof mass suspended by springs to a frame is maintained in a steady-state oscillatory motion along the drive axis. Rotation of the frame in the plane formed by the drive and sense axes produces, along the sense axis, a Coriolis acceleration that is proportional to the product of the drive velocity and the angular rate. If we express the drive oscillation as $x_d = x_{d0} \cos(\omega_d t)$ where x_{d0} and ω_d are respectively the amplitude and angular frequency of the drive oscillation, then the Coriolis acceleration due to an angular rate Ω is

$$a_c = 2\,\Omega\,\dot{x}_d$$
$$= \underbrace{-2\,\Omega\,\omega_d\,x_{d0}\sin(\omega_d t)}_{a_{c0}} \tag{5.1}$$

where a_{c0} is the amplitude of the oscillatory Coriolis acceleration. The angular rate Ω is inferred from measuring the Coriolis acceleration a_c.

To first order, each axis of a vibratory gyroscope is a second-order system. Vacuum packaging results in highly under-damped resonance modes. The resonance mode along the drive axis is referred to as the drive mode and the one along the sense axis is referred to as the sense mode. The two resonance frequencies are usually mismatched intentionally. Figure 5.2 shows a schematic illustration of the frequency response along the drive and sense axes. The drive oscillation normally occurs at the drive resonance frequency to benefit from the amplification by the quality factor of the drive mode. Consequently, the Coriolis acceleration is also centered at the drive resonance.

5.2.2 Electronic Interface

Figure 5.3 shows a simple generalized model of a gyroscope with the electronic interface necessary to produce the final output. An oscillator establishes the above mentioned drive oscillation at the drive resonance frequency, and the Coriolis readout interface detects and

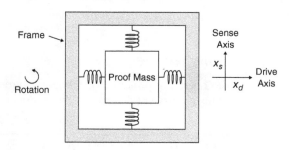

Figure 5.1   Vibratory gyroscope operating principle

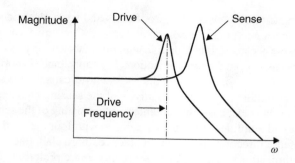

Figure 5.2   Vibratory gyroscope frequency characteristics

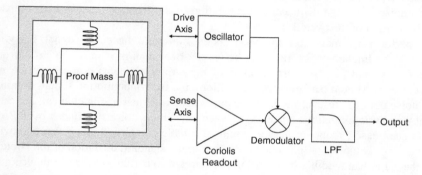

Figure 5.3   Simplified model of gyroscope with the necessary electronic interface

amplifies the Coriolis acceleration. A demodulator demodulates the angular rate signal from the Coriolis acceleration, and a low-pass filter removes, from the final output, artifacts of the demodulation and other unwanted signals outside the desired frequency band.

The high quality factors achievable with vacuum packaging greatly relax the oscillator power requirements. The demodulator and low-pass filter contribute marginally to the overall interface power dissipation since they handle already amplified signals and thus are not noise limited. This leaves the readout interface which detects Coriolis accelerations with extremely high precision as the dominant source of power dissipation. Many applications require a digital output resulting in additional power dissipation in the analog-to-digital conversion. The readout interface, therefore, holds the key to substantial reductions in the overall electronic interface power dissipation.

5.2.3   Readout Interface

The readout interface senses the Coriolis acceleration indirectly by detecting the motion the Coriolis acceleration induces on the proof mass along the sense axis. The induced motion is oscillatory and is of the form $x_s = x_{s0} \sin(\omega_d t + \phi_s)$ where x_{s0} is the amplitude of the motion and ϕ_s is the phase lag of the sense axis response at the drive frequency. The motion is detected by measuring the capacitance between the proof mass and fixed electrodes. The capacitance varies with, and is thus a good indicator of, displacement. Capacitive sensing is attractive for low cost inertial sensors because it is compatible with most fabrication processes and the capacitive interface can be easily used for force actuation. The capacitance is normally implemented with transverse comb fingers for maximum displacement-to-capacitance sensitivity which is advantageous for maximizing the overall sensitivity of the sensing element.

Figure 5.4 illustrates the most basic readout interface consisting of the sensing element and a position sense front-end amplifier. A simple suspension with four beams that are compliant in both the drive and sense directions is shown; in practice, more elaborate solutions that enable the independent optimization of the drive and sense modes are preferred [1, 2]. The front-end amplifier converts the differential capacitance between the proof mass and the fixed electrodes into voltage or current. Unfortunately, the amplifier's output will, invariably, be corrupted by electronic noise which we model by an equivalent source in series with the input of the amplifier. This noise limits the displacement resolution of the front-end and directly impacts the power dissipation of the overall interface.

A key performance metric for any sensor is the minimum detectable signal. For vibratory gyroscopes, the fundamental limit is set by (1) the Brownian motion of the gas surrounding the proof mass and (2) the thermal noise of the circuit elements comprising the readout interface. For maximum performance (resolution), the Brownian motion should dominate the system noise floor to preserve the intrinsic performance of the sensing element, but often and especially in systems that are operated in vacuum, circuit noise dominates over Brownian noise. In such cases, reducing the minimum detectable signal within a given bandwidth by a factor of two requires a proportional reduction in the standard deviation of the circuit noise within the same bandwidth which, in turn, requires a four-fold increase in device currents. Improving the angular rate resolution without the associated increase in power dissipation is thus a major challenge.

Since, due to manufacturing tolerances, the drive and sense springs are imperfectly orthogonal, some of the drive oscillation leaks directly into the sense axis, resulting in large undesired sense axis oscillatory motion that is in quadrature with the desired Coriolis acceleration induced motion. The demodulation signal from the oscillator circuit normally has substantial phase noise which the so called quadrature error can mix down reciprocally,

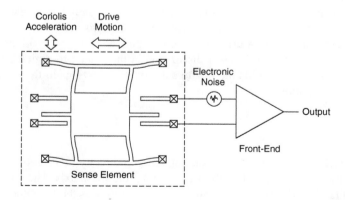

Figure 5.4   Basic Coriolis readout interface (drive details omitted)

raising the overall interface noise floor beyond that set by the front-end. Fortunately, most but not all of the quadrature error can be nulled using special quadrature nulling electrodes [3]. The residual error can be rejected during the demodulation process by using an appropriately phased demodulation signal since the error is in quadrature with the desired signal. Achieving a high degree of rejection requires a well defined phase relationship between the quadrature error and the drive oscillation from which the demodulation signal is normally derived. Ensuring that the phase relationship is well defined is the second significant challenge for the readout interface.

There is also the Coriolis offset which comes from leakage of the drive force into the sense axis due to misalignment of the drive combs. This error is minimized by vacuum packaging since the increased quality factors enable the use of smaller drive forces which result in smaller forces feeding through to the sense axis [4].

Other challenges include obtaining a wide enough signal bandwidth and ensuring that the overall gain (scale factor) is stable over fabrication tolerances and ambient variations. A wide bandwidth is necessary especially in control applications such as vehicle stability control where sensors with minimum phase lag are required.

5.2.4   Improving Readout Interface Power Efficiency

In a vibratory gyroscope, rotation is converted to Coriolis acceleration that is detected by measuring the consequent motion of the proof mass. A "rate grade" resolution of $0.1°/s/\sqrt{Hz}$ translates into a displacement resolution on the order of $100 \; fm/\sqrt{Hz}$ in typical gyroscope designs. Current state-of-the art interfaces resolve $60 \; fm/\sqrt{Hz}$ while dissipating 30 mW [2]. Applications such as image stabilization in cameras and vehicle stability control require an order of magnitude better angular rate resolution for similar or lower power dissipation. Unfortunately, power dissipation in a noise limited readout interface is, to first order, inversely proportional to the square of the displacement resolution. Thus, while $0.01°/s/\sqrt{Hz}$ can be achieved through traditional means by simply resolving $10 \; fm/\sqrt{Hz}$, the 1 W of power required makes such a noise floor impractical in the target applications. Essentially, the readout interface power efficiency must be improved to enable the use of high-resolution angular rate sensors in power constrained applications. Increasing the signal-to-noise ratio (SNR) passively requires increasing the sensing element's angular rate-to-sense motion sensitivity ($\Delta x_{s0}/\Delta\Omega$) so that the same angular rate produces a larger sense motion amplitude.

The angular rate-to-sense motion sensitivity can be expressed as the product of two factors:

$$\frac{\Delta x_{s0}}{\Delta\Omega} = \left(\frac{\Delta a_{c0}}{\Delta\Omega}\right)\left(\frac{\Delta x_{s0}}{\Delta a_{c0}}\right). \tag{5.2}$$

The first factor is the angular rate-to-Coriolis acceleration sensitivity which indicates the amplitude of the Coriolis acceleration produced by a given angular rate. This factor is normally maximized by a large drive oscillation amplitude. The second factor is the Coriolis acceleration-to-sense motion sensitivity which indicates the sense motion amplitude resulting from a given Coriolis acceleration amplitude. The drive and sense resonance frequencies are normally mismatched either by design or due to fabrication tolerances and ambient variations. However, if they were perfectly matched, the Coriolis acceleration would be centered at the sense mode frequency, and because of the consequent amplification by the sense mode quality factor, the same Coriolis acceleration, and consequently the same angular rate, would produce a much larger sense motion [5]. Continuing with the previous example and assuming a ten fold increase in Coriolis acceleration-to-sense motion sensitivity, an angular rate resolution of $0.01°/s/\sqrt{Hz}$ would require a displacement resolution on the order of $100 \text{ fm}/\sqrt{Hz}$ rather than the more stringent $10 \text{ fm}/\sqrt{Hz}$, which translates into two orders of magnitude power dissipation reduction over the original example.

Based on the foregoing discussion, we propose to exploit the free mechanical amplification provided by the sense resonance to greatly relax the noise requirements, and therefore substantially reduce the power dissipation, of the readout interface.

5.2.5   Exploiting the Sense Resonance

Matching the drive and sense modes, or so-called mode-matching, increases sense displacements by the sense mode quality factor and thereby relaxes the noise requirements of the front-end, but also brings several problems, chief among which are an extremely narrow sense bandwidth due to the high quality factor, and increased gain variation and phase uncertainty due to fabrication tolerances and ambient variations. The bandwidth is given by

$$f_{BW} = \frac{f_s}{2Q_s} \qquad (5.3)$$

where f_s and Q_s are the frequency and quality factor of the sense resonance. With mode-matching, the frequency of the sense resonance is equal to that of the drive. The drive frequency and sense mode quality factor, typically on the order of 15 kHz and 1000 respectively, result in bandwidths on the order of 7.5 Hz which is in stark contrast to the 50 Hz required by automotive and consumer applications. The 7.5 Hz 3-dB bandwidth is moreover poorly controlled due to the normally substantial variation of the quality factor with the ambient. The variation of quality factor also results in gain variation. Figure 5.5 illustrates this problem. Also, the invariably limited accuracy and bandwidth of any practical mode-matching scheme will result in a small residual frequency mismatch. Especially considering the process and ambient variations of the residual mismatch, the very abrupt phase change near the sense resonance results in substantial phase uncertainty which exacerbates the task of rejecting quadrature error. Figure 5.6 illustrates this problem. Due to these difficulties, many gyroscope implementations avoid mode-matching and instead operate away from the sense resonance, obtaining a larger bandwidth and better defined gain and phase at the expense of sensitivity [4, 6]. A practical readout interface exploiting the sense resonance must overcome the problems arising from mode-matching in a way that neither interferes with gyroscope performance nor negates the power advantage derived from mode-matching.

Feedback is widely used in electronics to obtain precise characteristics, for example precise gains, from imprecise elements. It has been used in sensors to improve bandwidth, dynamic range, linearity, and drift [7, 8]. Especially in high-Q vibratory gyroscopes with matched modes, feedback is imperative to ensure proper operation. Figure 5.7 shows the sensing element enclosed in a force feedback loop. A compensator and a force transducer are added to the basic *open-loop* interface to form a *closed-loop* interface. Based on the motion sensed by the front-end, the compensator produces an estimate of the Coriolis force which the force transducer applies with opposite polarity on the proof mass to null the sense motion. Perfect nulling

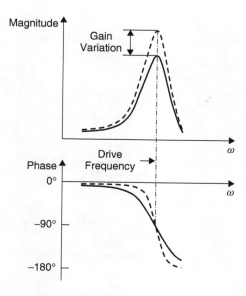

Figure 5.5 Gain (scale factor) variation with quality factor

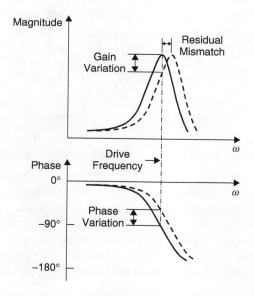

Figure 5.6 Gain and phase variation with residual mismatch

of the proof mass motion implies that the feedback force is exactly equal and opposite to the Coriolis force. While this is impossible to achieve over all frequencies, in practice, adequate nulling is possible within a limited frequency band where the force feedback open-loop gain is sufficiently high. Within that frequency band, the output of the closed-loop interface is an accurate representation of the Coriolis acceleration. Figure 5.8 compares the frequency responses

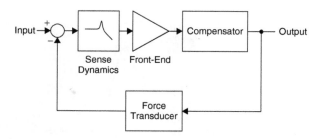

Figure 5.7   Basic force feedback loop

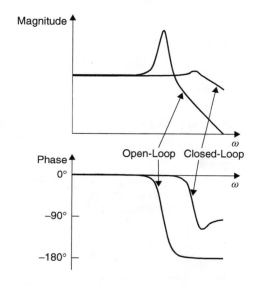

Figure 5.8   Illustrative example of sensor and closed-loop frequency responses

of the open-loop sensor and that of a closed-loop interface that has a high open-loop gain over a frequency range that extends beyond the resonance of the sensing element. Electronic circuits implementing the compensator provide the necessary open-loop gain. Regardless of the variations of the sensor parameters, the closed-loop response remains flat and stable over a much wider frequency range. Thus, the traditional tradeoff of mechanical sensitivity for larger bandwidth and better defined gain and phase is unnecessary.

5.3   Mode Matching

Maximizing the signal-to-noise ratio (SNR) improvement afforded by mode-matching requires the frequency matching error to be less than the reciprocal of the sense mode quality factor. For example, a sense mode quality factor of 1000 requires less than 0.1% matching error. Process tolerances and ambient variations limit the minimum matching error achievable with low-cost manufacturing to about 2% [4], mandating resonance frequency calibration.

One way to perform the calibration is to fully characterize the dependence of frequency matching on physical parameters such as temperature and then use the data to calibrate the sense resonance frequency at runtime. The high cost of fully characterizing the sensing element at the factory puts this technique at odds with the cost constraints of MEMS gyroscope applications. The alternative approach is to continuously monitor sensor properties that vary with frequency matching. Previously proposed calibration schemes of this type determine frequency matching by monitoring sensor properties such as gain and phase lag [1, 9]. Unfortunately, those properties are not easily measurable when the sensing element is part of a force feedback loop. Since force feedback is imperative to ensure proper operation of a mode matched gyroscope, we need to develop a way to measure the relevant sensor properties in a way that is compatible with closed-loop sensing.

5.3.1 Estimating the Mismatch

Figure 5.9 models the dynamics of the sense axis as a lumped spring-mass-damper system. The system from the force input to the displacement output has the transfer function

$$H_s(s) = \frac{1}{m_s \, s^2 + b_s \, s + k_s} \tag{5.4}$$

where m_s and b_s are respectively the mass and damping factor, and k_s is the variable stiffness which we aim to observe and ultimately control. The task of mode-matching is to force k_s to approach the optimal stiffness $k_{s,opt} = m_s \omega_d^2$ at which the sense resonance frequency equals the drive frequency. This requires the monitoring of the deviation of the actual stiffness from the optimal value.

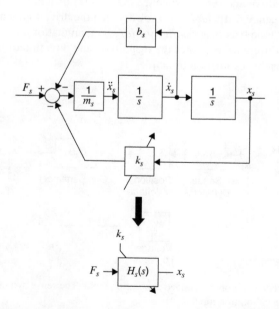

Figure 5.9 Second-order sense dynamics with variable stiffness

The feedback path of a stable closed-loop system determines the closed-loop response provided that the open-loop gain is much greater than unity. We exploit this property to isolate the characteristics of the sensing element from the rest of the feedback loop by choosing a calibration input that places only the sense dynamics in the feedback path. Figure 5.10 shows the force feedback loop with the added calibration input. We have replaced the front-end and the force transducer by position-to-voltage gain K_{x-v} and voltage-to-force gain K_{v-f} respectively. The transfer function from the calibration input to the output assuming a high open-loop gain is

$$G_{cal}(s) \approx \frac{1}{K_{v-f}\, K_{x-v}\, H_s(s)} \propto \frac{1}{H_s(s)}$$

$$\approx \frac{1}{K_{v-f}\, K_{x-v}}(m_s\, s^2 + b_s\, s + k_s). \qquad (5.5)$$

The gain terms K_{v-f} and K_{x-v} affect only the static gain of $G_{cal}(s)$ and not the location of the complex zeros. Figure 5.11 compares the frequency responses of $H_s(s)$ and $G_{cal}(s)$. The *notch* and 90° phase *lead* of $G_{cal}(s)$ exactly mirror the *peak* and 90° phase *lag* of $H_s(s)$ at resonance making $G_{cal}(s)$ an excellent, albeit inverse, proxy for $H_s(s)$. In a sense, $G_{cal}(s)$ is preferable over $H_s(s)$ because it avoids the high-Q poles in $H_s(s)$ that severely limit the tracking bandwidth of conventional open-loop sensing based frequency calibration techniques [9].

One possible way to use the calibration input to estimate the frequency mismatch is to monitor the phase shift from the calibration input to the output using a pilot tone at the drive frequency. Unfortunately, this approach is problematic because the tone would invariably interfere with the Coriolis signal. We overcome this problem by using *two* pilot tones that are referenced to the drive frequency and located outside the desired signal band with one tone above and the other below the desired signal band. We adjust the tones to *equalize* their output amplitudes when the drive and sense resonance frequencies match. If after the adjustment the sense resonance frequency drifts *higher* (or *lower*) than the drive frequency, the amplitude of the higher frequency tone becomes *smaller* (or *larger*) than that of the lower frequency tone. Thus, the amplitude difference indicates the magnitude and direction of frequency mismatch. Figure 5.12 illustrates this estimation principle.

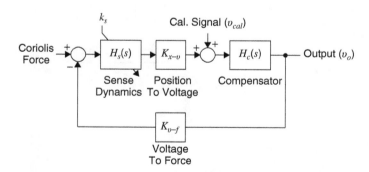

Figure 5.10   Force feedback loop with added calibration input. The sense dynamics are in the feedback path with respect to the calibration input

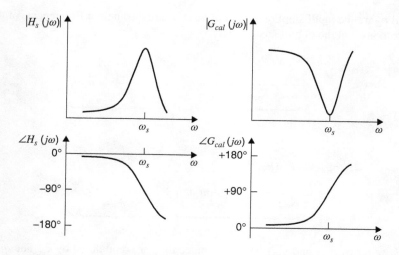

Figure 5.11   Comparison of the frequency responses of $H_s(s)$ and $G_{cal}(s)$

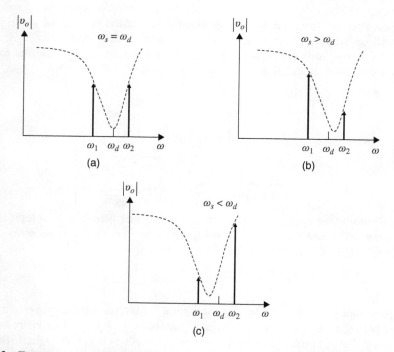

Figure 5.12   Frequency mismatch estimation principle. The dashed lines indicate $|G_{cal}(j\omega)|$. (a) The amplitudes match when the drive and sense frequencies match. The higher frequency tone becomes (b) smaller when the sense resonance frequency drifts higher, and (c) larger when the sense resonance frequency drifts lower

If v_1 and v_2 are the input amplitudes and ω_1 and ω_2 are the angular frequencies of the tones, then the responses at the output are

$$v_{o1} = G_{cal}(j\omega_1)v_1$$

$$= \underbrace{\frac{v_1}{K_{v-f}\,K_{x-v}}\left(k_s - m_s\omega_1^2\right)}_{v_{o1,I}} + \underbrace{j\frac{v_1}{K_{v-f}\,K_{x-v}}\,b_s\omega_1}_{v_{o1,Q}} \tag{5.6}$$

and

$$v_{o2} = G_{cal}(j\omega_2)v_2$$

$$= \underbrace{\frac{v_2}{K_{v-f}\,K_{x-v}}\left(k_s - m_s\omega_2^2\right)}_{v_{o2,I}} + \underbrace{j\frac{v_2}{K_{v-f}\,K_{x-v}}\,b_s\omega_2}_{v_{o2,Q}}. \tag{5.7}$$

The *in-phase* terms $v_{o1,I}$ and $v_{o2,I}$ are useful since they are modulated by k_s. The *quadrature* terms $v_{o1,Q}$ and $v_{o2,Q}$ are useless and are rejected by synchronously demodulating the in-phase terms. Their rejection makes this approach insensitive to damping factor variations to first order. Another welcome feature of synchronous demodulation is that it preserves the signs of the in-phase terms. This, combined with the phase inversion beyond the sense resonance frequency, allows amplitude differencing to be realized by simply summing $v_{o1,I}$ and $v_{o2,I}$.

Figure 5.13 shows the very simple realization of the estimator. Signals similar to the pilot tones are used in the demodulation. The final error signal is

$$v_{err} = v_{o1,I} + v_{o2,I}$$

$$= \underbrace{\frac{v_1 + v_2}{K_{v-f}\,K_{x-v}}}_{\text{estimator gain}}\left[k_s - m_s\underbrace{\left(\frac{\omega_1^2}{1 + \frac{v_2}{v_1}} + \frac{\omega_2^2}{1 + \frac{v_1}{v_2}}\right)}_{\text{reference stiffness}}\right]. \tag{5.8}$$

We adjust the pilot tone parameters as previously mentioned such that the reference stiffness is equal to the optimal stiffness $k_{s,opt}$. If we fix the tone frequencies to $\omega_1 = \omega_d - \omega_{cal}$ and $\omega_2 = \omega_d + \omega_{cal}$, then the amplitudes must satisfy

$$\frac{v_2}{v_1} = \frac{2\omega_d - \omega_{cal}}{2\omega_d + \omega_{cal}}. \tag{5.9}$$

The unequal amplitudes account for the logarithmic nature of frequency behavior and the asymmetry between the low and high frequency responses of $G_{cal}(s)$. The constraint results in an error signal that is exactly proportional to the difference between the actual stiffness and the optimal stiffness, i.e. $v_{err} = K_e(k_s - k_{s,opt})$ where K_e is the estimator gain.

5.3.2 Tuning Out the Mismatch

Figure 5.14 shows a simplified model of a balanced transverse comb electrostatic actuator. The proof mass is grounded and the fixed electrodes are biased at V_{tune}. This actuator configuration

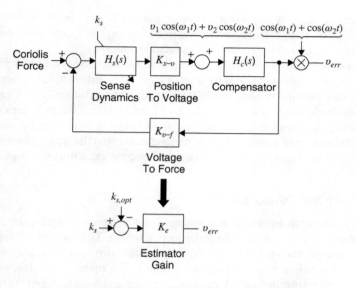

Figure 5.13 Estimator realization and linearized model

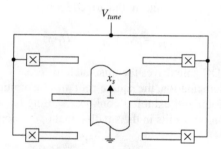

Figure 5.14 Voltage tunable spring implemented by a balanced transverse comb electrostatic actuator

implements a voltage tunable spring in the transverse direction with stiffness [1]

$$k_e = -\frac{C_{tune}}{x_g^2} V_{tune}^2 \tag{5.10}$$

where x_g and C_{tune} are respectively the gap and net capacitance between the proof mass and the fixed electrodes in the transverse direction when the proof mass is undeflected. The voltage tunable spring combines with other springs suspending the sense axis to yield the net stiffness

$$k_s = k_m + k_e$$

$$= k_m - \frac{C_{tune}}{x_g^2} V_{tune}^2 \tag{5.11}$$

where k_m is the combined stiffness of the other springs consisting mainly of flexures and parasitic springs from *electrostatic* force feedback and quadrature nulling. Since the tunable spring only softens k_s, it is important to design the flextures to be stiffer than the optimal stiffness by sufficient margin to accommodate force feedback and quadrature nulling induced spring softening in addition to process and ambient variations.

The position sense electrodes are normally realized by transverse combs and thus can double as tuning combs, eliminating the need for a set of electrodes dedicated to stiffness tuning only. Time multiplexing position sensing and stiffness tuning at a sufficiently high rate is one way to share the electrodes. In this case, the effective electrostatic stiffness is scaled by the duty factor of the stiffness tuning phase. From the point of view of minimizing power dissipation, however, a dedicated set of tuning combs is preferable to avoid the typically substantial power penalty associated with charging and discharging the sense capacitors at a high rate.

5.3.3   Closing the Tuning Loop

The frequency mismatch estimator and voltage tunable spring comprise the necessary elements to implement automatic resonance frequency tuning. The only remaining element is a controller to close the tuning loop. The controller should drive the frequency mismatch estimate to zero and remain stable at all operating points. The square dependence of the tunable stiffness on voltage results in signal dependent loop gain and must be taken into account in the controller design. Figure 5.15 shows one way to implement the tuning loop. The loop includes an explicit square-root function to counter the square dependence of stiffness on voltage. The open-loop transfer function of the resulting linearized loop is

$$G_{tune} = K_e \, V_{ref} \frac{C_{tune}}{x_g^2} H_f(s). \tag{5.12}$$

A loop filter with infinite DC gain drives the mismatch to zero.

To simplify system implementation, the square-root may be omitted as shown in Figure 5.16 with the penalty that the loop will exhibit a nonlinear settling behavior. The uncompensated square nonlinearity in the loop results in the small-signal open-loop transfer function

$$G_{tune} = 2 \, K_e \, V_{tune} \frac{C_{tune}}{x_g^2} H_f(s) \tag{5.13}$$

which depends on the bias point V_{tune}.

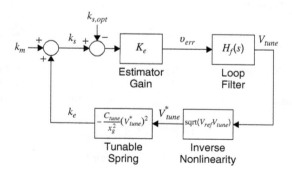

Figure 5.15   Tuning loop with nonlinearity compensation

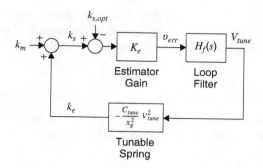

Figure 5.16   Tuning loop without nonlinearity compensation

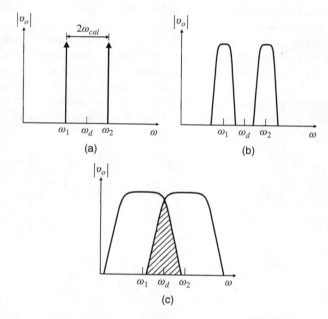

Figure 5.17   Possible spectra of pilot tones modulated by frequency mismatch. (a) Static mismatch is fully recoverable. (b) Dynamic mismatch with variation bandwidth less than ω_{cal} is fully recoverable. (c) Dynamic mismatch with variation bandwidth greater than ω_{cal} resulting in overlapping spectral components is not fully recoverable

A fundamental property of the proposed estimator is that the mismatch information is modulated onto carriers at $\omega_d \pm \omega_{cal}$. If the mismatch is not constant but time varying, then the modulated signals occupy a non-zero bandwidth. Full recovery of the modulated information is possible provided that their spectral components do not alias. Figure 5.17 illustrates the various possible cases. A unity gain or gain crossover frequency of much less than ω_{cal} allows the loop filter, which can be a simple integrator, enough margin to provide adequate anti-aliasing filtering at ω_{cal} to prevent aliasing and ensure proper loop operation. A higher-order loop filter could provide better attenuation, but the consequent increased phase lag would limit the potential tracking bandwidth improvement. A tracking bandwidth of about 25 Hz or settling time constant of about 6 ms is possible for $\omega_{cal} = 2\pi \times 250$ Hz.

5.3.4 Practical Considerations

We now turn to issues concerning practical signal synthesis, demodulation, and filtering, and the effects of finite force feedback open-loop gain and potential interference from large inertial forces.

5.3.4.1 Practical Signal Synthesis, Demodulation, and Filtering

As shown in Figure 5.18, an offset before the loop filter is indistinguishable from the actual error signal and is consequently a source of systematic frequency offset. Digital implementation of calibration signal synthesis, demodulation, and loop filtering in the experimental prototype avoids the substantial offsets that are possible in analog implementations.

Even with digital implementation, generating a calibration signal with the precise amplitude ratio given by (5.9) is inconvenient. Using tones with equal amplitudes is far more convenient and allows the calibration signal to be used directly for demodulation, leading to simpler system implementation as shown in Figure 5.19. With $v_1 = v_2$ and the tone frequencies defined previously, (5.8) becomes

$$v_{err} \propto k_s - m_s(\omega_d^2 + \omega_{cal}^2) \approx k_s - m_s\omega_d^2 \left(1 + \underbrace{\frac{1}{2}\frac{\omega_{cal}^2}{\omega_d^2}}_{\text{offset}}\right)^2 \tag{5.14}$$

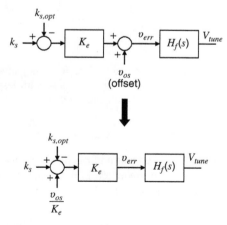

Figure 5.18 Problem of calibration signal, demodulator, and loop filter offset. The lumped offset before the loop filter appears as an equivalent stiffness offset

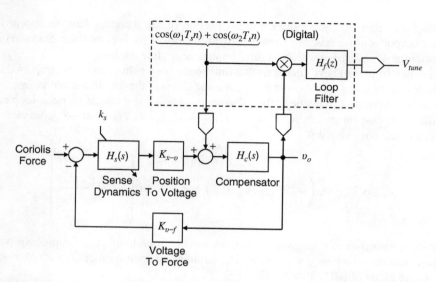

Figure 5.19 Practical estimator with digitally synthesized equal amplitude tones and digitally implemented demodulator and loop filter

where the approximation assumes that $\omega_{cal} \ll \omega_d$ and we have omitted the gain factor for simplicity. Basically, using equal amplitudes introduces a frequency offset that forces the sense resonance frequency to be slightly higher than the drive frequency. Fortunately, the error is negligible relative to signal bandwidth if the tones are located just outside the desired signal band as the following example illustrates. A bandwidth of 50 Hz is typical in consumer and automotive applications. Choosing $\omega_{cal} = 2\pi \times 250$ Hz places the pilot tones well outside the desired signal band and, with a drive frequency of 15 kHz, results in an offset of 0.013% or 2 Hz.

5.3.4.2 Finite Force Feedback Open-Loop Gain

Since an arbitrarily high open-loop gain is difficult to attain in practice, it is useful to quantify the impact of finite open-loop gain on estimator performance. With finite open-loop gain, the transfer function from the calibration input to the output becomes

$$G_{cal}(s) = \left(1 - \frac{1}{1 + T(s)}\right) \frac{1}{K_{v-f} \, K_{x-v} \, H_s(s)} \tag{5.15}$$

where $T(s) = K_{v-f} \, K_{x-v} \, H_s(s) H_c(s)$ is the open loop transfer function, and the in-phase output components become

$$v_{o1,I} = \left(1 - \frac{\Re\left\{1 + T(j\omega_1)\right\}}{|1 + T(j\omega_1)|^2}\right) \frac{v_1}{K_{v-f} \, K_{x-v}} (k_s - m_s\omega_1^2) \tag{5.16}$$

and

$$v_{o2,I} = \left(1 - \frac{\Re\left\{1 + T(j\omega_2)\right\}}{|1 + T(j\omega_2)|^2}\right) \frac{v_2}{K_{v-f} \, K_{x-v}} (k_s - m_s\omega_2^2). \tag{5.17}$$

The imaginary part of $T(j\omega)$ causes a small portion of the damping term to appear in the in-phase output components. We have neglected this effect for brevity since mode-matching implies a high-Q resonance which, in turn, implies negligible damping.

It is evident from the above equations that finite open-loop gain introduces errors in the tone amplitudes. Only a negligible estimator gain error arises if the amplitude errors in $v_{o1,\mathrm{I}}$ and $v_{o2,\mathrm{I}}$ match, otherwise a frequency offset also arises. Assuming a minimum open-loop gain of $T_{\min}$, the worst case mismatch occurs when $T(j\omega_1) = T_{\min}$ and $T(j\omega_2) = -T_{\min}$ or vice versa. In this case, the error signal is

$$v_{err} \propto k_s - m_s \left(\omega_d^2 + \frac{2}{T_{\min}} \omega_d \omega_{cal} \right) \approx k_s - m_s \omega_d^2 \left(1 + \underbrace{\frac{1}{T_{\min}} \frac{\omega_{cal}}{\omega_d}}_{\text{offset}} \right)^2. \tag{5.18}$$

Fortunately, the offset is negligible for any reasonable open-loop gain. Continuing with the previous example where $\omega_{cal} = 2\pi \times 250$ Hz, a minimum open-loop gain of 40 dB results in a worst case offset of 0.017% or 2.5 Hz.

5.3.4.3   Interference from Large Inertial Forces

Since there is no filter to limit the bandwidth of Coriolis and other inertial forces that appear on the sense axis, spectral components of those forces around $\omega_d \pm \omega_{cal}$ can interfere with the pilot tones and produce signal dependent frequency offset. The use of a tuning fork structure largely rejects the linear acceleration component leaving the Coriolis acceleration component. In the following analysis, we quantify the worst case error that the Coriolis acceleration component can contribute.

The Coriolis acceleration can be expressed as

$$a_c = 2\,\Omega\,\dot{x}_d + \dot{\Omega}\,x_d. \tag{5.19}$$

The $\dot{\Omega}\,x_d$ term captures an often neglected higher order effect that is important in the following analysis. The worst case interference occurs when the angular rate is sinusoidally varying at ω_{cal} in which case the angular rate can be expressed as

$$\Omega = \Omega_0\,\cos(\omega_{cal}t + \phi_\Omega). \tag{5.20}$$

where Ω_0 is the amplitude and ϕ_Ω is the phase which can assume any value. If the drive axis oscillates according to $x_d = x_{d0}\,\cos(\omega_d t)$, then the Coriolis acceleration resulting from the angular rate is

$$a_c = \underbrace{-2\,\Omega_0\,\omega_d\,x_{d0}\,\cos(\omega_{cal}t + \phi_\Omega)\,\sin(\omega_d t)}_{2\,\Omega\,\dot{x}_d} - \underbrace{\Omega_0\,\omega_{cal}\,x_{d0}\,\sin(\omega_{cal}t + \phi_\Omega)\,\cos(\omega_d t)}_{\dot{\Omega}\,x_d}. \tag{5.21}$$

The acceleration appears at the output of the force feedback loop scaled by m_s/K_{v-f} (see Figure 5.10). The $2\,\Omega\,\dot{x}_d$ term dominates by far since it is multiplied by ω_d while the $\dot{\Omega}\,x_d$ term is multiplied by the much smaller ω_{cal}. It is therefore important to generate the calibration and demodulation signals by using a sinusoid at ω_{cal} to modulate the amplitude of a carrier that is *in phase* with the drive displacement (and thus *in quadrature* with the drive velocity) to enable the rejection of the dominant $2\,\Omega\,\dot{x}_d$ term. After demodulation, the $\Omega\,x_d$ term remains and the error signal becomes

$$v_{err} = K_e(k_s - m_s\omega_d^2) - \underbrace{\frac{m_s}{K_{v-f}}\Omega_0\,\omega_{cal}\,x_{d0}}_{\text{worst case}}. \tag{5.22}$$

The consequent offset error is minimized by maximizing the estimator gain K_e which requires the use of large amplitude pilot tones. The amplitudes can not be arbitrarily large, however, since the resulting output signals must live within the force feedback loop's limited output range. As we have already seen in Figure 5.12, the output amplitudes of the tones vary substantially with frequency mismatch. Since the amplitudes at worst case frequency mismatch can be substantially higher than with perfect matching, the amplitudes should be small enough to avoid overloading the output during startup when the system has the worst case frequency mismatch. As frequency matching improves, the amplitudes, and consequently the estimator gain, may be increased to minimize the impact of Coriolis interference. It is important to reciprocally lower some other gain factor while increasing estimator gain to maintain an optimal tracking bandwidth. If the pilot tones are maximized, then for full-scale Coriolis acceleration sinusoidally varying at the worst frequency and with the worst phase, the resulting fractional matching error is on the order of $\omega_{cal}^2/\omega_d^2$, which is similar to the magnitude of the error coming from both finite force feedback open-loop gain and the use of equal amplitude tones.

5.4 Force Feedback

5.4.1 *Mode-Matching Consideration*

Figure 5.20 shows how to drive the proof mass and position sense electrodes during the force feedback phase to realize differential actuation. The proof mass is grounded, and the top and bottom electrodes are biased at V_{bias} and driven differential by the feedback voltage v_{fb}. In another approach, the top and bottom electrodes are biased at V_{bias} and $-V_{bias}$ respectively and the proof mass is driven by v_{fb}. Both approaches produce similar results. However, the first approach is preferable since it requires only one bias voltage. In any case, the feedback

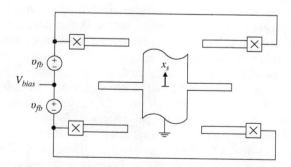

Figure 5.20 Schematic diagram of the sense combs doubling as differential actuator

force applied on the proof mass for displacements that are small relative to the gap is [1]

$$F_{fb} = \underbrace{2\frac{C_{s0}}{x_g}V_{bias}\upsilon_{fb}}_{K_{\upsilon-f}} + \underbrace{2\frac{C_{s0}}{x_g^2}\left(V_{bias}^2 + \upsilon_{fb}^2\right)x_s}_{\substack{\text{signal dependent} \\ \text{stiffness}}} \tag{5.23}$$

where x_g is the nominal gap and C_{s0} is the nominal sense capacitance between the proof mass and each pair of connected electrodes. In addition to the desired voltage controlled force with a voltage-to-force gain of $K_{\upsilon-f}$, the transducer produces an unwanted stiffness term that also depends on the feedback voltage. During normal operation, the tuning loop forces the pilot tones used for resonance frequency calibration to have equal amplitudes at the output of the force feedback loop. Neglecting other signals that may be present and assuming that proportional feedback is used, the output signal will be of the form $\cos[(\omega_d - \omega_{cal})t] - \cos[(\omega_d + \omega_{cal})t] = 2\sin(\omega_{cal}t)\sin(\omega_d t)$. The feedback voltage is derived from the output and thus can be expressed as $\upsilon_{fb} = |\upsilon_{fb}|\sin(\omega_{cal}t)\sin(\omega_d t)$. The square of this voltage modulates the signal dependent stiffness term resulting in spectral components at DC in addition to $2\omega_{cal}$, $2\omega_d$, and $2\omega_d \pm 2\omega_{cal}$. While the DC component will be removed by the tuning loop, the AC components, all of which are beyond the tracking bandwidth of the tuning loop, will remain. Additional signals in the output, the Coriolis force for example, exacerbate the problem. Left unaddressed, the parasitic tuning of stiffness by the feedback voltage would result in about 1% dynamic variation of the sense resonance in the experimental prototype.

Bang-bang control, a feedback control strategy in which the feedback voltage is restricted to just two levels, say $\pm V_{bias}$, overcomes this problem. With bang-bang control, the feedback voltage toggles between V_{bias} and $-V_{bias}$ in such a way that its time-average approximates the feedback voltage under proportional control. The technique converts the dynamic frequency variation into a static error since, regardless of the spectral content of the feedback voltage, the square is constant, V_{bias}^2. The resulting static error is removed by the tuning loop.

5.4.2 Preliminary System Architecture and Model for Stability Analysis

Figure 5.21 summarizes the preliminary system architecture. To realize the above mentioned bang-bang control, a single-bit quantizer placed after the compensator restricts the feedback voltage to just two levels resulting in an architecture akin to a $\Sigma\Delta$ modulator with the noise shaping realized by the sensing element complemented by the compensator [10–15]. The inherent analog-to-digital conversion obviates the need for a dedicated high-resolution A/D following the output, and the single-bit output facilitates implementation of the mode-matching algorithm by reducing the demodulator to a simple multiplexer that either keeps or inverts the sign of the demodulation signal. The boxcar filter captures the behavior of the power-efficient position-sensing front-end proposed in [16]. The impulse response of the feedback DAC accounts for the time-multiplexing of force feedback onto the position sense electrodes. The delay to the beginning of the pulse accounts for the time it takes the compensator and quantizer to process the position signal and produce the next output. As required by the mode-matching algorithm, a DAC for injecting the calibration signal is included.

A major design goal is ensuring that the system is stable in the sense that the digital output is free of large limit cycles and is a faithful representation of the Coriolis force. The presence

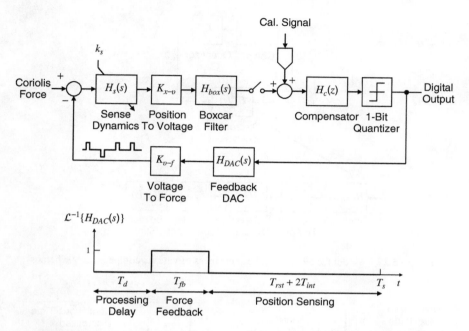

Figure 5.21 Preliminary system architecture. The time $T_{rst} + 2T_{int}$ accounts for the reset, error integration, and error and signal integration phases of position sensing

of the quantizer results in a complex system behavior that is hard to analyze directly. The describing function model, a widely used approximation in which the nonlinear element is replaced with a signal-dependent gain and an additive noise source [17], captures sufficient detail of the nonlinear behavior under certain conditions to yield valuable insight into the nature of instability in the modulator [18]. Figure 5.22 shows the describing function model for evaluating the robustness of various compensation schemes. To further facilitate the analysis, we have also replaced the electromechanical chain with its impulse invariant discrete-time equivalent [19].

5.4.3   Accommodating Parasitic Resonances

While the linear model above does not identify sufficient conditions for stability, it has been found via simulation that the lack of phase margin in the model is a sufficient condition for instability. We use this powerful capability in the following analysis to evaluate the robustness of various compensation schemes.

Although practical gyroscopes typically have countless resonance modes across a wide range of frequencies, we consider for simplicity a hypothetical sensor with only one parasitic resonance at 300 kHz in addition to a main resonance at 15 kHz. Figure 5.23 shows the frequency response of the sensor along with that of the discretized electromechanical chain for a sampling rate of 480 kHz. Since collocated control ensures the presence of a phase restoring anti-resonance between successive resonances, the continuous-time phase response does not

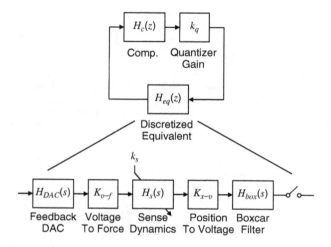

Figure 5.22 Model for evaluating the robustness of various compensation schemes

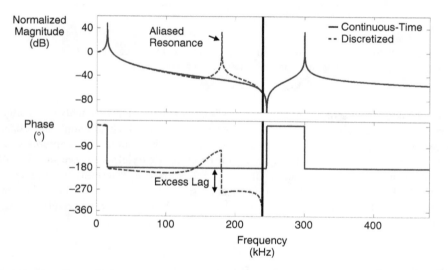

Figure 5.23 Frequency responses of a sensor with a parasitic resonance at 300 kHz that aliases down to 180 kHz with increased excess phase lag in the discretized frequency response

cross the $-180°$ threshold. However, the parasitic resonance, being in the second Nyquist zone, aliases down with a very large excess phase lag since signals in even Nyquist zones alias with inverted phase. Unfortunately, increasing the sampling rate to bring all resonances below the Nyquist frequency is both impractical and ineffective. It is impractical because the sense and parasitic capacitances and the wiring resistance of real sensors impose time constants that limit the maximum sampling rate. It is ineffective because the processing delay together with other delays in the electromechanical chain introduce additional phase lag that pushes the discretized phase response well below $-180°$ even in the absence of parasitic resonances. The very large excess phase lag in the example is therefore a fairly common occurrence. We now evaluate the abilities of various compensation schemes to accommodate parasitic resonances with such excess phase lag.

5.4.3.1 Traditional Lead Compensation

Readout interfaces employing $\Sigma\Delta$ force feedback must reject the quantization noise in the desired signal band to a very high degree to satisfy stringent resolution requirements. Because the frequency shaping provided by the sensing element alone is always insufficient to reject quantization noise to a level below that set by the position sense front-end [19], the widely used second-order modulator, the architecture in which the sensing element is the sole provider of frequency shaping and the compensator only supplies phase lead, always has a degraded SNR. The degradation can be quite substantial, for example 20 dB in [15]. It is possible to realize a second-order architecture that avoids the SNR degradation by using multi-bit rather than single-bit quantization. Among the disadvantages of this approach are the difficulty of applying the required multi-bit force-feedback in a way that is compatible with the previously mentioned bang-bang control, the increased complexity of the demodulator used by the mode-matching algorithm, and the increased complexity of the decimator used to process the oversampled multi-bit output. Additional frequency shaping provided by the compensator in the fourth-order modulator reported in [14] eliminates the SNR degradation with minimal added complexity. We consider only that fourth-order architecture since it preserves the SNR improvements with only a very modest increase in the overall system complexity.

The compensator used in the fourth-order architecture is of the form

$$H_c(z) = \underbrace{\frac{z+a}{z}}_{\substack{\text{lead} \\ \text{comp.}}} \underbrace{\frac{z^2 + b_1 z + b_2}{z^2 + cz + 1}}_{\substack{\text{frequency} \\ \text{shaping}}}. \tag{5.24}$$

The pair of imaginary poles controlled by c provides the above mentioned frequency shaping, and the pair of complex zeros controlled by b_1 and b_2 compensates the phase lag of the imaginary poles. The zero at a provides phase lead at high frequencies to compensate the phase lag of the discretized system. The pole at the origin, unavoidable since a physically realizable system cannot have more zeros than poles, unfortunately negates the phase lead provided by the zero, limiting the available phase lead near the Nyquist frequency. Figure 5.24 shows the resulting open-loop frequency response for typical coefficient values. As one might expect, the system is stable without the parasitic resonance. With the parasitic resonance, however, the system possesses three unity gain frequencies, the last of which is characterized by a large negative phase margin. Unfortunately, the system cannot be stabilized by simply lowering the overall gain since doing so introduces negative phase margin at a different frequency. Even after lowering the gain to a point where the available in-band loop gain is far too low to be useful, the system remains unstable. Lacking the large phase lead needed to accommodate high-Q parasitic resonances, compensators of this kind are inadequate for practical vacuum packaged gyroscopes.

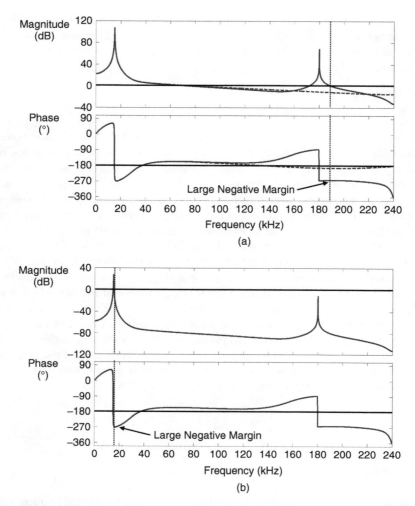

Figure 5.24 Open-loop frequency response of a fourth-order modulator with a parasitic resonance. (a) The system is unstable since there is no phase margin at the third unity gain frequency. (b) Lowering the gain introduces negative phase margin at a different frequency and therefore fails to stabilize the system. The dashed lines indicate the ideal response

5.4.3.2 Positive Feedback Technique

A block whose output is simply the negative of the input is normally thought of as introducing 180° of phase *lag* at all frequencies. Equally valid is thinking of the block as introducing 180° of phase *lead* at all frequencies since $e^{\pm j\pi} = -1$. The 180° of phase lead is free to be exploited provided that the consequent positive feedback is handled with care. To prevent the positive feedback from leading to instability, the open-loop DC gain must be set below unity and a lag compensator must be included to provide adequate phase margin at the first unity gain frequency. Note that we have expanded the definition of phase margin to mean the minimum margin to $\pm 180°$, not just the traditionally used $-180°$. An open-loop DC gain below unity, effectively resulting in the absence of force feedback at DC, is permissible in this application since the Coriolis force is away from DC. With these adjustments and the inclusion of frequency shaping, the form of the compensator becomes

$$H_c(z) = -\underbrace{\frac{z}{z+a}}_{\substack{\text{lag} \\ \text{compensator}}} \underbrace{\frac{z^2 + b_1 z + b_2}{z^2 + cz + 1}}_{\substack{\text{frequency} \\ \text{shaping}}}. \qquad (5.25)$$

The minus sign provides the *automatic* $180°$ of phase lead. The pole at a provides the above mentioned phase margin at the first unity-gain frequency. The zero at the origin is added to cancel the phase lag contributed by the pole at high frequencies since potential parasitic resonances at those frequencies, having substantial phase lag themselves, require no additional lag compensation. In similarity to the previously considered compensator, the pair of complex zeros compensates the phase lag of the pair of imaginary poles included to provide the necessary frequency shaping. Figure 5.25 shows the resulting open-loop frequency response for typical coefficient values. The compensator provides ample phase margins for the parasitic resonance. The margin at low frequencies, though small, is enough for stability. Figure 5.26 shows the root locus of the system. Except for the real pole of the compensator which exits the unit circle for open-loop DC gains greater than unity, all the closed loop poles stay within the unit circle at all gains. Setting the open-loop DC gain below unity with some safety margin guarantees stability. The guarantee of stability for all open-loop DC gains below unity implies that the modulator will always recover from an overload condition since overload only reduces the open-loop DC gain. With these assurances of stability, parasitic resonances can be safely neglected.

A major drawback of this compensation scheme is that the open-loop gain cannot be increased arbitrarily by including yet additional imaginary pole pairs since the entire phase space from $+180°$ to $-180°$ has already been consumed by the phase lag coming from the

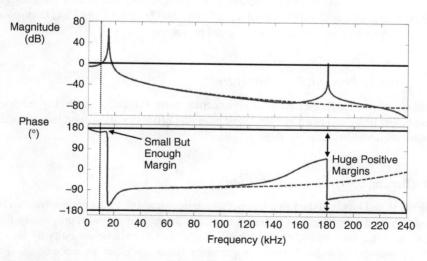

Figure 5.25 Open-loop response of the positive feedback compensated loop

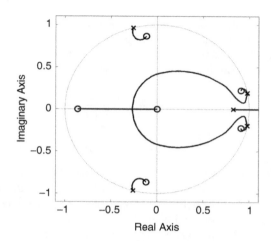

Figure 5.26 Root locus of the positive feedback compensated loop. The high frequency imaginary poles are due to the parasitic resonance. One of the two pairs of low frequency imaginary poles comes from the sensor. The other pair and the real pole come from the compensator

imaginary poles of the sensor and the compensator (see Figure 5.25). Since the accuracy of the mode-matching algorithm depends on the open-loop gain at the pilot tone frequencies, it is important to verify that the open-loop gain resulting from the use of this scheme is sufficient. The maximum achievable open-loop gain at a frequency offset Δf from the drive frequency is given by

$$G_{\max} = \frac{d}{4} \left(\frac{f_d}{\Delta f} \right)^2 \tag{5.26}$$

where d is a correction factor with a typical value of 0.25 to account for the less than unity DC gain and the in-band gain reduction coming from the real pole and pair of complex zeros of the compensator. The achievable gain of 47 dB at a 250 Hz offset from 15 kHz surpasses the open-loop gain requirement in the experimental prototype.

5.4.4   Positive Feedback Architecture

A practical positive feedback based $\Sigma\Delta$ architecture must be capable of forcing the open-loop DC gain below unity and must be tolerant of potential offsets in the sensor. In this section, we derive one such architecture.

5.4.4.1   Setting the Open-Loop DC Gain

Scaling the signal levels in the force feedback loop does nothing to the open-loop gain because the quantizer gain simply adjusts to keep it constant. Since, as mentioned previously, the quantizer gain is signal dependent, it is possible to force the DC gain below unity by injecting an appropriate amount of dither before the quantizer. Since the input variance of the quantizer increases while the output remains constant, the quantizer gain decreases, lowering with it the open-loop gain. A pseudo-random binary sequence is a good dither signal since it also helps to remove the tonal behavior of the modulator and is easily generated using a linear feedback shift register. The sequence does not degrade the overall interface noise floor since, like the quantization noise, it appears at the output frequency-shaped by both the sensor and the compensator. The dither signal can also be injected before the compensator, in which case it is important to inject dither that has undergone frequency-shaping to make up for the loss of frequency shaping by the compensator. Figure 5.27 shows the alternative solutions. In the experimental prototype, the dither is injected before the compensator to reuse the DAC used to inject the calibration signal.

In the interest of reducing analog complexity, the calibration signal can be band-pass $\Sigma\Delta$ modulated, then injected using a coarse DAC as shown in Figure 5.28 since the consequent truncation noise outside the desired signal band is acceptable. In fact, the truncation noise, having been shaped away from the desired signal band, can double as the dither, obviating the need for an additional frequency shaped dither. This, however, requires careful calibration of the coarse DAC gain to ensure that the truncation noise provides just the right amount of dithering. Because of the difficulty posed by MEMS and IC process tolerances, we avoid this solution in the experimental prototype and instead add a frequency shaped dither signal whose magnitude is digitally adjusted to the correct value, and reduce the truncation noise so that the process variation of the coarse DAC gain results in only a minor variation in the

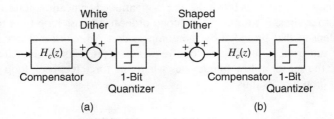

Figure 5.27 Lowering quantizer gain by injecting (a) white dither before the quantizer, and (b) shaped dither before the compensator

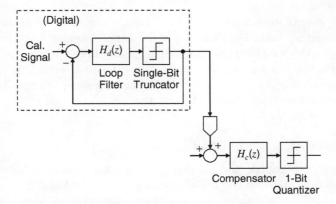

Figure 5.28 Using a digital $\Sigma\Delta$ modulator to reduce the analog complexity of the DAC, with the truncation noise doubling as dither

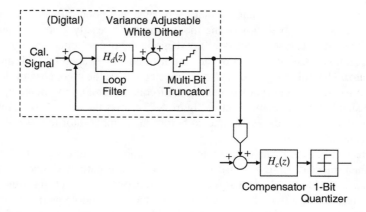

Figure 5.29   A more robust way to generate dither while minimizing analog complexity. The frequency shaped dither is realized by reusing the $\Sigma\Delta$ modulator to *Delta* modulate the variance adjustable white dither

total dither. Figure 5.29 shows this solution. The frequency shaped dither is realized by reusing the $\Sigma\Delta$ modulator to Δ modulate a white dither signal whose variance is adjustable. Multi-bit truncation is necessary here to achieve the above mentioned truncation noise reduction.

The modulator requires, at minimum, a second-order band-pass loop filter to provide noise shaping equivalent to that provided by the compensator. An additional integrator, included to reject dither and truncation noise at low frequencies, prevents the injection of too much disturbance into the sensor at low frequencies where the force feedback loop gain is too low. The form of the loop filter is thus

$$H_d(z) = \underbrace{\frac{1}{z-1}}_{\substack{\text{low-}\\\text{frequency}\\\text{shaping}}} \underbrace{\frac{z^2 + d_1 z + d_2}{z^2 + cz + 1}}_{\substack{\text{in-band}\\\text{shaping}}}. \tag{5.27}$$

where d_1 and d_2 control the pair of complex zeros that compensate the phase lag introduced by the pair of imaginary poles. The frequency of the pair of imaginary poles is controlled by c and therefore coincides with the frequency-shaping poles of the compensator. The loop filter is implemented entirely in the forward path with feed-forward summation to minimize the in-band input-to-output phase lag since the mode-matching algorithm employs synchronous demodulation and is therefore sensitive to phase error.

5.4.4.2   Accommodating Sensor Offset

Due to fabrication tolerances and packaging stress, the sensor typically suffers from non-idealities such as non-zero nominal displacement from the balanced position and mismatch between the differential sense and parasitic capacitances. These non-idealities manifest as a DC or slowly drifting offset that often exceeds the full-scale measurement range of the sensor [11]. Having substantial loop gain at DC, traditional negative feedback loops easily

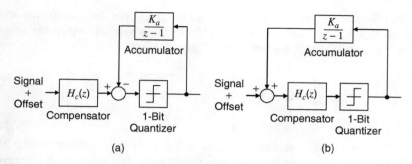

Figure 5.30 Accommodating sensor offset by applying offset compensation before (a) the quantizer, and (b) the compensator

accommodate the offset. The loss of feedback at DC in the positive feedback architecture results in the accumulation of offset before the quantizer, resulting in the departure of the quantizer gain from the desired value and consequent degraded operation. Fortunately, the problem is easily solved by a slow regulation loop that subtracts out the offset before the quantizer. Alternatively, the offset compensation signal can be applied before the compensator. In this case, the signal is added rather than subtracted since the compensator already performs sign inversion. Figure 5.30 shows the two possible solutions. Regulating the DC value of the output of the quantizer is permissible in this application since the Coriolis force is away from DC. We implement the second approach as shown in Figure 5.31 to reuse the already available system blocks. With this, we arrive at a system architecture that minimizes analog complexity while providing all the functions necessary for digital implementation of the mode-matching algorithm. Above all, the architecture is robust against parasitic resonances.

5.4.4.3 System Design

Figure 5.32 shows the analytical model for design. Arriving at a fully optimized design is quite challenging due to the nesting of the digital $\Sigma\Delta$ loop within the offset compensation loop, and interaction of the electronic loops with the main electromechanical loop. To make the process more tractable, we proceed incrementally starting with only the main electromechanical loop, adding the other loops as we proceed. The following is a design procedure that has been found to enable rapid specification of all coefficients. Of course, the analytical results must always be verified and refined via simulation.

- Design the compensator to provide adequate phase margins for the discretized electromechanical chain with K_q adjusted to provide about 6 dB of gain margin at DC. Over design of phase margin should be avoided since it requires the complex zeros and real pole of the compensator to move closer to the unit circle which penalizes the in-band gain.
- Design the digital $\Sigma\Delta$ modulator with the noise notches placed at both DC and the drive frequency. The white dither should be nominally sized to dominate over the truncation noise so that the process variation of the coarse DAC gain is easily accommodated by digitally resizing the white dither as mentioned previously. The number of truncation levels is as yet unimportant provided that it is enough to fully accommodate the truncation noise and dither.

- Close the offset compensation loop and choose the product of the accumulator and coarse DAC gains $K_a K_d$. The overall open-loop transfer function with the loop broken immediately after the quantizer (the spot marked by x in Figure 5.32) is

$$G_{open}(z) = K_q H_c(z) \left[H_{eq}(z) - \frac{\overbrace{K_a K_d}}{z-1} \frac{K_t H_d(z)}{1 + K_t H_d(z)} \right]. \qquad (5.28)$$

Figure 5.33 shows how increasing values of $K_a K_d$ affect the overall open-loop response. The offset compensation path introduces one more unity-gain point at low frequencies for a total of three unity-gain frequencies (with parasitic resonances excluded). This additional unity-gain frequency is critical since the phase also crosses over in that vicinity. Provided that doing so does not adversely impact phase margin, the product of the gains should be increased to maximize the bandwidth of the offset compensation loop to minimize the settling time during startup.
- While keeping the product $K_a K_d$ constant, determine the value of K_d that provides just the right amount of dithering to force K_q to the value selected in the first step.
- Select the number of truncation levels that fully accommodates the calibration signal, the dither, and the worst case offset compensation signal.

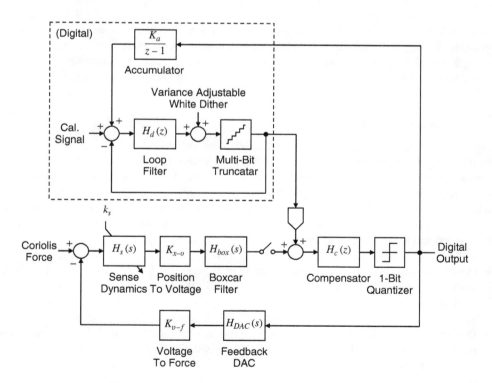

Figure 5.31 Final system architecture

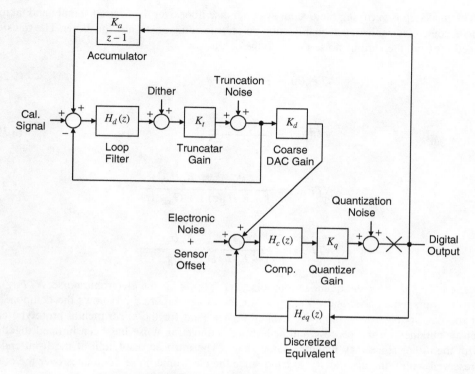

Figure 5.32   Analytical model of final architecture

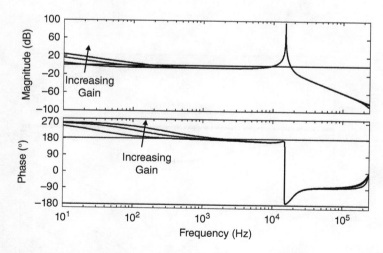

Figure 5.33   Effect of increasing the static gain of the offset compensation loop

The final step is verifying the overall system noise floor over the desired signal band taking into account electronic noise, quantization noise, truncation noise, and the dither. The transfer functions from the various noise inputs to the output are

$$NTF_q(z) = \frac{1}{1 + G_{open}(z)} \tag{5.29}$$

$$NTF_e(z) = \frac{K_q H_c(z)}{1 + G_{open}(z)} \tag{5.30}$$

$$NTF_t(z) = \frac{K_d}{1 + K_t H_d(z)} \frac{K_q H_c(z)}{1 + G_{open}(z)} \tag{5.31}$$

and

$$NTF_d(z) = \frac{K_d K_t}{1 + K_t H_d(z)} \frac{K_q H_c(z)}{1 + G_{open}(z)} \tag{5.32}$$

where $NTF_q(z)$ is for the quantization noise, $NTF_e(z)$ is for the electronic noise, $NTF_t(z)$ is for the truncation noise, and $NTF_d(z)$ is for the dither. Figure 5.34 shows the components of the output spectrum around the desired signal band for the experimental prototype and sensor obtained by the process outlined above. Truncation noise has been lumped together with the dither since they have the same shape. The intrinsic resolution of the front-end is preserved within the desired signal band since the electronic noise dominates over the other noise sources except for the Brownian noise, included for reference. The electronic noise is slightly over designed to prevent degradation of the overall SNR when small mode-matching errors exist.

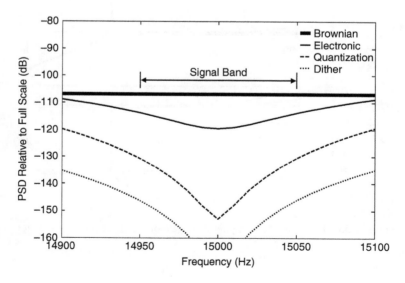

Figure 5.34   In-band output spectrum

5.5 Experimental Prototype

The techniques developed in the previous sections have been applied towards the implementation of an experimental readout interface in a 0.35 µm CMOS process. The intended sensor has a drive resonance frequency of about 15 kHz and a Brownian noise floor of about $0.004°/s/\sqrt{Hz}$. The required bandwidth is 50 Hz (100 Hz double-sided). The operating frequency is locked to 32 times the sense element's drive resonance frequency of nominally 15 kHz.

5.5.1 Implementation

Figure 5.35 shows the overall interface. The sense/feedback switch time multiplexes the same set of electrodes between position sensing and feedback to implement collocated sensing and actuation. A digital estimator injects out-of-band pilot tones before the front-end to monitor the mismatch between the drive and sense resonance frequencies. The estimate feeds into an accumulator that generates a voltage used to electrostatically tune the sense resonance frequency. Eleven bit precision is needed to achieve the required tuning accuracy. The DAC is implemented with a 1-bit $\Sigma\Delta$ modulator followed by a switched-capacitor integrator that serves as the accumulator and doubles as reconstruction filter. Leak and offset in the integrator result in a systematic mode mismatch. A digital PI filter with infinite DC gain rejects this error. The modulated calibration, dither, and offset compensation signals are applied using a 3-bit DAC. By injecting them before rather than after the front-end amplifier, the large displacements are subtracted before the amplifier. The resulting smaller signal sensed by the front-end is advantageous for minimizing the adverse effects of jitter, drift, transconductance variation, and differential pair nonlinearity. The following subsections elaborate on the circuit details of key blocks.

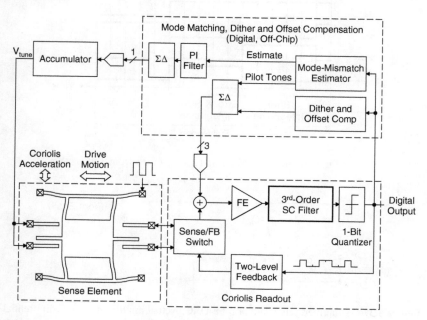

Figure 5.35 Interface block diagram

5.5.1.1 Front-End and 3-Bit DAC

Figure 5.36 shows a schematic of the front-end with its timing diagram. The sense electrodes are connected to feedback voltages through feedback switches (omitted) during the feedback phase and to the front-end amplifier through the sense switches during the sense phase. The output of the amplifier is connected directly to the following switched-capacitor filter through the CDS switches during the two integration phases and reset to ground when the front-end is inactive. The output of the 3-bit DAC is capacitively coupled to the input of the amplifier. The coupling capacitor is only 70 fF, negligible compared to the 8 pF of combined sense and parasitic capacitances.

Figure 5.37 shows the 3-bit DAC. It is shown as single-ended for simplicity, but the actual implementation is differential. It consists of seven unit elements that, depending on the input code, are connected to either ground or the proof mass node. When the pulse is applied on the

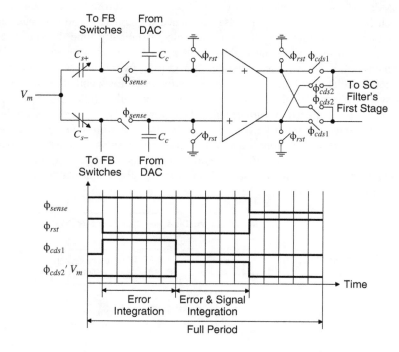

Figure 5.36 Schematic and timing of the front-end

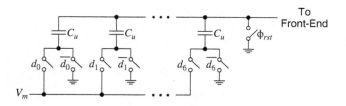

Figure 5.37 Simplified schematic of 3-bit DAC. Actual implementation is differential

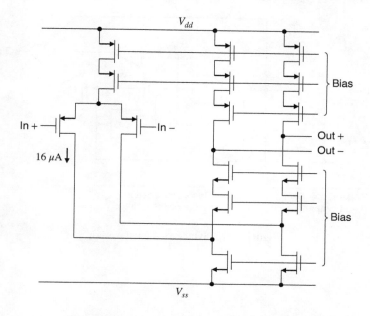

Figure 5.38 Front-end OTA schematic diagram

proof mass, a voltage proportional to the capacitance imbalance between the sense capacitors, together with a voltage dependent on the input code of the DAC, develops at the input of the amplifier. The proof mass and DAC are excited by the same voltage pulse to keep the displacement to voltage gain and the DAC gain ratiometric.

Figure 5.38 shows the transistor-level circuit diagram of the front-end OTA. A folded cascode with PMOS inputs is chosen to enable an input common-mode level of V_{ss} (since the sense electrode are reset in preparation for position sensing). The double cascodes provide high output impedance.

5.5.1.2 Compensator

The first step in implementing the compensator is the synthesis of the transfer function out of only unit delay and gain elements. Figure 5.39 shows a realization that places the delays strategically to minimize the settling path of the switched-capacitor integrator stages that ultimately implement the compensator. The resulting transfer function is

$$H_c(z) = -\frac{z}{z+a} \frac{z^2 + b_1 z + b_2}{z^2 + cz + 1} \tag{5.33}$$

where

$$a = k_1 - 1 \tag{5.34}$$

$$b_1 = k_2 + k_3 + k_4 - 2 \tag{5.35}$$

$$b_2 = 1 - k_3 \tag{5.36}$$

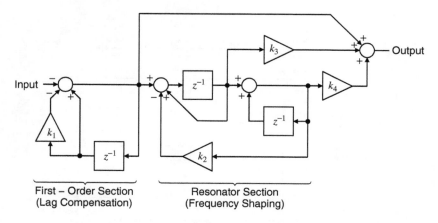

First – Order Section   Resonator Section
(Lag Compensation)   (Frequency Shaping)

Figure 5.39   Synthesized compensator

and

$$c = k_2 - 2. \qquad (5.37)$$

The above system of equations can be solved to find the gains that yield the desired compensator coefficients. The gains should be rational to enable accurate realization by simple capacitor ratios. This may require several iterations of the process of choosing the compensator coefficients so that, while meeting the other design goals, they also result in easily realizable gains.

Figure 5.40 shows the switched-capacitor circuit implementation of the compensator. Final summation of the signals from the resonator and the feed forward paths is realized passively to avoid additional power dissipation. The gain error introduced by the parasitic capacitances at the summation node is unproblematic since the comparator following the filter is sensitive only to the polarity of the signal. Input sampling capacitors are absent since the output current of the front-end during the two correlated double sampling phases integrates directly onto the integration capacitors of the first stage. The extra time during which the front-end is inactive allows the amplifier stages to settle and the comparator to reach a bit decision. This extra time is the source of the processing delay mentioned in the previous section. The amplifiers are one fifth scale versions of the fully differential folded double cascode OTA used in the front-end. The capacitor ratios are related to the gains above as follows:

$$k_1 = \frac{C_{fb1}}{C_{int1}} \qquad (5.38)$$

$$k_2 = \frac{C_{fb2}}{C_{int2}} \frac{C_{s2}}{C_{int3}} \qquad (5.39)$$

$$k_3 = \frac{C_{s1}}{C_{int2}} \frac{C_{ff2}}{C_{ff1}} \qquad (5.40)$$

$$k_4 = \frac{C_{s1}}{C_{int2}} \frac{C_{s2}}{C_{int3}} \frac{C_{s3}}{C_{ff1}} \qquad (5.41)$$

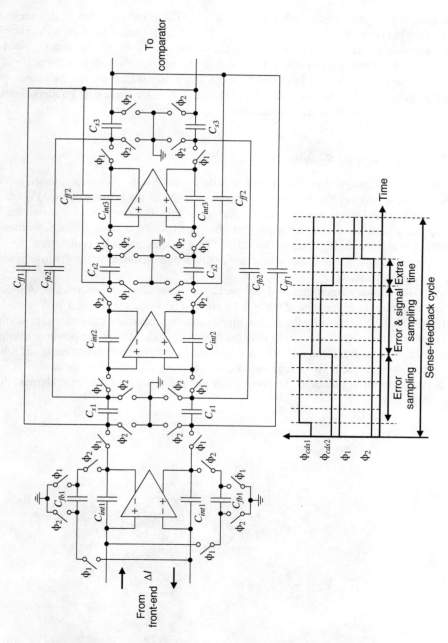

Figure 5.40 Switched-capacitor circuit implementation of the compensator

The gain of the front-end depends on the front-end OTA transconductance, the integration time, and the load capacitor which, in this case, is C_{int1}. C_{int1} is chosen large enough to keep the signal swing within the supply. In this design, $G_m = 160$ uS, $T_{int} = 0.65$ us, and $C_{int1} = 1$ pF, resulting in a front-end gain of about 40 dB. This gain is large enough that the noise of the filter is negligible but is not too large that the OTA and sensor offsets exceed the output swing of the first integrator. The values of the rest of the capacitors are chosen as a tradeoff between matching, and power consumption while satisfying the ratios above.

5.5.2   Experimental Results

The interface was designed and fabricated in a 0.35 μm CMOS process and tested with the gyroscope presented in [20]. Figure 5.41 shows an SEM of the sense element. Figure 5.42 shows the measured frequency response of the sense axis. Besides the main resonance near 15 kHz, many parasitic resonance modes can be found across a wide frequency range, the major ones being around 95 kHz and 300 kHz. These modes, normally problematic for loops employing traditional lead compensation, are easily accommodated by the positive feedback compensation scheme.

Figure 5.43 shows a micrograph of the packaged sense element and readout ASIC. The interface occupies an active area of 0.8×0.4 mm^2 and consumes less than 1 mW from 3.3 V and 12 V. The 12 V is used by the high-voltage switched-capacitor integrator (accumulator) that generates the electrostatic tuning voltage. Approximately 20% of the power is dissipated in the position sense front-end and 10% is dissipated in the switched-capacitor filter. Another 10% is dissipated in the high-voltage switched-capacitor integrator, and about 40% is due to CV^2 losses incurred in switching the proof mass and sense nodes of the sense element during

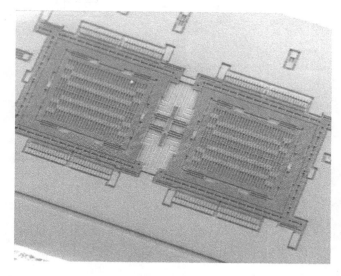

Figure 5.41   SEM of the sense element. The sense element consists of two mechanically coupled sensing structures

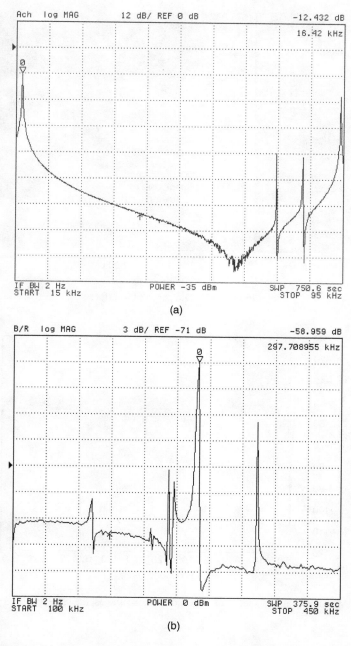

Figure 5.42 Measured frequency response of the sense axis. (a) From 15 kHz to 95 kHz. (b) From 100 kHz to 450 kHz

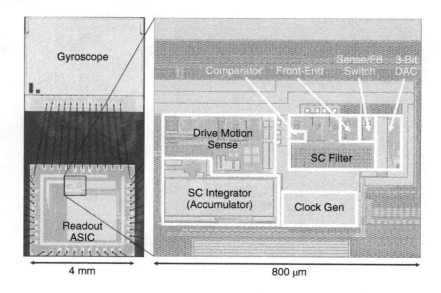

Figure 5.43   Die photo of interface

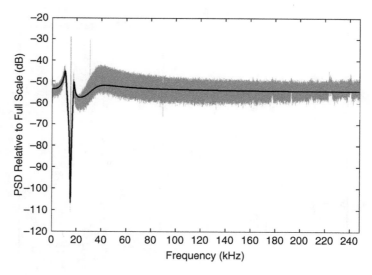

Figure 5.44   Measured output spectrum. The solid dark line is the analytically predicted output spectrum

the various phases of each sampling period. Additional circuits (not included in the 1 mW) are a conventional switched-capacitor charge integrator front-end and buffers to detect the drive motion of the gyroscope. The digital blocks, including the digital $\Sigma\Delta$ modulators, the digital PI filter, and the calibration signal synthesizer and demodulator, were implemented in a Xilinx FPGA. The packaged gyroscope and readout ASIC were mounted on a test board that includes

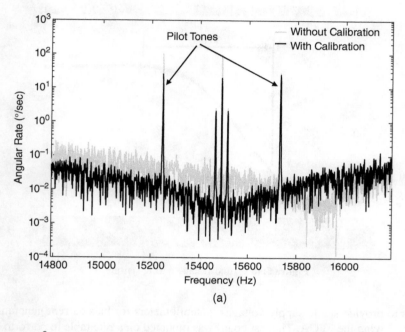

(a)

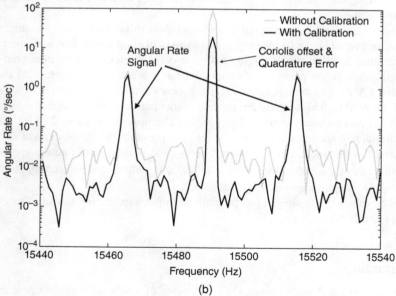

(b)

Figure 5.45 Measured output spectrum showing (a) the misplaced noise notch in the uncalibrated system, and (b) the improvement in the in-band noise floor with calibration

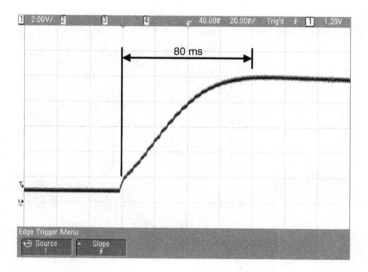

Figure 5.46 Tune voltage during startup

regulators to provide stable supply voltages, potentiometers for bias current generation, and buffers for driving the FPGA. The test board was mounted on a rate table to perform angular rate measurements.

Figure 5.44 compares the measured output spectrum to the analytically predicted output spectrum. The overall shape of the output spectrum is in good agreement with the prediction of the describing function model. Figure 5.45 shows the output spectrum measured with and without calibration in the presence of an angular rate sinusoidally varying at 25 Hz with an amplitude of $5.3°/s$. The sinusoidal rate signal appears amplitude modulated at the drive frequency of 15.49 kHz. The spectral component coincident with the drive frequency is due to Coriolis offset and quadrature error. The spectral components at about 250 Hz offset from the drive frequency are the pilot tones. Calibration prevents the misplacement of the noise notch inherent in the uncalibrated system. The in-band portion of the output spectrum indicates a reduction of the noise floor from $0.04°/s/\sqrt{Hz}$ when mode-matching is disabled to $0.004°/s/\sqrt{Hz}$ when mode-matching is enabled.

Figure 5.46 shows the measured electrostatic tuning voltage during startup. The calibration loop settles within 80 ms.

5.6 Summary

Matching the drive and sense resonance frequencies of a vibratory gyroscope enables substantial improvements in the power efficiency of the readout interface, but implies certain architectural choices. We have presented the architecture and circuits used to exploit the sense resonance to achieve more than 30 dB improvement in the Coriolis acceleration noise floor over traditional solutions dissipating the same amount of power.

The system architecture developed here uses background calibration to match the drive and sense resonance frequencies beyond fabrication tolerances to enable the full exploitation of the sense resonance. It uses force feedback to overcome challenges such as limited sense bandwidth and poor scale factor stability brought about by mode-matching. It uses bang-bang control to prevent the feedback voltage from inadvertently tuning the sense resonance, and uses a positive feedback compensation technique to ensure that the force feedback loop remains stable and robust against the parasitic resonance modes of the sense element. The result is the first experimentally verified 1 mW gyroscope readout interface with a $0.004°/s/\sqrt{Hz}$ noise floor over a 50 Hz band.

References

[1] W. A. Clark, *Micromachined Vibratory Rate Gyroscopes*. PhD thesis, Electrical Engineering and Computer Sciences, University of California, Berkeley, 1997.

[2] J. A. Geen, S. J. Sherman, J. F. Chang, and S. R. Lewis, "Single-chip surface micromachined integrated gyroscope with 50°/h Allan deviation," *IEEE Journal of Solid-State Circuits*, vol. 37, pp. 1860–1866, Dec. 2002.

[3] W. A. Clark, R. T. Howe, and R. Horowitz, "Surface micromachined Z-axis vibratory rate gyroscope," in *Technical Digest of the Solid-State Sensor and Actuator Workshop* (Hilton Head Island, SC), pp. 283–287, 1996.

[4] M. S. Weinberg and A. Kourepenis, "Error sources in in-plane silicon tuning-fork MEMS gyroscopes," *Journal of Microelectromechanical Systems*, vol. 15, pp. 479–491, June 2006.

[5] M. W. Putty and K. Najafi, "A micromachined vibrating ring gyroscope," in *Technical Digest of the Solid-State Sensor and Actuator Workshop* (Hilton Head Island, SC), pp. 213–220, 1994.

[6] J. Choi, K. Minami, and M. Esashi, "Silicon resonant angular rate sensor by reactive ion etching," in *Technical Digest of the 13th Sensor Symposium* (Tokyo, Japan), pp. 177–180, 1995.

[7] W. Yun, R. T. Howe, and P. R. Gray, "Surface micromachined, digitally force-balanced accelerometer with integrated CMOS detection circuitry," in *Technical Digest of the Solid-State Sensor and Actuator Workshop* (Hilton Head Island, SC), pp. 126–131, 1992.

[8] Analog Devices, One Technology Way, Norwood, MA 02062, *ADXL50: Monolithic accelerometer with signal conditioning*.

[9] A. Sharma, M. Zaman, and F. Ayazi, "A 0.2°/hr mirco-gyroscope with automatic CMOS mode matching," in *ISSCC Digest of Technical Papers* (San Fransisco, CA), pp. 386–387, 2007.

[10] T. Smith, O. Nys, M. Chevroulet, Y. DeCoulon, and M. Degrauwe, "A 15b electromechanical sigma-delta converter for acceleration measurements," in *ISSCC Digest of Technical Papers* (San Fransisco, CA), pp. 160–161, 1994.

[11] M. Lemkin and B. E. Boser, "A three-axis micromachined accelerometer with a CMOS position-sense interface and digital offset-trim electronics," *IEEE Journal of Solid-State Circuits*, vol. 34, pp. 456–468, Apr. 1999.

[12] X. Jiang, J. I. Seeger, M. Kraft, and B. E. Boser, "A monolithic surface micromachined Z-axis gyroscope with digital output," in *Symposium on VLSI Circuits Digest of Technical Papers* (Honolulu, HI), pp. 16–19, 2000.

[13] H. Kulah, A. Salian, N. Yazdi, and K. Najafi, "A 5V closed-loop second-order sigma-delta micro-g micro accelerometer," in *Technical Digest of the Solid-State Sensor and Actuator Workshop* (Hilton Head Island, SC), pp. 219–222, 2002.

[14] V. P. Petkov and B. E. Boser, "A fourth-order $\Sigma\Delta$ interface for micromachined inertial sensors," *IEEE Journal of Solid-State Circuits*, vol. 40, pp. 1602–1609, Aug. 2005.

[15] H. Kulah, J. Chae, N. Yazdi, and K. Najafi, "Noise analysis and characterization of a sigma-delta capacitive microaccelerometer," *IEEE Journal of Solid-State Circuits*, vol. 41, pp. 352–361, Feb. 2006.

[16] C. D. Ezekwe and B. E. Boser, "A mode-matching $\Sigma\Delta$ closed-loop vibratory gyroscope readout interface with a 0.004°/s/$\sqrt{hz}$ noise floor over a 50 hz band," *IEEE Journal of Solid-State Circuits*, vol. 43, pp. 3039–3048, Dec. 2008.

[17] J.-J. E. Slotine and W. Li, *Applied Nonlinear Control*. Prentice Hall, 1991.

[18] S. R. Norsworthy, R. Schreier, and G. C. Temes, eds., *Delta-Sigma Data Converters: Theory, Design, and Simulation*. IEEE Press, 1997.

[19] X. Jiang, *Capacitive Position-Sensing Interface for Micromachined Inertial Sensors*. PhD thesis, Electrical Engineering and Computer Sciences, University of California, Berkeley, 2003.

[20] U.-M. Gómez, B. Kuhlmann, J. Classen, W. Bauer, C. Lang, M. Veith, E. Esch, J. Frey, F. Grabmaier, K. Offterdinger, T. Raab, H.-J. Faisst, R. Willig, and R. Neul, "New surface micromachined angular rate sensor for vehicle stabilizing systems in automotive applications," in *Proceedings of the 13th International Conference on Solid State Sensors, Actuators and Microsystems* (Seoul, Korea), pp. 184–187, 2005.

6

Introduction to CMOS-Based DNA Microarrays

Roland Thewes

Technische Universität Berlin (Berlin Institute of Technology), Berlin, Germany

6.1 Introduction

Electronic and in particular CMOS-based microarrays used for detection of biomolecules such as DNA or proteins have gained huge interest in recent years since they promise to provide advantages compared to state-of-the-art commercially available tools using optical readout principles. Integration of the transducer together with signal-processing circuitry in closest proximity on a solid-state chip supports signal integrity and thus translates into robustness, leads to provision of signals in the same domain (i.e. electrically) as they are further post-processed, and may open the way to decrease overall system costs and to increase flexibility concerning application scenarios. In this chapter an overview of CMOS-based microarrays is given from an engineer's point of view, covering the general operation principle of microarrays, functionalization techniques, and different detection principles. Special emphasis is put on CMOS integration and the related processing issues, as well as on CMOS circuit-design requirements. In the latter context, a number of examples published in the literature are considered.

This chapter is organized as follows: In Section 6.2, the basic operation principle as well as applications of DNA microarrays are introduced. After that, in Section 6.3, functionalization techniques – that is, techniques which merge chip and biomolecules and make a chip a bio chip – are briefly discussed. Section 6.4 considers requirements and challenges related to CMOS integration of specific bio-compatible materials demanded by the transducer principle. Sections 6.5 and 6.6 comment on various readout techniques. There, Section 6.5 focuses on electrochemical detection principles, required potentiostatic setup, and examples for related readout circuitry, and Section 6.6 discusses different non-electrochemical techniques. A few considerations on packaging and assembly are provided in Section 6.7, and in Section 6.8 summarizing remarks are given.

Smart Sensor Systems: Emerging Technologies and Applications, First Edition.
Edited by Gerard Meijer, Michiel Pertijs and Kofi Makinwa.
© 2014 John Wiley & Sons, Ltd. Published 2014 by John Wiley & Sons, Ltd.

6.2 Basic Operation Principle and Application of DNA Microarrays

Before focusing our discussion on DNA microarrays, we briefly review a few important micro-biological terms and definitions.

The term *genome* is used as a synonym for the entirety of all *genes* of an organism. In terms of classical genetics, a gene describes hereditary disposition; in terms of molecular genetics – which is more suitable in our context – a gene is a functional intercept of the *Desoxyribonucleic Acid (DNA)*. The DNA is a *Nucleic Acid* and carries – as all types of nucleic acids do – the full genetic information. Besides DNA, which *conserves and transfers* the genetic information by replication, a second important type of nucleic acids is represented by the *Ribonucleic Acid (RNA)*, which *expresses* the genetic information. *Proteins* are essential for cell operation, provide significant contributions to the entire cell metabolism, and fulfill various specific functions.

Within the context of this chapter, for simplicity we restrict ourselves to the discussion of DNA microarrays. As far as electrical engineering and technology issues are concerned, all statements concerning DNA microarrays can be applied to protein microarrays as well, since related technical boundary conditions and engineering challenges apply in a similar way.

In the following, first a few important properties of DNA molecules are summarized. As schematically depicted in Figure 6.1a, the DNA usually consists of a double strand with base pairs in between the two single strands. The backbone of a single strand (Figure 6.1b) is a chain of phosphor acid rests and pentoses. The bases of the base pairs of the double strand are bound by the pentose complex.

DNA base pairs consist of four different bases (Figure 6.1c), <u>A</u>denine, <u>G</u>uanine, <u>T</u>hymine, and <u>C</u>ytosine. These bases build complementary bindings, Adenine binds with Thymine (A-T), and Guanine with Cytosine (G-C). The mass of one such nucleotide pair is approximately 1.1×10^{-21} g.

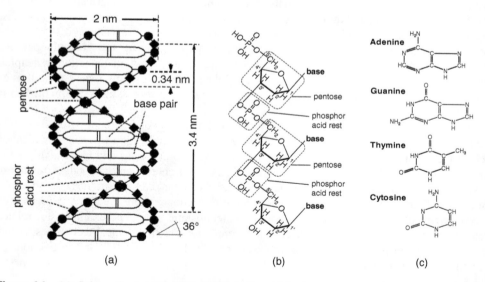

(a)   (b)   (c)

Figure 6.1 (a): Schematic plot of a DNA double helix with basic geometric parameters. (b): Chemical structure of the backbone (of a single strand). (c): Chemical structure of the four DNA bases Adenine, Guanine, Thymine, and Cytosine

A number of characteristic geometrical parameters is shown in Figure 6.1a as well. Moreover, it is worth mentioning that the entire DNA molecule is negatively charged, and that the length of the human DNA is of the order of 3.2×10^9 base pairs.

DNA microarrays using hybridization assays allow the investigation of a given sample concerning the presence or absence or also the quantitative amount of specific user-defined DNA sequences within this sample with high parallelism and high throughput [1–9]. In other words, this means they are only capable of detecting what they are "taught" to detect. In order to avoid any misunderstanding it should be noted that this approach must be clearly distinguished from sequencing [10–13] techniques which are used to determine the order of the bases of unknown DNA strands without any pre-information.

Today's most important applications fields of microarray-base hybridization assays are genome research, drug development, and medical diagnosis, whereas in particular the latter area is believed to provide high growth rates in the future. Important parameters which characterize DNA microarrays (and related assays) include the number of test sites, sensitivity, dynamic range, and specificity. Depending on the particular application, the relative importance of these parameters varies: for example, for the first two mentioned application areas, usually high density arrays and high dynamic range are requested, whereas for diagnostic applications often low to medium numbers of test sites are sufficient but high specificity is demanded [1–9].

The basic setup and operation principle of DNA microarrays are schematically depicted in Figure 6.2. The microarray itself is a slide or a chip typically made of glass, polymer material, or silicon, with an active area in the order of square millimeters to square centimeters. Within that area, single-stranded DNA receptor molecules, frequently referred to as probe molecules, are immobilized at predefined positions (Figure 6.2a). They typically consist of 16–40 bases.

In Figures 6.2b' and b'', two different sites within an array are considered. For simplicity, single strands with only five bases are depicted in this schematic illustration. The different bases are sketched as different symbols. As shown in Figures 6.2c' and c'', during the measurement phase, first the entire chip is flooded with a sample containing the ligand or target molecules. Note that these molecules can be up to two orders of magnitude longer

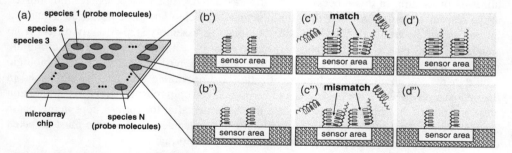

Figure 6.2 Schematic representation of the basic operation principle of DNA microarrays. (a) Microarray chip with a number of test sites with different species of probe molecules. (b') and (b''): Two test sites with different probe molecules. For simplicity, probe molecules are shown here with only five bases. The different bases are schematically emphasized by different symbols. (c'), (c''): Hybridization phase. A sample containing target molecules to be detected is applied to the whole chip. Hybridization occurs in the case of matching DNA strands (c'). In the case of mismatching molecules (c'') chemical binding does not occur. (d'), (d''): Situation after washing step (see Plate 1)

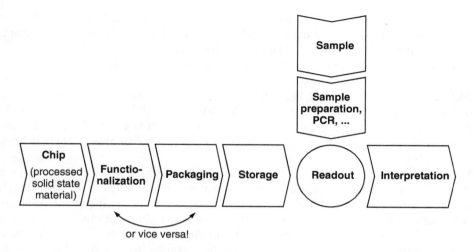

Figure 6.3   Entire operation chain of a DNA microarray

than the probe molecules. In the case of complementary sequences of probe and target molecules, this match leads to hybridization (Figure 6.2c'). If probe and target molecules mismatch (Figure 6.2c''), this binding process does not occur. Finally, after a washing step, double-stranded DNA is obtained at the positions with matching strands (Figure 6.2d'), and single-stranded DNA remains at the mismatch sites (Figure 6.2d''). Since the probe molecules and their positions are known and well-defined, the amount of double-stranded DNA at a respective position reveals the concentration of the related target in the sample. Optical and electronic techniques to distinguish between sites with single- and double-stranded DNA or to quantitatively evaluate the amount of double-stranded DNA are discussed later in this chapter.

The entire application chain of microarrays is schematically summarized in Figure 6.3: Starting with a chip made of a solid-state material, this chip must first undergo a function-alization procedure: In this process, single-stranded DNA probe molecules are immobilized at the respective positions. After that, the functionalized chip has to be packaged. Note that in the case of electronic or CMOS-based solutions this assembly step strongly differs from standard CMOS packaging concepts, as the chip must have an electronic and a microfluidic interface, and as the materials used for the latter interface must be compatible with biologic requirements. As an alternative, functionalization and packaging may also be performed in the opposite order. Then, the packaged and functionalized chip must be stored, as application and operation of the chip is usually not directly performed after the former steps in this chain are completed. As the chip already carries biological molecules, special care must be taken concerning temperature and perhaps also humidity while storing it.

We now consider the second branch dealing with the sample to be investigated: Full blood or other DNA carrying media cannot be applied directly but first has to undergo a sample preparation phase. There, the DNA in the sample is isolated, cut, and in many cases the amount of target material is also amplified by using specific biotechnological tools [14].

Chip and pre-processed samples then meet during the application phase (cf. Figures 6.2c and d) in a reader unit which handles the sample, operates and reads out the chip. Finally, the resulting data are interpreted using the methods and insights from the field of bioinformatics.

6.3   Functionalization

For functionalization, various approaches are used. In Figure 6.4a the relationship is depicted between different functionalization techniques and related probe molecule synthesis techniques as a function of the amount of sites per chip. Using a corresponding horizontal axes to Figure 6.4a, Figure 6.4b shows related DNA microarray application areas as a function of site density. Considering the functionalization techniques mentioned in Figure 6.4a, today, the two most important techniques are spotting of off-chip synthesized probe molecules [15, 16] using microspotters and optically-controlled in-situ growth of probe molecules [17, 18] directly on-chip.

Today's microspotters are capable of handling volumes below 1 nl and realizing pitches between the sites of the order of 100 μm. Thus, this technique is adequate for low- and medium-density arrays which are frequently suggested for diagnostic applications.

Aiming for very high-density arrays ($\geq 100,000$ sites) as, for example, used for drug development purposes, in-situ growth is the only way to realize the requested number of sites within a reasonable area. For this purpose, the market leader in microarrays, Affymetrix, uses a lithography-based mask technique (to some extent similar to that known from the semiconductor manufacturing world). There, the probe molecules are synthesized base-by-base on-chip. Ligation of a base at the strands under construction is triggered or blocked by the presence or absence of light at the respective sites. Thus, the required mask count is approximately equal to $4 (=$ number of different bases, i.e. Adenine, Guanine, Thymine, and Cytosine, respectively$) \times$ length of the probe molecules (cf. Section 6.2).

Another technique for the low- and medium-density range suggested by Nanogen uses off-chip synthesized probe molecules and directs these to their target positions by means of

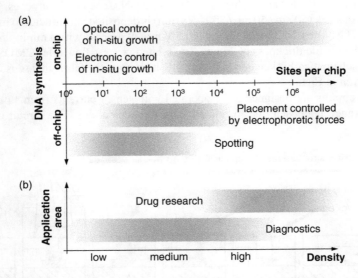

Figure 6.4   (a) Overview diagram showing the relationship between different functionalization techniques and related probe molecule synthesis techniques as a function of the amount of sites per chip. (b) Density-related DNA microarray application areas with the plots in (a) and (b) having corresponding horizontal axes

electrophoretic forces [19, 20]. For this purpose, some logic circuitry is needed to provide the required voltages to electrodes located at the respective sensor sites.

Techniques have also been demonstrated for an electronically-controlled on-chip in-situ synthesis [21, 22]. By means of CMOS circuitry, during the functionalization process electrical signals are applied to the test sites equipped with noble metal electrodes controlling the related biochemical reactions. Devices with more than 10,000 test sites have successfully been demonstrated.

6.4   CMOS Integration

The interaction of solid-state CMOS chips with the wet world of biology usually requires introduction of extra processing steps to extend a given standard CMOS technology to provide the transducers, related materials, and a biocompatible passivation. For example, in the case of electrochemical principles operated on CMOS chips – or in the case of chips using electronically controlled functionalization as discussed above – noble metal electrodes are demanded. In particular, often gold is requested as this material provides a well-known and frequently used system for electrochemical purposes. Processing of such materials directly within a CMOS production line is often impossible, since they may cause contamination problems which have a significant impact on performance and yield of the CMOS devices.

For that reason, the concept of CMOS post-processing must be applied. However, also in that case care must be taken that post-processing does not deteriorate the performance of the CMOS process and related devices, in particular when sensitive analog circuitry is realized. An example is discussed in the following.

For the sensor principle discussed later in Section 6.5.1.2, interdigitated gold electrodes are used as sensor elements for array chips for electronic DNA detection purposes. For this purpose, a Ti/Pt/Au stack (50 nm / 50 nm / 300–500 nm) is deposited and structured using a lift-off process [23]. The basic CMOS technology is a 5 V n-well process with a minimum gate length of 0.5 μm and an oxide thickness of 15 nm. The related process flow after CMOS passivation with Si_3N_4 is schematically sketched in Figure 6.5; a photo of the fabricated sensor and a cross section are given in Figure 6.6.

Simple test circuits are designed to be operated with sensor currents from 1 pA to 100 nA. This current range is chosen due to related sensor specifications. The circuit consists of a

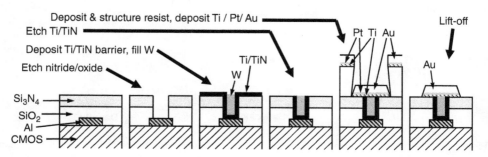

Figure 6.5   Schematic post-CMOS process flow used to provide interdigitated gold electrodes on CMOS [23]

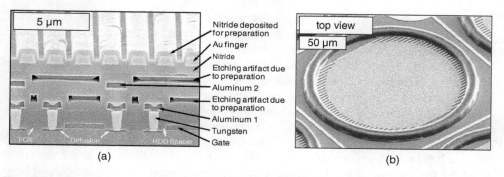

(a)   (b)

Figure 6.6 SEM photographs of the extended CMOS technology used in [23]. (a): Cross section with Au sensor electrodes and CMOS elements after the complete process run. Note that the nitride layer on top of the sensor electrodes is only used for preparation purposes. (b): Bird view of a sensor with interdigitated gold electrodes. Width and spacing of the sensor electrodes are 1 μm

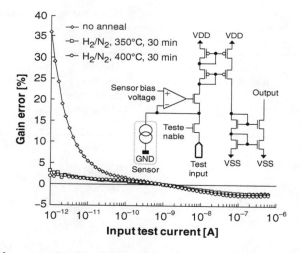

Figure 6.7 Error of current gain of the circuit shown in the inset as a function of the input test current normalized to the gain at test current of 1 nA for different annealing options after gold processing

regulation loop to control the bias voltage of the electrodes, whose current is recorded and amplified by a factor of approximately 100 using two current mirrors in series. The simplified circuit diagram of one branch is shown in the inset of Figure 6.7 [23]. The circuit can be characterized using a test/calibration input.

Figure 6.7 shows the measured gain as a function of the input current (average value of all test sites from an array chip with 128 positions). Data are shown with and without additional annealing steps after the gold process module is performed. If no annealing step is applied, a severe deviation of the measured gain is obtained for input currents below 10 pA. This effect coincides with an increase of the gate oxide interface state density, values above 2×10^{11} cm^{-2} are obtained (cf. Table 6.1). These excessive values translate into an increased inverse subthreshold slope of the transistors, worsened off-state characteristics, increased junction-to-substrate or junction-to-well leakage currents [24], and thus deteriorate the transfer characteristics for low currents. Application of forming gas annealing steps (N_2, H_2 at 400°C/350°C, 30 min) after gold processing significantly reduces the gate oxide interface state density again and leads to reasonable transfer characteristics.

In addition to the CMOS process front-end parameters considered so far, the characteristics of the gold electrodes with and without annealing must be investigated. Measured resistance data of gold and top aluminum lines, and of the related via connections are given in Table 6.1. The data without annealing step and with annealing at 350°C are similar for all parameters. At 400°C, a 20% increase of the gold resistance occurs. The SEM photos in Figure 6.8, moreover, reveal that this increase at 400°C coincides with a rearrangement of grains and deformations within the gold layer. For the Au-process annealed at 350°C, however, the SEM picture looks the same as the photo obtained without annealing step. Consequently, annealing at 350°C is chosen as a process window where both CMOS frontend device and backend electrode properties are optimized (Table 6.1).

Table 6.1 Sheet resistance of gold and aluminum-2 lines, resistance of the related via connections, and transistor gate oxide interface state density, without and with annealing steps at different temperatures after gold processing

	square resistance Au lines [mΩ/square]	resistance via holes (top Al to Au) [mΩ]	square resistance top Al lines [mΩ/square]	gate oxide interface state density [1/cm²]	overall performance evaluation
CMOS only (i.e. without Au process)	–	–	–	$\sim 10^{10}$	–
CMOS + Au process, no anneal	48	370	79	$\sim 2 \times 10^{11}$ (!)	CMOS frontend deterioration
CMOS + Au process, N_2/H_2 anneal with 350°C, 30 min	51	360	76	$< 10^{10}$	**good**
CMOS + AU process, N_2/H_2 anneal with 400°C, 30 min	61(!)	340	74	$< 2 \times 10^9$	sensor backend deterioration

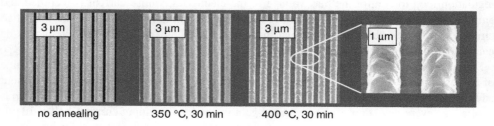

no annealing   350 °C, 30 min   400 °C, 30 min

Figure 6.8 SEM photos showing the Au sensor electrodes without and with annealing steps performed at different temperatures after Au processing [23]

6.5   Electrochemical Readout Techniques

Today's most widely used, state-of-the-art, and commercially available DNA microarray read-out technique is based on an optical detection principle [4, 5, 18]. As schematically depicted in Figure 6.9, there the target molecules are labeled with fluorescence molecules before the sample is applied to the chip. After hybridization and a subsequent washing step, the whole chip is illuminated or scanned with monochromatic light with a wavelength matched to the absorption profile of the marker molecules. A camera system with a blocking filter for the excitation wavelength takes an image of the array chip. The amount of fluorescent light emitted at a respective position reveals the amount of successful hybridization and double-stranded DNA at this position.

The motivation to develop fully electronic readout techniques is driven by the idea to provide systems with increased user friendliness, decreased system cost, and increased flexibility avoiding the relatively bulky and expensive optical readout equipment. This advantage is believed to also (economically) open the way for applications in new areas such as diagnostics and individualized medicine. In that context note that the chip price of a functionalized and packaged device in the case of a fully electronic system will have a similar price as compared to an optical chip; that is, the price of the naked chip alone does not decide on success or failure.

6.5.1   Detection Principles

Today, various electronic approaches have been suggested for DNA detection purposes. A number of these techniques are described in this and in the following section. As a significant number of electronic readout approaches uses electrochemical techniques [25], we treat these techniques in this section. The following section then comments on non-electrochemical approaches.

The family of electrochemical readout techniques can be further subdivided into three groups:

- **Labeling-based principles:**
  The DNA target molecules carry label molecules which enable and contribute to the electrochemical reaction;

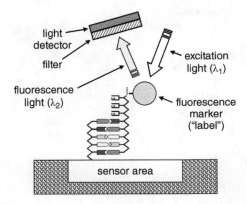

Figure 6.9   Schematic plot describing the principle of optical DNA detection (see Plate 2)

- **Quasi label-free principles:**
  The assay makes use of label molecules but a labeling step directly applied to the target molecule is avoided;
- **Label-free principles:**
  The use of label molecules is completely circumvented.

In the following, examples for each of these groups are provided. The most classical methods in the electrochemical domain are coulometry and cyclic voltammetry [25], which use the same electrical setup (Section 6.5.1.1), and redox cycling (Section 6.5.1.2). Labeling-free and so called quasi-label-free electrochemical approaches are briefly discussed in Sections 6.5.1.3 and 6.5.1.4, respectively. Finally, non-electrochemical methods, using or avoiding labels, are considered in Sections 6.6.1 and 6.6.2, respectively.

6.5.1.1   Coulometry and Cyclic Voltammetry

The basic setup for both methods is schematically depicted in Figure 6.10a. It consists of a three-electrode system with a potentiostat, whose output and input are connected to a counter electrode and to a reference electrode, respectively. This potentiostatic setup is used to control the potential of the electrolyte and keep it under defined conditions.

The working electrode is functionalized and carries the probe molecules. In the case of successful hybridization target molecules are also present which are labeled with an electrochemically active marker molecule. If the potential of the working electrode is changed by a suitable amount depending on the electrochemical system (in practice typically a few 100 mV) with respect to the reference potential, oxidation or reduction of the electrochemical labels and a charge transition from label to electrode or vice versa are achieved.

In the case of coulometric approaches, under practical measurement configurations the total amount of charge is measured through integration within a given time window. The total charge achieved from the labels, Q_{label}, amounts to

$$Q_{label} = z \times q \times D_{probe} \times A_{we} \tag{6.1}$$

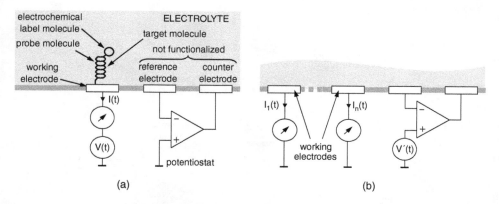

Figure 6.10   (a): Basic setup for detection systems based on Coulometric or Cyclic Voltammetry readout using a three-electrode configuration. (b): Electrically equivalent configuration suitable for array operation with parallel readout

with z being the amount of electrons per oxidation/reduction and label molecule (typically a low integer number, frequently 1), q being the elementary charge ($= 1.6 \times 10^{-19}$ As), D_{probe} being the density of probe molecules on the test sites (typical values are $(10 \text{ nm})^{-2}$ or slightly below), and A_{we} being the working electrode area, respectively. For example, for a circular electrode with 100 μm diameter and $D_{probe} = (10 \text{ nm})^{-2}$, $Q_{label} \approx 12.6$ pAs is obtained. It is obvious that an electrically equivalent situation is achieved, if the potential change is not applied to the working electrode but through the potentiostat (with inverted polarity as compared to the setup in Figure 6.10a). This configuration is particularly advantageous in the case of arrays with many working electrodes operated and read out in parallel [26]. The related setup is shown in Figure 6.10b.

As the electrode-to-electrolyte interface acts as capacitance, whose capacitance in the case of a blank electrode is given through the capacitance of the Helmholtz double-layer (of the order of 10 ... 40 μF/cm²), as depicted in Figure 6.11 a displacement current occurs through this interface when the electrode-to-electrolyte voltage is changed (a more detailed electrode equivalent circuit is discussed in the context with Figure 6.17 in Section 6.5.2). This displacement current also contributes to the final electrical measurement result and represents an offset signal, whose integrated value translates into the charge

$$Q_{double-layer} = V_{step} \times C'_{we} \times A_{we} \tag{6.2}$$

with V_{step} being the amplitude of the voltage step and C'_{we} the working electrode double-layer capacitance per area, respectively. Calculating this value for a blank electrode with typical values for V_{step} (e.g. 200 ... 250 mV) and the other parameters as given above, it is found that this offset value is more than one order of magnitude bigger as the value calculated for Q_{label} in our example.

Consequently, this offset would significantly impact accuracy and reliability of the measurement result. Thus, special biochemical procedures are applied to the electrode after the immobilization process – for example, growing a dense meadow of "blocking molecules" in between the linker molecules which bind the probe molecules to the electrode – which do not hinder the charge transfer between label and electrode, but decrease the electrode capacitance by a factor of the order 10 as compared to a non-biochemically post-processed electrode.

The two methods, Coulometry and Cyclic Voltammetry, differ concerning the applied signals and the related time scales. In the case of coulometric principles (Figures 6.12a and b), a rapid voltage step (slew rate of order 1 V/μs) of suitable magnitude and polarity is applied to the

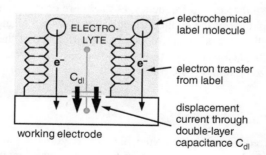

Figure 6.11 Contributions to charge flow through working electrode from electrochemically active label and double-layer capacitance

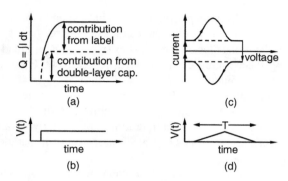

Figure 6.12   (a) and (b): Integrated current as a function of time in the case of coulometric readout (a) and signal applied to the electrode (b). Typical time scales of the entire time axes shown are of order up to a few milliseconds, slew rates of the signal applied to the electrode of order 1 V/μs. The bold line in the upper plot represents the entire signal, whereas the dashed line depicts the contribution of the parasitic electrode-to-electrolyte capacitance (displacement current, not to scale). (c) and (d): Measured current as a function of voltage in the case of voltammetric readout (c) and signal applied to the electrode as a function of time (d). Typical time scales of the entire time axis shown here are of order (tens of) milliseconds up to seconds. The bold line in the upper plot represents the entire signal, whereas the dashed line depicts the contribution of the parasitic electrode-to-electrolyte capacitance (displacement current)

electrode, and oxidation or reduction of the electrochemical labels and a charge transition from label to electrode or vice versa are rapidly achieved. Charges originating from electrochemical agents in the bulk liquid contribute later to the charge flow through the electrode as their time constants are limited by diffusion processes in the bulk. The evaluated signal is usually the integrated charge within a time window of the order of a few microseconds to maximally milliseconds after the voltage step is applied.

In the case where the contribution from electrochemical agents in the bulk liquid is negligible, cyclic voltammetry principles are applicable (Figure 6.12c and d). There, a triangular voltage with frequencies from below 1 Hz up to several 100 Hz is applied and the electrode current is measured. As electrochemical reactions do not occur at a sharp onset, the current first increases after a certain voltage is reached until a maximum current is reached, and then decreases again until the electrochemical reaction is complete and only the contribution from the displacement current is still present (Figure 6.12c). Note that the current levels, that is, both the current associated with the electrochemical reaction as well as the displacement current, are proportional to the applied frequency, as the entire charge transfer (under condition that base line and amplitude of the applied triangle signal are not changed) is a constant. As a parameter, which is finally evaluated, frequently the difference of the two peak currents is taken, whereas in practical cases the two peaks do not necessarily coincide on the voltage axis as shown in the idealized plot in Figure 6.12c [25].

6.5.1.2   Redox-Cycling

The electrochemical redox-cycling principle [27–30] requires a four electrode system as depicted in Figure 6.13. A single sensor (Figure 6.13a) consists of interdigitated noble-metal (generator and collector) electrodes. Probe molecules are immobilized on both of them

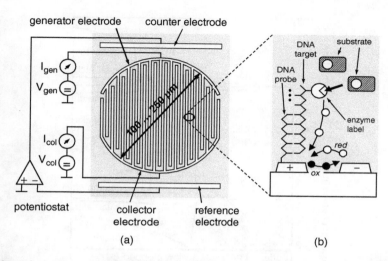

Figure 6.13   Schematic plot showing the redox-cycling sensor principle and the sensor layout (a): Single sensor consisting of interdigitated gold electrodes and potentiostat circuit with counter and reference electrodes. (b): Blow-up of a tilted sensor cross-section showing two neighboring working electrodes after successful hybridization. For simplicity, probe and target molecule are only shown on one of the electrodes.

(note that the blowup in Figure 6.13b shows DNA strands on one of these only for simplicity). The target molecules in the sample are tagged by an enzyme label (e.g. Alkaline Phosphatase). After hybridization and washing phases, another chemical substrate (e.g. para-Aminophenylphosphate) is added to the sample electrolyte. The enzyme label, which itself is not electrochemically active, cleaves the substrate so that now an electrochemically active species (in our example: para-Aminophenol) is generated at those positions where double-stranded DNA is available.

By applying simultaneously an oxidation and a reduction potential to the sensor electrodes (typically of the order $+/-$ few 100 mV with respect to the reference potential), this species (example: para-Aminophenol) is oxidized (example: to Quinoneimine) at the one electrode, and reduced to the former state at the other one. The activity of these electrochemically redox-active compounds translates into an electron current at the electrodes.

As compared to principles where the electrochemically active molecule is directly linked to the DNA target molecule this method allows provision of a higher amount of charge per sensor site as this is not limited by the amount of available label molecules. Moreover, it produces a quasi DC current, and electrode potentials remain at constant voltages during the measurement, so that displacement currents through the double-layer capacitance do not play a role here.

As a certain amount of the electrochemical players involved in this principle also diffuses away from the respective site where they are generated, a potentiostatic setup is also applied here to control the potential of the electrolyte.

Figure 6.14 shows measured data of two sites within an array, one with matching strands and the other one with mismatching random sequences. While the substrate is pumped over the chip (hatched area in Figure 6.14), the electrical current detected at the sensor electrodes is a function of the contribution of the initial generation rate of the enzyme labels, and of the

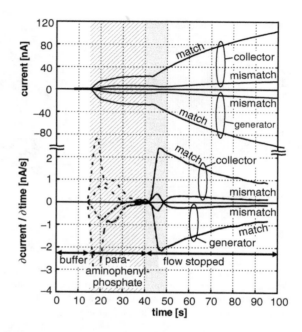

Figure 6.14 Measured redox-cycling sensor currents and their derivatives with respect to time for a position with matching strands and another position with mismatching random sequences within an array. Sensor site diameter = 250 μm, electrode width and spacing = 1 μm

redox-cycling related contribution at the sensor electrodes. After a certain time an equilibrium is achieved between the amount of electrochemically active species generated at a given site with matching strands and the amount washed away due to the continuous flow on the chip, so that the measured current increases no further.

As part of the electrochemically active species generated at a position with double-stranded DNA is also pumped to positions without hybridized target strands, in this phase we also see a certain amount of electrical current at such mismatch positions. This phenomenon can be understood as electrochemical crosstalk.

For the final measurement (gray area in Figure 6.14, time frame between 40 and 50 s) the pump is stopped. Now, the concentration of electrochemically active particles at a match position can further increase without being washed away, while the amount of electrochemically active particles at a mismatch position decreases due to diffusion processes. For that reason, often the derivatives of the sensor currents with respect to the measurement time are evaluated instead of the absolute values. As is clearly evident from the figure, concerning this parameter a strong increase is found for the match positions while in the case of the mismatch position not only a decrease but also a change of sign occurs. As can be seen in this example using sensor site diameters of 250 μm and electrode widths and spacings of 1 μm, the parameters to be electrically evaluated, that is, the currents of relevance and their related derivatives, are of the order +/− several tens of nA and +/− a few nA/s, respectively.

6.5.1.3 Quasi-Label-Free Electrochemical Approaches

In the case of Quasi-Label-Free approaches labeling of the target molecule is avoided. This simplifies the biochemical process during the sample preparation phase (cf. Figure 6.3). It does not mean, however, that the use of label molecules in the entire assay is generally avoided.

An example is given in the following [31, 32]: The assay schematically depicted in Figure 6.15 makes use of intercalators. Such molecules are captured during the hybridization phase in between double-stranded DNA, but do not bind to single strands. These intercalators can be used as carriers of label molecules, so that double-stranded DNA can be electrochemically detected using one of the measurement methods shown before. Moreover, they can also carry more than only one label molecule which allows provision of relatively high signals per sensor area. A disadvantage of this method is that direct contact of humans to intercalators poses a risk to health, so that security requirements are significantly higher in case such assays are used.

6.5.1.4 Label-Free Electrochemical Readout

Figure 6.16 provides an example of a technique using cyclic voltammetry and a three-electrode system as shown in Figure 6.10 while completely avoiding the use of label molecules [33]: Here, a redox reaction takes place within an electropolymer (polypyrrole) covering the working electrode. Hybridization and the related presence of double strands hinder the movement of the chloride counter ions needed for this reaction and thus decrease the measured redox current.

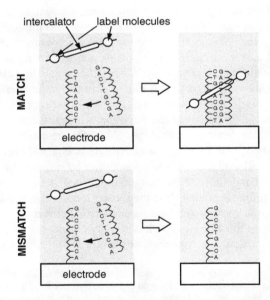

Figure 6.15 Schematic description of the behavior of intercalator molecules in hybridization assays for DNA detection [31, 32]

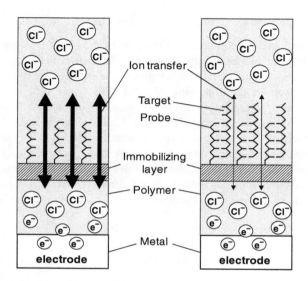

Figure 6.16   Schematic principle of an electropolymer redox reaction whose strength is a function of the amount of double-stranded DNA thus allowing completely label-free electrochemical DNA detection [33]

6.5.2   *Potentiometric Setup*

As already mentioned before, a potentiostatic setup is necessary in all detection methods considered so far [25]. In this section we will thus discuss the specific boundary conditions and requirements related to the operation of electronic DNA microarrays [34–36]. A few general guidelines are derived for circuit design and implementation for our purposes.

Stability is of course absolutely mandatory. A simplified equivalent circuit of a potentiostat operated in an array is shown in Figure 6.17. Each electrode is characterized by a capacitance, a resistance, and an electrode-to-electrolyte voltage drop. Resistance and voltage drop[1] depend on electrolyte and material properties of the electrode, whereas the values of the capacitances show a lower concentration dependence than the values of the resistances, and can be estimated with an accuracy of about one decade. The electrolyte is characterized by a resistive and a capacitive element as well, and the operational amplifier by standard opamp parameters such as open loop gain, gain bandwidth, phase margin, input capacitance, and output resistance.

An evaluation of stability is achieved, if the closed loop is cut at the opamp input and a transfer function using the above mentioned parameters is derived.

For this purpose it is in most cases sufficient to only consider the resistive components of the electrolyte and the capacitive components of the electrodes. After some mathematics, and

---

[1] A standard potentiostatic setup usually uses a reference electrode with negligible ion-concentration dependence concerning the voltage drop between reference electrode and electrolyte such as an Ag/AgCl reference electrode. In the case of assays where electrochemical concentrations do not vary much, however, CMOS-based arrays have also been published, where working electrode, counter electrode, and reference electrode use the same material, e.g. Au. As these electrodes do not fulfill the requirements of a "good" reference electrode with wide-range operating capability concerning ion concentrations in the proximity of that electrode, they are frequently referred to as "quasi reference electrodes".

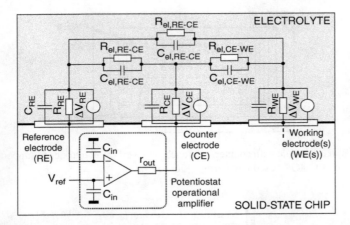

Figure 6.17 Simplified equivalent circuit of a potentiostat operated in an array. For simplicity only one working electrode is depicted, whereas in practice an entire ensemble of these distributed over the chip must be considered [26, 36]

assuming an opamp with sufficient phase margin at unity-gain frequency (e.g. 70°), we arrive at the following recommendations:

- The resistance between reference and counter electrode should be small, in particular much smaller than the resistance between reference and working electrodes.
- The counter electrode capacitance (driven by the potentiostat opamp) should be much bigger than the capacitance of the working electrodes.
- Moreover, capacitance of reference and counter electrode should be much higher than the opamp input capacitance.

Besides the standard CMOS circuit design parameters which determine the properties of the opamp, we can tune all these parameters by means of physical design, that is, by choosing adequate electrode areas and arrangements on the chip. A simple solution which provides the above mentioned recommended ratios is schematically depicted in Figure 6.18. The ratios of electrode-to-electrode resistances (through the electrolyte) and electrode capacitances are obvious.

Requirements concerning accuracy and input offset voltage are in this context relaxed. In order to understand this we consider Figure 6.19, where the strengths of an electrochemical reaction is schematically depicted as a function of the electrolyte-to-electrode voltage in arbitrary units. For small voltages, a reaction is completely suppressed. While increasing the voltage, we reach the onset of the target reaction, and after a further increase of this voltage, the reaction is completed, so that a plateau level is reached again in this plot. Thus, a voltage used in this context to guarantee a completely finalized reaction can be chosen up to a few tens of mV above the point of completed reaction. As the onset of a next electrochemical process occurs at significantly higher voltage (in the case of every suitable electrochemical assay at least a few 100 mV), a slight shift of the target operating point to higher or lower voltages is negligible.

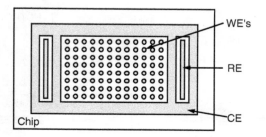

Figure 6.18   Layout example with counter, reference, and working electrodes to support stability of the potentiostatic loop of CMOS-based microarrays [26, 36]

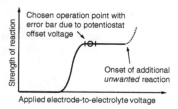

Figure 6.19   Schematic plot depicting the strength of an electrochemical reaction as a function of an applied electrode-to electrolyte voltage [26, 36]

A similar argument applies to the required open-loop gain: since for the same reasons as above a slight regulation error is acceptable, high open-loop gain (e.g. >> 60 dB) is not necessarily required here.

Last but not least, slew rate and gain bandwidth (GBW) shall be considered. Here, the requirements strongly depend on the related detection method. If, for example, a method with relatively fast transients is applied (such as coulometry) also the potentiostat must provide high slew rate and high GBW. On condition, however, that the detection method provides quasi DC signals (e.g. redox-cycling) or has to handle low-frequency AC signals only (e.g. cyclic voltammetry), also the related slew rate and gain-bandwidth requirements for the potentiostat opamp are relaxed.

Published circuits for such purposes often use well-known opamp topologies, e.g. two-stage opamps or opamps with AB output stage. If applicable concerning power consumption and gain-bandwidth, also standard folded cascodes are used as they allow stability to be easily achieved by simply operating them with large capacitive output loads. As the potentiostat drives a usually large counter electrode this condition is automatically fulfilled.

6.5.3   Readout Circuitry

In this section, examples concerning circuit design aspects and suitable circuit topologies are given for the detection methods discussed so far.

An obvious sensor site circuit for coulometric approaches is an in-sensor-site integrator (Figure 6.20a). A smart solution – also providing a benefit for small pixel electrode pitches – is

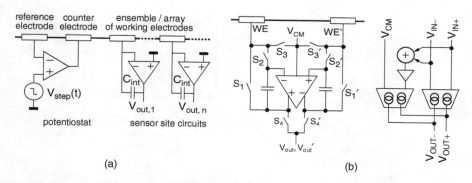

Figure 6.20 (a): Straight-forward solution for the design of electrochemical DNA arrays utilizing a couloumetric detection principle. (b): Solution using a fully-differential opamp approach with common mode input feedback regulation for readout of two sites [26]

used in an array with 384 pixels [26] fabricated on the basis of a 0.5 μm, 5 V, 3M + 2P CMOS technology extended by sensor electrodes made of gold.

The pixel circuit is shown in Figure 6.20b): During the measurement the working electrodes are held at a constant potential (here: V_{CM}) while the electrolyte voltage is changed through the potentiostat. The fully-differential pixel integrator and the related opamp use an input-referred common-mode feedback. This ensures that the working electrodes are held at the requested values during the integration phase and during the readout phase. Moreover, the voltages received at the two output nodes of the differential opamp have the same values as a standard single-ended solution with one integrator for each electrode would provide.

The purpose of switch S_3, connecting the working electrodes to V_{CM}, is to fix their potential at V_{CM} during the readout phase. Moreover, it reduces the leakage current of switch S_2 and makes it signal independent. The integrator op-amp itself uses a folded-cascode topology with approximately rail-to-rail output voltage swing.

Whereas the design discussed above carefully considers the effect of leakage in the pixel circuit during sequential readout of all sites, direct in pixel analog-to-digital conversion and digital data storing completely circumvents this problem. An example using a single-slope ADC applied to a redox-cycling chip with 128 sensor sites is given in Figure 6.21 [30].

For simplicity, only one of the two branches (generator and collector) is shown in the diagram. On the chip itself, a complimentary circuit is also used for the other branch. The voltage of the sensor electrode is controlled by a regulation loop via an operational amplifier and source follower transistor. A/D conversion is achieved by charging an integrating capacitor C_{int} by the sensor current. When the switching level of the comparator is reached, a reset pulse is generated and the capacitor is discharged by transistor M_{res} again. This circuit translates the current into a frequency which is approximately proportional to the sensor current and inversely proportional to the switching threshold voltage and to the capacitance value of C_{int}. For example, using a threshold of 1 V and $C_{int} = 140$ fF, frequencies are obtained between 7 Hz and 700 kHz for a sensor current range from 10^{-12} A to 10^{-7} A.

As the current obtained with the redox-cycling method provides a quasi-DC signal, the number of reset pulses within a given time window is counted with an in-sensor-site counter with

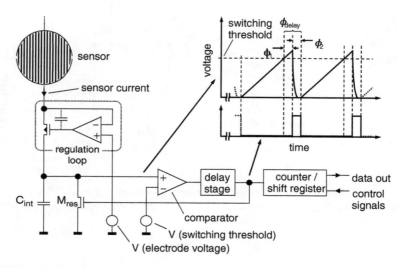

Figure 6.21   Pixel circuit used to directly A/D convert sensor currents origination from a redox-cycling type sensor array [30]

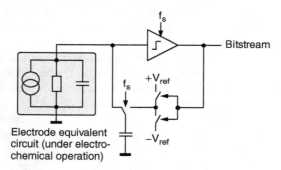

Figure 6.22   First-order delta-sigma modulator using the working electrode as integration electrode applied on a CMOS-based microarray utilizing label-free cyclovoltammetric readout [33]

24 stages. For readout, the counter circuit is converted into a shift register by a control signal and the A/D converted result is provided to the output of the chip.

This conversion principle and extensions of this principle have also been successfully applied in further works focusing on biomedical array applications [37].

A further interesting approach with direct A/D conversion is used on a chip with 576 sensor sites and 24 output channels for label-free cyclic voltammetry readout following the principle explained in Figure 6.22 [33]. The chip fabricated uses a channel-wise first order delta-sigma ADC and operates the sensor site electrodes as integration capacitance. A very high oversampling ratio is achieved due to the slow nature of the electrochemical signals (frequency of applied triangle voltage is of order 1 Hz here, so that an effective number of bits is achieved of approximately 11.

6.6 Further Readout Techniques

In this section non-electrochemical methods are reviewed. Again, we first consider labeling-based approaches and switch to completely label-free techniques later.

6.6.1 Labeling-Based Approaches

A number of approaches have been published using the idea of labeling the target molecules with gold nanoparticles and applying a silver precipitation step after the hybridization phase [38–40]. The basic idea is sketched in Figure 6.23: After the hybridization phase (Figure 6.23a) silver is applied to the sample bulk volume. The gold particles present at the sites with double-stranded DNA act as a seed layer for silver clusters which start to grow at these positions (Figure 6.23b). After further precipitation, a dense silver carpet is formed at the related positions (Figure 6.23c). However, as this carpet will further extend and eventually also cover positions with a lower or zero amount of double-stranded DNA, all assays based on this technique require consideration of the temporal development of the parameters chosen to detect this silver layer extension.

In order to measure the silver carpet extension at the considered sites, various techniques have been proposed:

- Conductivity measurement between electrodes separated by an isolating layer [38].
- AC parameter measurements between isolated electrodes [39].
- Optical attenuation (detected by a CMOS imager chip or in completely pure optical setups) [40].

In the first case, the sensor site consists of a pair of noble metal electrodes electrically separated from each other by an isolating layer in between the electrodes which is also the location of probe molecule immobilization. During the precipitation phase, a conductive silver layer

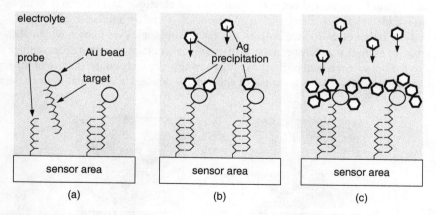

Figure 6.23 Schematic plot depicting the basic principle of detection methods based on gold bead labeling and subsequent silver precipitation. (a): Hybridization phase; (b): Silver precipitation and start of silver clusters growth; (c): Formation of a dense silver carpet after further precipitation

between the electrodes is formed which leads to a sharp decrease of the ohmic resistance over several orders of magnitude between the electrodes. As emphasized before, the discrimination between match and mismatch requires the time behavior of this parameter to be considered. Measurements reported in the literature apply silver enhancement times up to one hour, so that precise time discrimination can be easily achieved.

In the second case, neighboring or interdigitated electrodes are used which are covered by a dielectric. As the silver precipitation and the formation of a metallic layer at the interface of this dielectric provides other electrical parameters as compared to a dielectric interface to the electrolyte or to a functionalized area without silver layer, AC parameters such as impedance change or other RF parameters can be evaluated here.

In the case of optical or semi-optical techniques, the entire chip is illuminated and the decrease of the optical transmission through the active area is measured due to formation of the silver layer. For this purpose, completely optical setups as well as CMOS chips with photodiodes (similar to standard CMOS imager chips) are used.

Another labeling-based technique uses magnetic beads as markers of the target molecules [41, 42]. There, the sensor sites must provide a device whose electrical parameters change dependent on the presence of magnetic matter. Recently, CMOS arrays have been demonstrated with integrated GMR sensors, whose resistance is read out [42].

6.6.2 Label-Free Approaches

The methods discussed here make direct use of electrical or physical properties of DNA or other biomolecules.

The basic idea to detect hybridization dependent impedance changes has been investigated by a number of research groups who also published about CMOS array chips designed for this purpose (e.g. [43–45]). The basic idea is sketched in Figure 6.24. In many cases, the simple lumped-element equivalent circuit consisting of a capacitor in parallel to a resistor (whose value would be infinite in the ideal case) is sufficient. Various feasibility studies of this method report hybridization-driven capacitance decreases. The capacitance – or the entire impedance – can be measured between a noble metal electrode and the electrolyte, between interdigitated electrodes, or also by using electrodes covered by a dielectric.

Whereas a number of design techniques and circuit topologies are known which allow us to characterize capacitance and impedance with excellent accuracy, a certain disadvantage of this method is that the achievement of proper results strongly depends on the quality of the layer of immobilized molecules. If this layer is not sufficiently dense, for example, in the case where

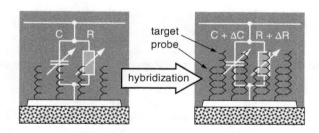

Figure 6.24 Schematic plot showing the basic idea behind impedance-based DNA detection

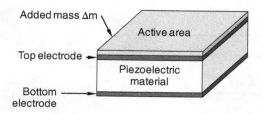

Figure 6.25 Basic device structure of FBAR sensors

it has a few pinholes, these areas may provide a contribution to the impedance of the entire electrode whose impedance is significantly lower as compared to the portion of the electrode with perfect functionalization. As a consequence, such pinholes may more or less shunt the effect obtained on the entire well-functionalized portion of the electrode and thus decrease the reliability of this method.

Gravimetric sensors consist of electro-mechanical oscillators. There, the oscillation frequency depends on the electrical properties and on the mechanical properties, and in particular also the mass of the resonator device. If additional mass is attached to their surface – in our context through hybridization of DNA – the oscillation frequency decreases. Note that this family of sensors is tolerant against pinholes in the receptor molecule layer, as non-well-functionalized areas of the sensor interface simply do not contribute to the sensor's response but do not shunt the sensor's signal as in the case of impedance related approaches.

A cantilever-based solution, as described in [46], fabricated by means of an extended CMOS technology and applied to biomolecule detection is a representative of this class of sensors. The gravimetric principle as such is known and has been successfully applied for a wide range of other purposes as well, also using discrete and large-area quartzes with frequencies in the MHz range. Following the Sauerbrey equation [47], the mass resolution increases in proportion to the resonator frequency. Film bulk acoustic wave resonator (FBAR) technology (Figure 6.25) provides thin-film piezoelectric devices whose frequency is in the low GHz range [48]. It is thus a promising candidate for high sensitivity sensors.

Due to their high operation frequency, short distances between sensor and amplifying circuit are a must. In the case where the FBAR technology cannot directly be fabricated on the same chips as the CMOS technology used, flip-chip technology provides an acceptable solution [49]. An example is sketched in Figure 6.26. The FBAR itself consists of a piezoelectric AlN layer sandwiched between two metal electrodes. In order to avoid a loss of acoustic energy into the substrate, an acoustic Bragg mirror is formed by several layers (W, SiO_2) with alternating low and high acoustic impedances within the substrate.

An important challenge for the operation of such devices and the related design of a suitable oscillator is the reduction of the quality factor in fluids like water or electrolytes used in related assays (Figure 6.27). The effect of the related mechanical attenuation is comparable to the effect of ohmic resistances in the electrical domain. Thus, many circuits used for piezo-oscillators are not applicable here as they rely on high quality factors as achieved in air.

Figure 6.28 presents an oscillator concept successfully applied for FBAR-in-water operation [49]. The FBAR is operated in a voltage-divider configuration in series with capacitance C_0. In the same figure, the transfer function of the FBAR – C_0 voltage divider is depicted revealing a 30° phase shift and a gain peak of −2 dB at resonance frequency. In order to meet

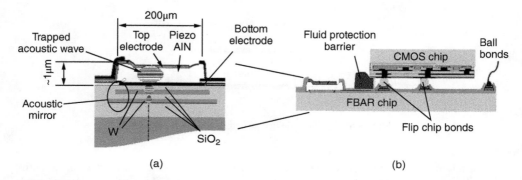

(a)   (b)

Figure 6.26   AlN-based FBAR device with acoustic Bragg mirror (a) and flip-chip stack with FBAR chip and standard CMOS chip in 130 nm technology (b) [48, 49]

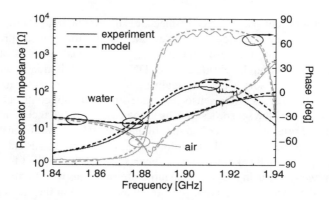

Figure 6.27   FBAR resonator impedance and phase in air and in water as a function of frequency [48, 49]

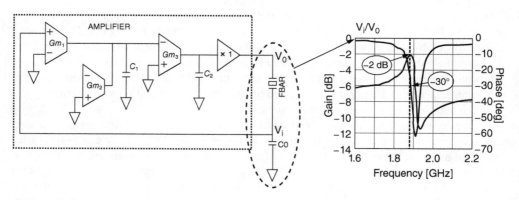

Figure 6.28   Oscillator concept applied for FBAR-in-water operation and transfer function of FBAR − C_0 voltage divider [49]

the Barkhausen criterion for oscillation and to have a gain margin to consider possible device parameter variations and their impact on the overall gain, the amplifier is designed to provide the remaining 330° and 5 dB gain. To precisely meet this gain – phase relationship at resonance frequency in all process/temperature corners the amplifier is built as a Gm-C filter, which is tunable by changing the transconductor bias current. The transfer function is process independent but only relies on the matching of Gm_1, Gm_2, and Gm_3, and C_1 and C_2.

6.7 Remarks on Packaging and Assembly

Packaging and assembly of electronic biochips require provision of a fluidic in addition to an electrical interface which clearly differs from the boundary conditions and packaging concepts known from the standard CMOS world. The solution must be reliable and should not be a show-stopping cost driver.

Concerning in-package (micro-) fluidics, a number of requirements must be considered, such as laminar flow and the avoidance of bubbles (or provision of bubble traps at predefined positions within the package). A detailed requirement catalogue will always depends on technical details concerning detection method, assay, application, and so on.

A number of packaging proposals for such applications have already been discussed in the literature. They range from assembly solutions for very simple open systems where PCBs act as carrier for sealed CMOS chips and electrical interface to reader units (e.g. [44]), to complex chip-card-like systems, in part not only providing an electrical and a fluidic interface to the outer world but also carrying reservoirs with chemical compounds in dried form so that sample preparation can, in addition, be completely performed within a single disposable unit [22, 50].

6.8 Concluding Remarks and Outlook

In this chapter, a brief overview has been given of CMOS-based DNA microarrays. Extended CMOS processing issues, various functionalization and detection techniques, related electrical characteristics, and suitable circuit design approaches have been discussed in this context.

Summarizing today's status, a number of approaches have been published and proven feasible. Commercial and technical success of pure electronic approaches require the entire system (including assembly, packaging, storage, microfluidics, software, target application and related assays) to be considered from the user's point of view. Whereas today's market is clearly dominated by optical systems, commercialization and further development of electronic and CMOS-based solutions as well as development of appropriate business models are on-going: Although around the turn of the millennium and shortly after a number of publications on CMOS-based biomolecule detection chips came from representatives from the semiconductor industry, it has turned out that today the volume of CMOS chips behind many bio-sensing or bio-interfacing applications is too small to translate into an attractive commercial scenario for a company whose business is mainly based on selling processed silicon area. On the other hand, a number of examples where CMOS devices interact with biology have also shown that such applications indeed have the potential to create ethical and economical value, when driven by system houses or specialized application-oriented companies. To name two prominent and commercially available examples, for example, consider cochlea implants [51, 52] or devices for Deep Brain Stimulation (DBS) [53, 54] where the volume of devices per year is clearly below 100,000. Making use of foundries which also offer low volume production on the basis of sufficiently advanced CMOS processes, and creating complete systems around such chips, provision of CMOS-based bio-interfacing devices and systems as mentioned above has been made available for the benefit of user and provider.

Thus today, similar scenarios are considered as a possible pathway to make CMOS-based microarrays also commercially available and successful.

References

[1] E.M. Southern, "An improved method for transferring nucleotides from electrophoresis strips to thin layers of ion-exchange cellulose", *Anals of Biochemistry*, November, vol. 62, no.1, pp. 317–318, 1974.

[2] "The Chipping Forecast", *Nature Genetics Supplement*, vol. 21, Jan. 1999.

[3] http://www.nature.com/ng/chips_interstitial.html

[4] *DNA Microarrays: A Practical Approach*, M. Schena ed., Oxford University Press Inc., Oxford, UK, 2000.

[5] *Microarray Biochip Technology*, M. Schena ed., Eaton Publishing, Natick, MA 01760, 2000.

[6] D. Meldrum, "Automation for genomics, sequencers, microarrays, and future trends", *Genome Research*, vol. 10, pp. 1288–1303, 2000.

[7] F. Bier and J.P. Fürste, "Nucleic acid based sensors", in *Frontiers in Biosensorics I*, F. Scheller, F. Schubert and J. Fedrowitz ed., Birkhäuser Verlag Basel/Switzerland, 1997.

[8] T. Vo-Dinh and B. Cullum, "Biosensors and biochips: advances in biological and medical diagnostics", *Fresenius J. Anals of Chemistry*, vol. 366, pp. 540–551, 2000.

[9] P. Hegde, R. Qi, K. Abernathy, C. Gay, S. Dharap, R. Gaspard, J.E. Hughes, E. Snesrud, N. Lee, and J. Quackenbush, "A concise guide to cDNA microarray analysis", *Biotechniques*, September, vol. 29, no.3, pp. 548–556, 2000.

[10] F. Sanger, S. Nicklen, and A.R. Coulson, "DNA sequencing with chain-terminating inhibitors"; Proceedings of the National Academy of Sciences, pp. 5463–5467,1977.

[11] M. Margulies et al., "Genome sequencing in microfabricated high-density picolitre reactors", *Nature* vol. 437, pp. 376–380, 2005.

[12] J. M. Rothberg et al., "An integrated semiconductor device enabling non-optical genome sequencing", *Nature* vol. 475, pp. 348–352, doi:10.1038/nature10242, 2011.

[13] J. Shendurel and H. Ji, "Next-generation DNA sequencing", *Nature Biotechnology*, vol. 26, pp. 1134–1145, doi:10.1038/nbt1486, 2008.

[14] K. Mullis, F. Falcomer, S. Scharf, R. Snikl, G. Horn, and H. Erlich, "Specific enzymatic amplification of DNA in vitro: the polymerase chain reaction", *Cold Spring Harbor Symposia on Quantitative Biology*, vol. 51 Pt 1, pp. 263–273, 1986.

[15] E. Zubritsky, "Spotting a microarray system", Anals of Chemistry, pp. 761A–767A, December 1, vol. 72, no.23, 2000.

[16] V.G. Cheung, M. Morley, F. Aguilar, A. Massimi, R. Kucherlapati, and G. Childs, "Making and reading microarrays", *Nature Genetics*, pp. 15–19, January, vol. 21(1 Suppl), 1999.

[17] S.P.A. Fodor, R.P. Rava, X.C. Huang, A.C. Pease, C.P. Holmes, and C.L. Adams, "Multiplexed biochemical assays with biological chips", *Nature*, vol. 364, pp. 555–556, 1993.

[18] http://www.affymetrix.com.

[19] M.J. Heller, A. Holmsen, R.G. Sosnowski, and J. O'Connell, "Active microelectronic arrays for DNA hybridization analysis" in *DNA Microarrays: A Practical Approach*, M. Schena ed., Oxford University Press Inc., Oxford, UK, pp. 167–185, 2000.

[20] M.J. Heller, "An active microelectronics device for multiplex DNA analysis", *IEEE Engineering in Medicine and Biology Magazine*, pp. 100–104, 1996.

[21] K. Dill, D. Montgomery, W. Wang, and J. Tsai, "Antigen detection using microelectrode array microchips", *Analytica Chimica Acta*, vol. 444, pp. 69–78, 2001.

[22] http://www.combimatrix.com.

[23] F. Hofmann, A. Frey, B. Holzapfl, M. Schienle, C. Paulus, P. Schindler-Bauer, D. Kuhlmeier, J. Krause, R. Hintsche, E. Nebling, J. Albers, W. Gumbrecht, K. Plehnert, G. Eckstein, and R. Thewes, "Fully electronic DNA detection on a CMOS chip: device and process issues", Tech. Dig. International Electron Device Meeting (IEDM), pp. 488–491, 2002.

[24] S.M. Sze and K.K. Ng, *Physics of Semiconductor Devices*, John Wiley & Sons, 2007.

[25] A.J. Bard and L.R. Faulkner, *Electrochemical Methods*, John Wiley & Sons, 2001.

[26] M. Augustyniak, C. Paulus, R. Brederlow, N. Persike, G. Hartwich, D. Schmitt-Landsiedel, and R. Thewes, "A 24x16 CMOS-based chronocoulometric DNA mircoarray", Tech. Dig. International Solid-State Circuit Conference (ISSCC), pp. 46–47, 2006.

[27] A.J. Bard, J.A. Crayston, G.P. Kittlesen, T.V. Shea, and M.S. Wrighton, "Digital simulation of the measured response of reversible redox couples at microelectrode arrays: consequences arising from closely spaced ultramicroelectrodes", Anals of Chemistry, vol. 58, pp. 2321–2331, 1986.

[28] R. Hintsche, M. Paeschke, A. Uhlig, and R. Seitz, "Microbiosensors using electrodes made in Si-technology", in *Frontiers in Biosensorics I*, F. Scheller, F. Schubert, and J. Fedrowitz ed., Birkhäuser Verlag Basel/Switzerland, 1997.

[29] M. Paeschke, U. Wollenberger, C. Köhler, T. Lisec, U. Schnakenberg, and R. Hintsche, "Properties of interdigital electrode arrays with different geometries", *Analytica Chimica Acta*, vol. 305, pp. 126–136, 1995.

[30] M. Schienle, A. Frey, F. Hofmann, B. Holzapfl, C. Paulus, P. Schindler-Bauer, and R. Thewes, "A fully electronic DNA sensor with 128 positions and in-pixel A/D conversion", *IEEE J. Solid-State Circuits*, pp. 2438–2445, 2004.

[31] N. Gemma, S.-I. O'uchi, H. Funaki, J. Okada, and S. Hongo, "CMOS integrated DNA chip for quantitative DNA analysis", Tech. Dig. International Solid-State Circuit Conference (ISSCC), pp. 560–561, 2006.

[32] http://dna-chip.toshiba.co.jp/eng/

[33] F. Heer, M. Keller, G. Yu, J. Janata, M. Josowicz, and A. Hierlemann, "CMOS electrochemical DNA-detection array with on-chip ADC", Tech. Dig. International Solid-State Circuit Conference (ISSCC), pp. 122–123, 2008.

[34] A. Frey, M. Jenkner, M. Schienle, C. Paulus, B. Holzapfl, P. Schindler-Bauer, F. Hofmann, D. Kuhlmeier, J. Krause, J. Albers, W. Gumbrecht, D. Schmitt-Landsiedel, and R. Thewes, "Design of an integrated potentiostat circuit for CMOS bio sensorchips", Proceedings of International Symposium on Circuits and Systems (ISCAS), 2003, pp. V9–V12.

[35] R. Kakerrow, H. Kappert, E. Spiegel, and Y. Manoli, "Low power single chip CMOS potentiostat", Proceedings of Transducers 95, Eurosensors IX, 1995, vol. 1, pp. 142–145.

[36] R. Thewes, "CMOS bio sensors – specifications, extended CMOS processing, circuit design, and system level aspects", Tutorial Short Course International Solid-State Circuit Conference (ISSCC), 2006.

[37] M.H. Nazari, H. Mazhab-Jafari, L. Lian; A. Guenther, and R. Genov, "192-channel CMOS neurochemical microarray", Proceedings of Custom Integrated Circuits Conference (CICC), pp. 1–4, 2010.

[38] M. Xue, J. Li, W. Xu, Z. Lu, K.L. Wang, P.K. Ko, and M. Chan, "A self-assembly conductive device for direct DNA identification in integrated microarray based system", Tech. Dig. International Electron Device Meeting (IEDM), pp. 207–210, 2002.

[39] L. Moreno-Hagelsieb, G. Laurent, R. Pampin, D. Flandre, J.-P. Raskin, B. Foultier, and J. Remacle, "On-chip RF detection of DNA hybridization based on interdigitated Al/Al_2O_3 capacitors", Proceedings of European Solid-State Device Research Conference (ESSDERC), pp. 125–128, 2006.

[40] J. Li, C. Xu, Y. Wang, H. Peng, Z. Lu, and M. Chan, "A highly manufacturable nano-metallic particle based DNA micro-array", Tech. Dig. International Electron Device Meeting (IEDM), pp. 1005–1008, 2004.

[41] G. Li, V. Joshi, R.L. White, S.X. Wang, J.T. Kemp, C. Webb, R.W. Davis, and S. Sun, "Detection of single micron-sized magnetic bead and magnetic nanoparticle using spin valve sensors for biological applications", *Journal of Applied Physics*, vol. 93, pp. 7557–7559, 2003.

[42] S.-J. Han1, L. Xu, H. Yu, R.J. Wilson, R.L. White, N. Pourmand, and S.X. Wang, "CMOS Integrated DNA Microarray Based on GMR Sensors", Tech. Dig. International Electron Device Meeting (IEDM), pp. 719–722, 2006.

[43] C. Guiducci, C. Stagni, G. Zuccheri, A. Bogliolo, L. Benini, B. Samori, and B. Ricco,"A Biosensor for direct detection of DNA sequences based on capacitance measurements", Proceedings of European Solid-State Device Research Conference (ESSDERC), pp. 479–482, 2002.

[44] C. Stagni degli Esposti, C. Guiducci, C. Paulus, M. Schienle, M. Augustyniak, G. Zuccheri, B. Samori, L. Benini, B. Ricco, and R. Thewes, "Fully electronic CMOS DNA detection array based on capacitance measurement with on-chip analog-to-digital conversion", *IEEE Journal of Solid-State Circuits*, pp. 2956–2964, 2006.

[45] A. Manickam, A. Chevalier, M. McDermott, A.D. Ellington, and A. Hassibi, "A CMOS electrochemical impedance spectroscopy biosensor array for label-free biomolecular detection", Tech. Dig. International Solid-State Circuit Conference (ISSCC), pp. 130–131, 2010.

[46] Y. Li, C. Vancura, C. Hagleitner, J. Lichtenberg, O. Brand, and H. Baltes, "Very high Q-factor in water achieved by monolithic, resonant cantilever sensor with fully integrated feedback", Proceedings of IEEE Sensors, pp. 244–245, 2003.

[47] G. Sauerbrey, Günter, "Verwendung von Schwingquarzen zur Wägung dünner Schichten und zur Mikrowägung", *Zeitschrift für Physik* vol. 155, no.2: pp. 206–222, 1959.

[48] R. Brederlow, S. Zauner, A.L. Scholtz, K. Aufinger, W. Simbürger, C. Paulus, A. Martin, M. Fritz, H.-J. Timme, H. Heiss, S. Marksteiner, L. Elbrecht, R. Aigner, and R. Thewes, "Biochemical sensors based on bulk acoustic wave resonators", Tech. Dig. International Electron Device Meeting (IEDM), pp. 992–994, 2003.

[49] M. Augustyniak, W. Weber, G. Beer, H. Mulatz, L. Elbrecht, H.-J. Timme, M. Tiebout, W. Simbürger, C. Paulus, B. Eversmann, D. Schmitt-Landsiedel, R. Thewes, and R. Brederlow, "An integrated gravimetric FBAR circuit for operation in liquids using a flip chip extended 0.13 μm CMOS technology", Tech. Dig. International Solid-State Circuit Conference (ISSCC), pp. 422–423, 2007.

[50] G. McMahon, *Analytical Instrumentation: A Guide to Laboratory, Portable and Miniaturized Instruments*, John Wiley & Sons Ltd, 2007.

[51] K. Wise and K. Najafi, "Fully-implantable auditory prostheses: restoring hearing to the profoundly deaf", Tech. Dig. International Electron Device Meeting (IEDM), pp. 499 – 502, 2002.

[52] K. Wise, "Wireless integrated microsystems: Coming breakthroughs in health care", Tech. Dig. International Electron Device Meeting (IEDM), pp. 1–8, 2006.

[53] S. Oesterle, P. Gerrish, and P. Cong, "New interfaces to the body through implantable-system integration", Tech. Dig. International Solid-State Circuit Conference (ISSCC), pp. 9–14, 2011.

[54] T. Denison, P. Cong, and P. Afshar, "Exploring smart sensors for neural interfacing", Chapter 8 in current book.

7

CMOS Image Sensors

Albert Theuwissen

Electronic Instrumentation Laboratory, Harvest Imaging, Bree, Belgium and Delft University of Technology, Delft, The Netherlands

Over the last decade, CMOS image-sensor technology made huge progress. Not only was the imagers' performance drastically improved, but also their commercial success boomed after the introduction of mobile phones with an on-board camera. Many scientists and marketing specialists predicted 15 years ago that CMOS image sensors were going to completely take over from CCD imagers, in the same way as CCD imagers did in the mid 1980s when they took over the imaging business from tubes [1].

Although CMOS has a strong position in imaging today, it did not rule out the business of CCDs. On the other hand, the CMOS-push drastically increased the overall imaging market due to the fact that CMOS image sensors created new applications areas and they boosted the performance of CCD imagers as well.

This chapter describes the state-of-the-art of CMOS image sensors.

7.1   Impact of CMOS Scaling on Image Sensors

It is common knowledge that the scaling effects in CMOS technology allow the semiconductor industry to make smaller devices. This rule holds for CMOS imaging applications as well.

Figure 7.1 gives an overview of CMOS imager data published at IEDM and ISSCC of the last 15 years [2]. The bottom curve illustrates the CMOS scaling effects over the years, as described by the ITRS roadmap [3]. The second curve shows the technology node used to fabricate the reported CMOS image sensors, and the third curve illustrates the pixel size of the same devices. It should be clear that:

- CMOS image sensors use a technology node that is lagging behind the technology nodes of the ITRS. The reason for this is quite simple: very advanced CMOS processes used to fabricate digital circuits are not imaging friendly (issues with large leakage current, low light sensitivity, noise performance, ...);

Smart Sensor Systems: Emerging Technologies and Applications, First Edition.
Edited by Gerard Meijer, Michiel Pertijs and Kofi Makinwa.
© 2014 John Wiley & Sons, Ltd. Published 2014 by John Wiley & Sons, Ltd.

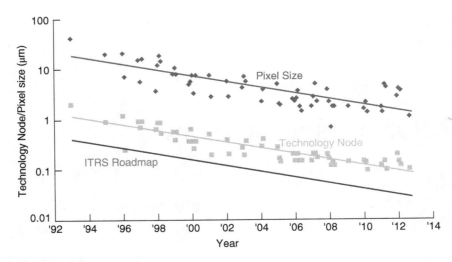

Figure 7.1 Evolution of pixel size, CMOS technology node used to fabricate the devices and the minimum dimension according to the ITRS Roadmap

- CMOS image sensor technology scales almost at the same pace as standard digital CMOS processes do;
- Pixel dimension scales with the technology node used, and the ratio is about a factor of 20.

Shrinking the pixel size for CMOS image sensors is a very important driver for the overall imaging business. It has a very large impact on various parameters of the complete camera system. For instance, if the pixel pitch of a CMOS image sensor is equal to p, the scaling factor for various parameters is (keeping the total pixel count unchanged):

pixel pitch	$\sim p,$
pixel area	$\sim p^2,$
chip area	$\sim p^2,$
chip cost	$\sim p^2,$
energy to read the sensor	$\sim p^2,$
lens volume	$\sim p^3,$
camera volume	$\sim p^3,$
camera weight	$\sim p^3.$

From this list it will be clear that there is a very strong driving force to shrink the pixel size as much as possible. Unfortunately, smaller pixels have a negative effect on the optical and electrical performance of the camera. For instance, the proportionality of the pixel/camera performance is:

depth of field	$\sim p^1$,
depth of focus	$\sim p^1$,
signal-to-noise	$\sim p^2$,
dynamic range	$\sim p^2$,

and they all become worse.

The effects of the shrinkage of a pixel can be summarized as follows: as long as a camera remains in its box, smaller pixels have only advantages. But once the camera is switched on, a smaller pixel has only disadvantages.

The market for consumer applications is asking for smaller pixel sizes at the same time that progress in CMOS technology is also offering the means to fabricate them. But as can be concluded from the table above, smaller pixels result in a weaker performance. It is a real challenge to improve the pixel design as well as the processing technology, at a pace that can counteract the loss of performance as the pixels shrink.

7.2 CMOS Pixel Architectures

In principle a CMOS image sensor has a very similar architecture as a digital memory, see Figure 7.2. It is composed of:

- an array of identical pixels, each having at least a photodiode and an addressing transistor, the number of pixels ranging from 330,000 for VGA-size imagers, to 100 M (or even more) for professional applications;
- a Y-addressing or scan register to address the sensor line-by-line, by activating the in-pixel addressing transistor;
- a X-addressing or scan register to address the pixels on one line, one after another;
- an output amplifier.

The structure of the pixels can be very simple: a combination of an n$^+$-p photodiode and an addressing transistor RS that acts as a switch, see Figure 7.3. The working principle can be understood as follows [4]:

- At the start of an exposure the photodiode is reverse biased to a high voltage (e.g. 3.3 V). This reset action is performed by means of circuitry present on the column bus (not shown in the figure). To allow the pixel to be reset, the row select (RS) needs to be active such that the pixel is connected to the column bus. Once the pixel is reset, the exposure can start and the RS switch will become inactive.
- During the exposure time, the n$^+$-region (cathode) of the photodiode is left floating. Imping-ing photons might get absorbed in the silicon and as a result of this action, electron-hole pairs can be generated. The present electrical field across the junction of the photodiode will sep-arate the two charge carriers. Electrons will move to the n$^+$ side of the photodiode and the hole will move to the p-substrate side of the photodiode. In this way, the reverse voltage across the photodiode will decrease.

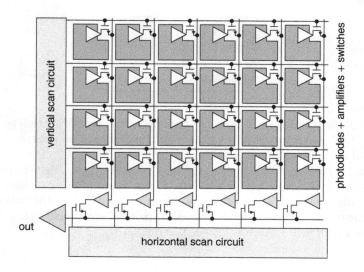

Figure 7.2   Schematic architecture of a 2-dimensional CMOS image sensor

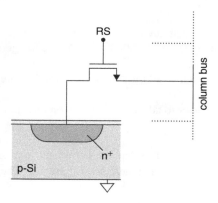

Figure 7.3   Passive CMOS pixel based on one in-pixel transistor, RS, used as the row-selection switch

- At the end of the exposure time the remaining voltage across the diode is measured, and its drop from the original value is a measure for the amount of photons falling on the photodiode during the exposure time. It should be clear that the measurement of the voltage across the diode requires an activation of the RS switch.
- To allow a new exposure cycle, the photodiode is reset again.

This so-called passive pixel is characterized by a large fill factor (ratio of diode area and total pixel area), but unfortunately, the pixel is suffering from a large noise level as well. The reasons for this are:

(a) the unavoidable presence of the kTC noise. kTC noise is introduced every time a capacitor is charged or discharged through a resistor, and this is what is happening in the photodiode. Every time the photodiode is being reset, its junction capacitance is charged to the reset voltage and kTC noise is introduced [k: Boltzmann's constant, T: absolute temperature, C is the junction capacitance of the photodiode];

(b) the mismatch between the small pixel capacitance and the large vertical bus capacitance will always result in relative low signal-voltage levels on the column bus, which are very susceptible to signal-to-noise issues.

A major improvement in the noise performance of the pixels was obtained by the introduction of the active pixel concept [5–7]: every pixel gets its own in-pixel amplifier, being a source-follower, see Figure 7.4. The pixel is composed out of the photodiode n⁺-p-substrate junction, the reset transistor RST, and addressing or row-select transistor RS and the driver of the source-follower SF (Figure 7.4). The current source of the source-follower is placed at the end of the column bus. The working principle of the active pixel sensor is basically the same as for the passive pixel sensor:

- The photodiode is reversely biased or reset by means of activating the RST switch.
- Impinging photons can be absorbed, which causes generation of electron-hole pairs. Under influence of the electrical field across the junction of the photodiode, the charge carriers will be separated and will move to the n⁺ side of the junction (for the electrons) or the p-substrate (for the holes). Consequently, the reverse voltage across the photodiode will be decreased.
- At the end of the exposure time, the pixel is addressed (by activating the RS switch) and the voltage change across the diode is measured by the source follower and "copied" on the column bus.

Next, the photodiode is reset again by means of re-activating the RST switch. This concept of active pixel sensor became very popular in the mid-1990s, it solved a lot of noise issues. Unfortunately the kTC noise component, introduced by resetting the photodiode, still remained.

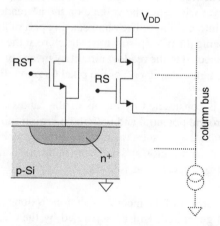

Figure 7.4 Active CMOS pixel based on an in-pixel amplifier. The transistors RST and RS are used for resetting and selection of the pixel

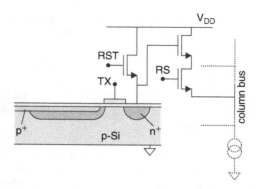

Figure 7.5 PPD CMOS pixel based on an in-pixel amplifier in combination with a pinned photodiode. RST, RS and TX are respectively the reset, row select and transfer transistor

To solve this kTC-noise issue, the so-called pinned photodiode pixel, also popular in CCD image sensors, was introduced. The pinned-photodiode pixel has the great advantage to allow Correlated-Double Sampling (CDS) to cancel the kTC noise of the reset action, $1/f$ noise of the source follower MOS transistor as well as the DC offset introduced by the source follower [8]. CDS in CMOS image sensors was demonstrated for the first time with a photogate Active Pixel Sensor or APS [9]. Having the CDS option in CMOS imagers like it was the case in CCDs, was a real breakthrough in allowing CMOS imagers to achieve higher performance.

The pinned-photodiode APS (or PPD), shown in Figure 7.5, can be seen as a logical improvement of the photogate APS, it combines the low-noise performance due to CDS with the high light sensitivity and low dark current of a photodiode [10, 11].

At the right side of Figure 7.5, one can recognize exactly the same structure as in the active pixel sensor. Although the structure is conceptually identical to the one shown in Figure 7.4, its functionality is different. The right part of Figure 7.5 is now acting only as the readout part of the pixel: the source-follower will sense the voltage on the n^+ readout node, the latter is used to convert a charge packet into a voltage. The photosensing part of the pixel is incorporated by the (pinned) photodiode, being the p^+-n-p-substrate structure at the left side of Figure 7.5. The (pinned) photodiode is connected to the readout circuit by means of an extra transfer gate, TX. With this pixel concept the light sensing part of the pixel (= photodiode) is separated from the readout node.

The pinned photodiode is constructed by means of the sandwich p^+-n^--p-substrate, with both p-regions biased at ground potential and with the n-region fully depleted. This results in a local potential maximum in the n-region (also called pinning voltage) that is fully and only determined by the doping of the n-region and the depth of the n-p-substrate and p^+-n junctions.

The pinned photodiode pixel operates as follows:

- Conversion into charge carriers of the incoming photons is done in the (pinned) photodiode. The electron-hole pairs generated, will be separated by the electrical field present in the n-region; electrons will be stored in this n-region, while the holes will move to the p^+-layer or the p-substrate.
- At the end of the exposure time, the n^+ readout node is reset by the reset transistor RST.

- A measurement of the output voltage after reset is done, this reference signal will contain the DC-offset of the source follower, $1/f$ noise of the source follower and kTC noise introduced by the reset action. The value of this first measurement is stored on a first capacitor present in the column circuitry.
- The photodiode is emptied by activating TX and transferring all charges from the photodiode to the n^+ readout node. The charge transfer is similar to the working principle of a CCD: transporting the complete charge packet from the pinned photodiode towards the floating diffusion.
- Next the charges contained in the pinned photodiode will be transferred to the readout node by activating TX. If all charges are removed from the pinned photodiode, the original pinning voltage of the pinned photodiode will be re-established.
- A measurement is done of the output voltage after transfer, this second signal will contain the photon-generated signal, but on top of that also the DC-offset of the source follower, also the $1/f$ noise of the source follower and also the kTC noise introduced by the reset action. This second measurement is stored on a second capacitor present in the column circuitry.
- The two measured values stored on the two capacitors are subtracted from each other in the analog domain (using Correlated Double Sampling, CDS) [8], and the major noise sources are cancelled out.

The completely depleted pinned-photodiode has several very attractive features:

- The kTC noise of the readout node can be completely cancelled by means of CDS.
- CDS has also a positive effect on the $1/f$ noise of the source follower, as well as on its residual off-set.
- The kTC noise of the photodiode itself is completely absent, because in the case of full depletion, the photodiode is completely emptied. Therefore, for the pinned photodiode, kTC noise does not exist.
- The light sensitivity is depending on the width of the depletion layer. Consequently, as compared to a classical photodiode, this sensitivity will be higher, because the depletion layer of a pinned-photodiode stretches almost to the Si-SiO$_2$ interface.
- Because of the double junction (p^+-n and n-p-substrate), the intrinsic charge-storage capacitance is higher, resulting in a larger dynamic range,
- The Si-SiO$_2$ interface is perfectly shielded by the p^+ layer and keeps the interface fully filled with holes, that makes the leakage or dark current extremely low.

Considering all these advantages, it will be clear that the pinned photodiode is the preferred choice for CMOS image sensor pixels. Nearly all products on the market these days make use of this pixel architecture, and it is the pinned photodiode that really boosted the introduction of CMOS image sensors into commercial products. Apparently history is repeating itself: also the CCD business really took off after the introduction of the pinned photodiode [12].

The active CMOS pixel with a pinned photodiode is implemented with four transistors and five interconnections in each pixel. This "complicated" architecture results in a relatively low fill factor. For that reason it is very hard to make pixels based on the PPD concept smaller than 2.5 μm. The in-pixel periphery consumes too much space.

A solution for the latter problem can be found in the "shared-pixel" concept: several neighbouring pixels share the same output circuitry [13, 14]. The basic idea is illustrated

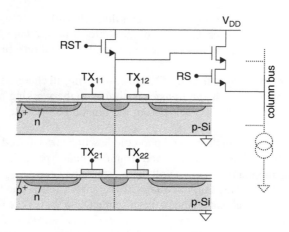

Figure 7.6 Shared pixel concept: 2×2 pinned photodiodes share the same in-pixel readout circuitry. RST and RS are the reset and row-select transistor, further selection of the individual pixels is done by means of the various transfer gates TX

in Figure 7.6: a group of 2 by 2 pixels have in common the source follower, the reset transistor, the addressing transistor and the readout node. Next to the listed components, the cluster of pixels has four pinned photodiodes and four transfer gates. The timing of pixels becomes a bit more complicated, but the shared pixel architecture is now characterized by eight interconnects and seven transistors, resulting in two interconnects and 1.75 transistors per photodiode. The positive effect on the fill factor should be clear. The price one has to pay for the shared pixel concept is an asymmetry in pixel design. The four individual pinned-photodiodes of a cluster as shown in Figure 7.6 are no longer perfectly identical to each other: within a square area, four pinned-photodiodes plus three transistors need to be placed. This results in a fixed-pattern noise component that needs to be corrected during the image-processing phase.

Lately reported image sensors with pixel sizes smaller than 2.5 µm and even down to 1.0 µm are all based on the shared pixel concept with pinned photodiodes.

7.3   Photon Shot Noise

Image sensors are characterized by many different noise sources, which can be categorized in temporal noise and spatial noise sources. Examples are:

- temporal noise: kTC noise, Johnson noise, flicker noise, RTS noise, dark current shot noise, photon shot noise, power supply noise, phase noise, quantization noise, ... ,
- spatial noise: dark fixed pattern, light fixed pattern, column fixed pattern, row fixed pattern, defect pixels, dead and sick pixels, scratches, ...

It is not the purpose of this chapter to study all these noise sources, only one important noise component will be discussed: the photon shot noise. This is the noise component due to the statistical variation in the amount of photons impinging the sensor during the exposure time. The latter is a stochastical process that can be described by Poisson statistics. If a pixel receives an amount of photons, equal to μ_{ph} during the exposure time, then this value μ_{ph} is the average value, that is also characterized by a noise component with standard deviation σ_{ph}, representing the photon shot noise. The relation between average value μ_{ph} and its associated noise σ_{ph}, is given by:

$$\sigma_{ph} = \sqrt{\mu_{ph}}. \tag{7.1}$$

After absorption of the incoming photons into the silicon, the flux of μ_{ph} photons results in μ_e electrons in every pixel, characterized by a noise component σ_e, connected by the same square root relation.

This ever-present photon shot noise component has a very interesting impact on the signal-to-noise behavior of an imaging system: in the case of a perfect noise-free imager in a perfect noise-free camera, the performance of the camera system is fully photon-shot-noise limited. The maximum signal-to-noise ratio $(S/N)_{MAX}$ is then given by:

$$\left(\frac{S}{N}\right)_{MAX} = \frac{\mu_e}{\sigma_e} = \frac{\mu_e}{\sqrt{\mu_e}} = \sqrt{\mu_e}, \tag{7.2}$$

or, the maximum signal-to-noise ratio is equal to the square root out of the signal value! This observation leads to an interesting rule of thumb: to make decent images for consumer applications, a minimum signal-to-noise ratio of 40 dB or more is needed, translated by means of the abovementioned formula into 10,000 electrons within every pixel. (This number tends to slowly go down due to extensive image processing and image-noise removal.)

On one hand, while the CMOS technology is shrinking further down, allowing for smaller pixels, the lower limit of the pixel size will no longer be determined by the minimum dimensions set by the CMOS technology, but it will be determined by the amount of electrons that can be stored in the pixel. On the other hand, a lower saturation level of electrons in a pixel will always result in a lower signal-to-noise ratio, because in best-case situations, the photon shot noise will be the dominant noise source.

7.4   Analog-to-Digital Converters for CMOS Image Sensors

It should be clear that in the era of digital imaging, most CMOS image sensors are provided with an analog-to-digital converter allowing the output signal to be accessible in the digital domain. Classical ADC architectures can be used in combination with the CMOS imager, for example, flash converter, sigma-delta converter, successive approximation, single-slope ADC, pipelined ADC, cyclic ADC, and so on. Only one particular architecture will be discussed in this chapter: the single-slope ADC. This concept is very appealing in the case where the CMOS imager is provided with an ADC for every column or even for every pixel. In particular, column-parallel conversion has some very interesting advantages for high-speed applications. Because in this case the sensor chip has as many ADCs as it has columns, and all these ADCs work fully in parallel [15–17].

The basic working principle of the single-slope ADC is illustrated in Figure 7.7. The analog input signal V_{IN} that needs to be converted is compared to an analog ramp signal V_{ramp}. A digital counter generates the latter. At the moment that the two voltages V_{IN} and V_{ramp} are equal to each other, the comparator changes state and latches the counter value into a memory. The data

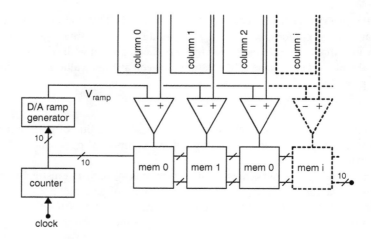

Figure 7.7   Basic architecture of a column-parallel single-slope analog-to-digital converter

stored into the memory will be the digital value corresponding to the analog input voltage V_{IN}. In the case of column-parallel conversion, the imager has at every column a comparator and a digital memory. The digital counter is common for all pixels on a single row.

After digitization, the output signal of the camera will have an extra quantization noise component σ_{ADC}, equal to:

$$\sigma_{ADC} = \frac{V_{LSB}}{\sqrt{12}},\qquad(7.3)$$

with V_{LSB} being the analog voltage of the least significant bit.

In relation to the photon shot noise, an interesting observation can be made: the noise floor in the output signal of an image sensor is always (best-case) determined by the photon shot noise. The latter will be small for small output signals of the sensor, but it will be large for large output signals of the sensor. In the case of a large output signal, the quantization error of the ADC does not have to be as low as it should be for smaller output signals. This idea allows an ADC converter with an adaptive quantization step: small for small signals, large for large signals. This idea can be relatively easily implemented by means of the single-slope ADC. In that case the ramp, generated originally by the digital counter, will be no longer linear with respect to the time, but can have a piece-wise-linear approach as shown in Figure 7.8 [18]. Next to the ramp itself, in Figure 7.8 also the photon shot noise is indicated as well as the quantization noise. It can be seen that when the quantization step is increased, the quantization noise is increased as well. But as long as it stays well below the photon-shot noise, it will not hamper the performance of the sensor. In this simple example (in which the quantization noise is kept a factor of 2 below the photon-shot noise), the ADC is changing from a single-slope to a so-called multi-slope ADC. In this way, its speed is increased by a factor of 3 without further increasing its power consumption.

Another way of increasing the speed of the single-slope ADC is changing the concept to a single-slope, multiple-ramp architecture. In this configuration, several ramps are running in parallel, they all have the same slope, but they differ from each other by a DC off-set [19]. Before starting the conversion, a coarse ADC action is performed, to assign every column of

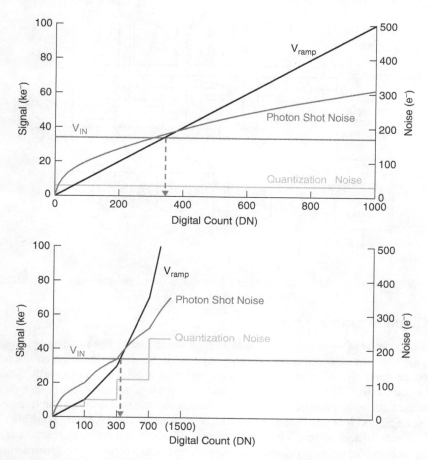

Figure 7.8 Ramp voltage for the single slope (top) and multi slope (bottom) ADC, in relation to the photon shot noise and quantization noise

the image sensor to a dedicated ramp. The coarse conversion is followed by a fine conversion cycle, and at the end, both results are combined. In Figure 7.9, the multiple-ramp concept is illustrated: at first the coarse action is taking place and its output is memorized in a 2-bit memory cell (two bits in this example with four parallel ramps).

These two bits not only represent the most significant bits of the digital words, but they also contain the information to which ramp the column needs to be assigned for the fine conversion. During the latter, the four parallel ramps are offered to all columns, but every column is checked against only one particular ramp. It should be clear that the increase in speed is about equal to the number of parallel ramps (neglecting the time needed to perform the coarse ADC).

This example of implementing a single-slope, multiple-ramp architecture column-parallel ADC greatly shows a key advantage of CMOS image sensors: implementation of additional analog and digital circuitry on the same chip as the imaging core. In the mean time the noise characteristics of the imager are taken into account to improve this on-chip circuitry as far as speed and power consumption are concerned.

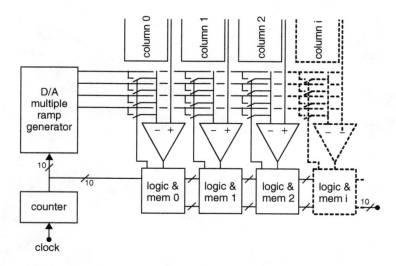

Figure 7.9   Basic architecture of a column-parallel single-slope, multi-ramp analog-to-digital converter

7.5   Light Sensitivity

The main purpose of an image sensor is to convert the incoming light into a measurable output signal. But with the shrinkage of the pixel size, light sensitivity is becoming an issue. The conversion of the incoming photons into electron-hole pairs is automatically done by the photo-electric effect of the silicon. Next, the electrons are separated from the holes by means of the electric field in the photodiodes. Collected charge carriers will decrease the reverse-biased diode voltage, and it is this decrease that can be measured. To make images of high quality and/or to make images at very low light levels it is of crucial importance to catch and convert as many as possible, not to say every, incoming photon. Unfortunately, the construction of the pixels does not allow all photons impinging the pixel to be used. The main reasons for this effect are:

- Part of the pixel is simply insensitive to incoming light, because part of the pixel will contain the readout electronics. The concept of the pixel limits the light sensitive area in comparison to the total pixel area (this ratio is also known as fill factor).
- Photons falling on the active pixel area not necessarily generate electron-hole pairs. The photons can be reflected at the silicon interface, can be reflected at one of the many interfaces in the multi-layered optical stack above the silicon, can be absorbed in one of these layers, or the photons even can pass through the silicon without absorption.
- Not every electron-hole pair generated by a photon will be collected. Charge carriers can recombine, or can get lost in the substrate.

Modern CMOS image-sensor technology has developed several techniques to overcome all the aforementioned sensitivity issues. Examples of these techniques are:

- Microlenses that are put on top of every single pixel, made out of deep UV photo-resist [20].

- Inner-lenses that are realized by means of the right combination of SiO_2 and Si_3N_4 layers during the fabrication process [21].
- Wave or light guides placed on top of the pixels [22].
- Back-side illumination [23].

The back-side illumination is a method already applied for professional CCDs, but recently it is introduced for small-pixel CMOS imagers as well. Bringing the photons from the backside has the following advantages:

- Nearly 100% fill factor, resulting in a very high quantum efficiency.
- Very low optical stack, resulting in a lower angular dependency of the light sensitivity and reduced optical cross-talk.
- The back-side (being the light sensitive part) can be processed fully independent of the front-side (being the CMOS circuitry part).

But the back-side technique comes with a certain price: the silicon wafer needs to be thinned down to just a few micrometers, the backside needs a very specific passivation technique, and new packaging methods have to be developed. Despite all these issues, back-side illuminated CMOS devices with pixels down to 1 μm are available on the market. Figure 7.10 shows an SEM cross section of such a back-side illuminated CMOS sensor with 1.65 μm pixel size.

Starting from the bottom part of the Figure 7.10 [23], one can recognize:

- The fourth metal layer, which is chosen to be the power line. Because the light is coming from the other side, the power line can be chosen thick and wide to prevent any voltage drops.
- Three metal layers which are used as the interconnect of the former front side.
- The silicon substrate, which is thinned down to a handful of microns.

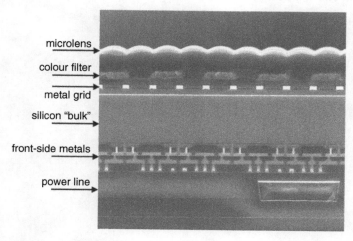

Figure 7.10 Cross section of a back-side illuminated CMOS image sensor. Reproduced by permission of IEEE [23]

- A metal grid used to prevent light falling "between" two color filter patches.
- The color filter array, although hardly visible, the cross section shows two different layers, e.g. red and green, or blue and green.
- Ultimately at the top of the array of microlenses. It may be surprising to find microlenses on top of a back-side illuminated sensor. But the microlenses have a double function: focusing the incoming rays away from the region that is covered with the metal grid, and focusing the incoming rays in the middle of the pixels to limit the crosstalk.

7.6 Dynamic Range

Another interesting issue that comes together with the pixel shrinkage is the reduction in dynamic range. The dynamic range of an imager is the ability of the device to detect details in high lights and low lights within the same image. In numbers, the dynamic range is defined by the largest signal that can be detected (= saturation level) and the minimum signal that can be detected (= noise floor in dark). But if the pixels get smaller, the amount of charge that can be stored in a pixel is reduced as well, so is the dynamic range.

Today, there are several potential solutions being proposed in the literature. One of the very first techniques implemented in products was the dual (or multiple) exposure. Every image is composed by means of two (or more) full-resolution exposures: a short exposure to capture the details in high lights and a long exposure to capture the details in low lights. Although this technique suffers from motion artifacts, it is pretty easy to implement, because it does not require any change in pixel design or lay-out.

Another approach can be found in the so-called LOFIC pixel design: Local Overflow with an In-Pixel Capacitor. This pixel extends its charge handling capacitance by adding an extra capacitor C_S as can be seen in Figure 7.11 [24]. The normal photon conversion takes place in the pinned photodiode, but in the case of a high light, the pinned photodiode will saturate and electrons will overflow across the transfer transistor TX onto the floating diffusion C_{FD}, and if that capacitor is completely "filled", further overflow can take place in C_S through the TS transistor. At the end of the exposure, first the charge stored on the C_S and C_{FD} capacitor is read

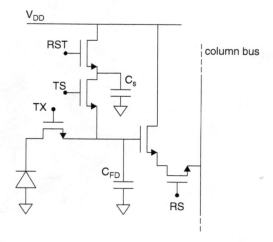

Figure 7.11 LOFIC pixel: local overflow with an in-pixel capacitor

out, next the floating diffusion is reset, the charge from the pinned photodiode is transferred and read. So the complete readout cycle is composed out of two read cycles. The extra capacitor serves as an extra storage capacitor in the image capture mode, and an extra capacitor in the signal readout mode. The LOFIC pixel is able to extend the dynamic range by an extra 60 dB.

The concept of this LOFIC pixel nicely illustrates one of the key advantages of CMOS over CCD imagers: the ability to integrate extra circuitry, even on pixel level. These kinds of solutions are not possible in CCDs.

7.7 Global Shutter

Another great example of extra integrated circuitry in the pixel is the global shutter CMOS pixel. Referring to Figure 7.2, it should be clear that the readout process of a CMOS image sensor is done row wise: row after row the pixels are being addressed, readout and reset. Every time the pixels are reset, a new exposure time start, every time the pixels are read, the exposure time ends. This way of reading a sensor is being known as a sensor with a rolling shutter. The drawback of the rolling shutter is the fact that the exposure time of every line starts at a different point in time and ends at a different point in time. This results in very annoying motion artifacts that are very difficult to compensate. For that reason a global shutter was developed in CMOS: all pixels start the exposure at the same time, and all pixels end the exposure at the same time, just like in a CCD. An example of such a global shutter pixel CMOS pixel is shown in Figure 7.12 [25]. The pixel is based on the 4T concept with a pinned photodiode. It works as follows:

- At the end of the exposure time, the floating diffusion nodes of all pixels are reset.
- The reset reference levels of all pixels are converted into a voltage by the floating diffusion capacitances, these voltages are stored on C_2 by activating switches V_{SAM1} and V_{SAM2}.
- After closing switches V_{SAM2}, the charges from all pinned photodiodes are transferred to the floating diffusions and sampled on C_1.
- Until this moment all pixels are operated fully in parallel, but from now on the readout is done line by line.
- First the voltage on C_2 is measured, this is the reference voltage after the pixels were reset.
- Next V_{SAM2} is made active, and the voltage on C_1 is shared on $C_1 + C_2$, in this way the video signal is measured.

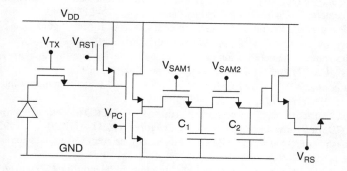

Figure 7.12   Global shutter CMOS pixel with in-pixel storage of the reset and video signal

Although the pixel is relatively complicated (two extra capacitors and four extra transistors), this global shutter implementation has the advantage of allowing correlated double sampling to cancel out the famous kTC noise.

7.8   Conclusion

CMOS image sensors made a huge technological progress over the last decade. The introduction of the pinned photodiode really boosted their success. Exploring the typical characteristics, needs and requirements of the imaging application can result in very attractive circuits and devices that increase the performance of the imagers. Because of the ever-shrinking dimensions in CMOS technology, further integration on column level and even on pixel level can make the imagers even smarter than they are already.

Acknowledgment

The author would like to thank all his imaging-community colleagues that contributed to the R&D of solid-state image sensors. A special thanks goes to the people who did a lot of pioneering work on CMOS imagers in the early 1990s. They not only introduced new technical, market and business opportunities, but they also challenged the CCDs as never before to increase their price-performance ratio.

References

[1] E.R. Fossum, "Active pixel sensors versus CCDs", *IEEE Workshop on CCDs & AIS*, June 9–11, 1993, Waterloo (ON).

[2] International Electron Devices Meeting and International Solid-State Circuits Conference, Digest of Technical Papers from 1990 to 2007.

[3] ITRS Roadmap, see also www.itrs.net

[4] G.P. Weckler, "Operation of p-n junction photodetectors in a photon flux integrating mode", *IEEE Journal of Solid-State Circuits*, pp. 65–73, September 1967.

[5] P. Noble, "Self-scanned silicon image detector arrays", *IEEE Trans. Electron Devices*, vol. 15, 1968, pp. 202–209.

[6] F. Andoh *et al.*, "A 250,000 pixel image sensor with FET amplification at each pixel for high-speed television cameras", *ISSCC Digest Technical Papers*, pp. 212–213, 1990.

[7] O. Yadid-Pecht *et al.*, "A random access photodiode array for intelligent image capture", *IEEE Transactions on Electron Devices*, pp. 1772–1780, 1991.

[8] M.H. White *et al.*, "Characterization of charge-coupled device line and area-array imaging at low light levels", *ISSCC Digest of Technical Papers*, pp. 134–135, 1973.

[9] S. Mendis *et al.*, "A 128 × 128 CMOS active pixel image sensor for highly integrated imaging systems", *IEDM Technical Digest*, pp. 583–586, 1993.

[10] P. Lee *et al.*, "An active pixel sensor fabricated using CMOS/CCD process technology", *1995 Workshop on CCDs and Advanced Image Sensors, Dana Point (CA)*, April 20–22, 1995.

[11] R.M. Guidash *et al.*, "A 0.6 μm CMOS pinned photodiode color imager technology", *IEDM Technical Digest*, pp. 927–929, 1997.

[12] N. Teranishi *et al.*, "No image lag photodiode structure in the interline CCD image sensor", *IEDM Technical Digest*, pp. 324–327, 1982.

[13] H. Takahashi *et al.*, "A 3.9 µm pixel pitch VGA format 10b digital image sensor with 1.5 transistor/pixel", *ISSCC Digest of Technical Papers*, pp. 108–109, 2004.

[14] M. Mori *et al.*, "A 1/4" 2M pixel CMOS image sensor with 1.75 transistors/pixel", *ISSCC Digest of Technical Papers*, pp. 110–111, 2004.

[15] K. Chen *et al.*, "PASIC – A processor-A/D converter sensor integrated circuit", *IEEE Int. Symp. Circuits and Systems*, pp. 1705–1708, 1990.

[16] S. Mendis *et al.*, "Design of a low-light-level image sensor with an on-chip sigma-delta analog-to-digital conversion", *Proc. SPIE*, vol. 1900, pp. 31–39, 1993.

[17] T. Sugiki *et al.*, "A 60 mW 10 bit CMOS image sensor with column-to-column FPN reduction", *ISSCC Digest of Technical Papers*, pp. 108–109, 2000.

[18] M.F. Snoeij *et al.*, "A low power column-parallel 12-bit ADC for CMOS imagers", *IEEE Workshop on CCDs & AIS*, pp. 169–172, Karuizawa (Japan), 2005.

[19] M.F. Snoeij *et al.*, "A CMOS image sensor with a column-level multi-ramp single-slope ADC", *ISSCC Digest of Technical Papers*, pp. 506–507, 2007.

[20] Y. Ishihara *et al.*, "A high photosensitivity IL-CCD image sensor with monolithic resin lens array", *IEEE IEDM*, pp. 497–500, 1983.

[21] Y. Sano *et al.*, "On-chip inner-layer lens technology for an improvement in photo-sensitive characteristics of a CCD image sensor", *The Journal of the Institute of Image Information and Television Engineers*, vol. 50, pp. 226–233, 1996.

[22] H. Watanabe *et al.*, "A 1.4 µm front-side illuminated image sensor with novel light guiding structure consisting of stacked lightpipes", *IEEE IEDM*, pp. 8.3.1–8.3.4, 2011.

[23] H. Wakabayashi *et al.*, "A 1/2.3-inch 10.3 Mpixel 50 frame/s back-illuminated CMOS image sensor", *ISSCC Digest Technical Papers*, pp. 410–411, 2010.

[24] N. Akahane *et al.*, "A sensitivity and linearity improvement of a 100 dB dynamic range CMOS image sensor using a lateral overflow integration capacitor", *Digest of Technical Papers VLSI Circuits*, pp. 62–65, 2005.

[25] X. Wang *et al.*, "A 2.2M CMOS image sensor for high speed machine vision applications", *SPIE Electronic Imaging*, vol. 7536, pp. 0M1-0M7, 2010.

8

Exploring Smart Sensors for Neural Interfacing*

Tim Denison, Peng Cong and Pedram Afshar
Medtronic Neuromodulation, Minneapolis, USA

8.1 Introduction

Neuromodulation aims to improve disease-state control with ongoing therapy adjustments that enhance therapeutic response while minimizing clinician and patient burden. Innovation in neuromodulation might be facilitated by modeling the interaction between device and the nervous system in a dynamic control framework. While advances in sensing [1, 2], therapy delivery [3, 4], and understanding the pathophysiology of disease states [5–7] have already aimed to improve device performance, dynamic control theory provides an alternative paradigm to advance the field of neuromodulation. As illustrated in Figure 8.1, a classic control paradigm consists of a "plant" (the nervous system), an actuator (neural stimulator), a sensor (clinical data collector), and a state estimator (assessment by the clinician, patient, caregiver or an automated algorithm). In this context, the actuator is any device or method that modulates the activity of a functional group of neurons. We call these "stimulators" for simplicity. Within a dynamic control framework, the detailed, desired function for each subcomponent is:

- Defining the patient's desired "state" using objective (preferably quantitative) criteria (Figure 8.1a). By "state", we mean the disease-relevant clinical condition of the patient. Some disease states are well correlated to biomarkers, such as the relationship between EKG and myocardial infarction. Many neurological disease states do not have well-correlated biomarkers and are difficult to discern, such as schizophrenia.
- Improving control through more sophisticated neurostimulation parameters (e.g., lead and electrode selection, field steering, selective stimulation, stimulation frequencies, amplitude, and pulse patterns). (Figure 8.1b).

* Elements of this work where published in reference [1]. The authors wish to thank Dr. Brian Litt at UPenn for helpful conversations on this subject.

Smart Sensor Systems: Emerging Technologies and Applications, First Edition.
Edited by Gerard Meijer, Michiel Pertijs and Kofi Makinwa.

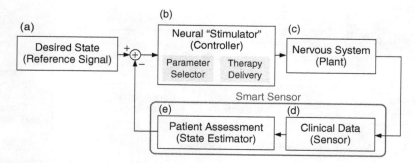

Figure 8.1 A dynamic control framework for analyzing neuromodulation systems. (a.) Desired neurological state is the input reference signal, (b.) the neural stimulator is the actuator, (c.) the nervous system is the plant, (d.) transducers and observations to collect clinical data are the sensors, and (e.) patient assessment is the state estimator. In current clinical practice, the difference between a physician's acute estimate of the patient state and the desired clinical state drives parameter changes in the device, with adjustments often limited to sparsely-sampled measurements. A role of smart sensors is to emulate and facilitate clinical judgment through quantitative measurements and algorithms

- Understanding and applying nervous-system disease pathophysiology as the foundation for control strategies, for example, how stimulation parameters affect the desired state (Figure 8.1c).
- Improving disease-state discernment by measuring relevant pathophysiological biomarkers (Figure 8.1d) and estimating the patient state (Figure 8.1e).

Currently, most stimulators operate in a so-called open-loop fashion, requiring an operator to change settings, such as voltage, frequency, or pulse width. In this case, clinical observations and tests serve as sensors to generate the data used by physicians to assess the patient. In practice, this is actually a closed-loop system, in which the clinician is the mechanism for providing feedback. A schematic representation of the dynamic system from a clinical flow perspective is shown in Figure 8.2. Merging combined biophysical sensors and state estimator will be defined as the "smart sensor" in this context and for the applications discussed in this chapter. The scope of "smart sensor" is illustrated both in Figure 8.1 and Figure 8.2.

When applying smart sensing technology, the control framework takes advantage of well-understood dynamic-control principles that aim to be clinically relevant and provide lessons for designing closed-loop neuromodulation systems. For example, enabling technologies should minimize time delay in the control loop by providing timely feedback and control automation from sensing and state estimation. Controlling delay helps to maintain stability by allowing the algorithm to respond at therapeutically relevant time scales: slow actuation may lead to over-damping that may fail to provide therapy in time, while fast actuation may lead to under-damping that risks driving system oscillation. It is also important to clearly understand the sensing-actuation interaction and minimize the impact of direct feed-through in the control loop, which might otherwise mask the observation of the true patient state. Control systems might also account for non-linearities and time-dependencies to improve performance during state transitions. Given the novelty of these designs, clinician oversight is required to initially optimize control parameters to ensure appropriate use of the therapy to obtain the desired clinical response without side effects. In practice, collaborations with clinicians are essential in defining the algorithms that ensure the sensor is actually "smart."

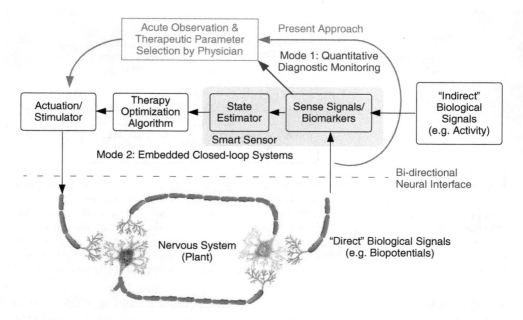

Figure 8.2   Modeling clinical flow of a patient from a feedback perspective. The pathways can include clinician and/or patient feedback-based observations, or be automated with embedded sensors and algorithms. In practice, most neuromodulation devices today are closed-loop, but the clinician and patient form the feedback mechanism. Technology can improve these systems through two modes. Mode 1 is the improvement of existing feedback paths with enhanced sensing of biomarkers. Mode 2 is implementation of a closed-loop system within the device. Both modes of operation employ smart sensing technology to facilitate quantified observation of patient state. Reprinted with permission from John Wiley & Sons [1]

In the sections that follow these concepts are explored in more detail. Section 8.2 discusses state-of-the-art challenges in each sub-block of a dynamic neural control framework from a broad neuromodulation perspective. To provide additional context for physiological control, Section 8.3 reviews early use cases from cardiac rhythm management that illustrate the application of closed-loop control for therapy enhancement. Translating to the neurological space, Section 8.4 presents a case study of control system methodologies employed in an implantable spinal cord stimulator. The device combines inertial sensing, stimulation and state estimation for real-time therapy titration based on a patient's posture and activity level. Section 8.5 discusses the early investigational application of smart sensors to the design and prototyping of closed-loop neural systems based on direct measurement of neural network activity. Section 8.6 briefly discusses the opportunity and challenges of extending these closed-loop methodologies to the broader neuromodulation space.

8.2   Technical Considerations for Designing a Dynamic Neural Control System

To be practical, the conceptual components of the feedback paradigm from Figures 8.1 and 8.2 must be consolidated in complete system, which often takes the form of a battery-powered,

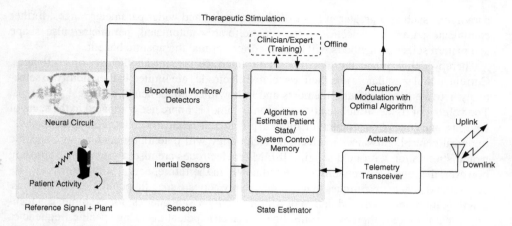

Figure 8.3 Mapping the abstracted feedback loops to a prototypical stimulator system. The key block diagrams of a functional stimulator are highlighted to illustrate the required technology. Reprinted with permission from John Wiley & Sons [1]

chronically implantable device. This mapping drives several technical considerations for designing a dynamic neural control system, including the application of smart sensing and feedback algorithms. One design architecture concept is illustrated in Figure 8.3, where the elements of the control loop are broken down into constituent parts for systematic analysis. The first step is to identify the potential sources for the key sub-elements from a broad neuromodulation perspective, which are then integrated together to form the complete system in the final system.

Desired Physiological and/or Clinical State (Reference Signal): The desired state, analogous to the reference signal in the control paradigm, represents the clinical outcome, which the clinician would like to achieve with optimal therapy. For example, a patient with Parkinson's disease might desire to be in the "on" state, meaning that symptoms are well controlled with medications and/or deep brain stimulation (DBS). An epileptic patient may desire homeostasis with no seizures (e.g., being free of the ictal neural state). For a chronic pain patient, the desired state is pain-free, but this might also be a more nuanced state such as presence of paresthesia in the affected dermatome. Precise definition of the desired state undergoes continuous refinement, as increasingly sensitive and specific biomarkers are identified and developed. In addition, the desired state may vary between patients, and usually changes over time for a given patient. Improved understanding of the mechanisms underlying neurological diseases should provide more objective measures for more accurately defining the desired state.

Neural Stimulator (Actuator): Current neural stimulators, effectively electric pulse generators, have an associated set of configuration parameters known as the stimulation settings (for example, frequency, amplitude, and pulse width; electrode anode/cathode; number of active electrodes; bipolar or referential channel selection). At present, the physiologic relationships between stimulation parameters and the nervous system are not well understood, leading to parameter selection procedures that can be both cumbersome and may not always optimize patient benefit. Increasingly complex therapy delivery paradigms, such as a greater number of electrodes [8] and wider parameter ranges, further complicate parameter selection. Efficient, preferably automated, parameter/pulse shape and pattern selection methods will be needed for optimal therapeutic benefit.

Therapy delivery and the electrode-tissue interface present another set of challenges. Current neural stimulators and their associated electrodes are limited in their ability to selectively activate specific neural elements and in how they affect neural electrical potentials. The volume of tissue stimulated (stimulation volume) is imprecise compared to the scale of neurons, and control over neural membrane potential is relatively coarse compared to physiological control. Advances in electrode technology will potentially enable more specific neural stimulation with field steering [8]. Microelectrodes are also a promising approach; however, the long-term reliabilities are unknown. In the future, neural stimulators may use cellular and genetic techniques that can selectively activate specific populations of neurons, enabling more nuanced modulation of neural activity. Optogenetics, for example, is a technique that uses gene therapy to control neural electric potentials using specific frequencies of light [5].

Nervous System (Plant): The representation and temporal dynamics of the disease state in the nervous system, or plant, are important in understanding the relationship between the neural stimulator and observed sensor data. However, dynamic system identification to describe the nervous system and its non-linear relationship to therapy remains a formidable challenge. Neural network models by Hahn and McIntyre [9] and Tass et al. [7] demonstrate examples of using physiology-based representations to help describe the dynamic effects of neural stimulation. Work by Holsheimer in the conventional spinal cord stimulation (SCS) field is another example of understanding stimulation mechanisms for optimizing therapy, in this case guiding stimulation for the desired activation of dorsal columns while avoiding stimulation in the dorsal roots that might lead to unwanted sensory side effects or pain [10].

Improved understanding of the nervous system at the cellular and network levels and increased computing power may improve the ability to characterize the plant and enable robust and precise neural stimulation and state observation strategies.

Quantified Clinical and Physiological Data Collection (Sensors): Sensors are critical components in the collection of physiological data from the patient. As sensors become smaller, more economical, and power-efficient [30] it is more likely that sensor-specific disease states can be defined. Chronic neural signals are a natural place to look for biomarkers of neurologic diseases. Work in areas such as Parkinson's disease indicates that the amplitude of neural activity in the beta band (10–30 Hz) in parts of the basal ganglia may be related to the degree of movement dysfunction. In these patients, beta band amplitude has been estimated to be in the order of $1 \mu V_{RMS}$ in the local field potential (LFP) [1], which is more than 100 times smaller than cardiac pacing signals. This poses significant technical challenges to developing sensors aimed at identifying these biomarkers and other important signals [8]. Other bioelectrical areas of interest, such as impedance and ECG, may correlate to physiological effects of diseases such as stress and have been previously used in closed-loop cardiac pacemakers. In addition to bioelectrical signals, it may be possible to infer disease state information from other physiologic markers (for example, limb movement, activity/posture, respirations) using sensors like inertial accelerometers, which are becoming ubiquitous in modern commercial devices such as cell phones and personal digital assistants.

Clinical Assessment of the Patient State (State Estimator): In neuromodulation, we define state estimation as the process of translating sensor information into an estimate of the patient state. In most of today's commercial neuromodulation systems, this estimation is performed via clinician or patient observation, and the clinician or patient then makes adjustments through the therapy programmer via telemetry as illustrated in Figure 8.2. With the goal of making the system "smarter," which in essence means to provide more information and be more automated, engineers and researchers are working toward embedding the state estimator with appropriate sensors to model and/or supplement clinical assessments. There are many ways to perform state estimation including using physiologic models and machine learning techniques (e.g., Kalman filters and support vector machines), each with their own trade-offs.

Several goals and challenges are important to understand in order to achieve adequate state estimation. First, sufficient information about the physiology and pathophysiology of the target neural network is needed to understand the relationship between sensed biomarkers and the disease state. This understanding is fundamental to adequately capture the complex relationships between sensor measurements and the desired patient state in order to deliver optimal therapy. Second, the process of data collection and algorithm deployment in the flow of patient care needs to be well characterized and committed to protocol for efficient care. Third, limited understanding of the true patient state complicates the process of state estimator validation and can confound algorithm validation in the clinic. A successful, practical state estimator requires solutions for all of these aspects.

This section provided a brief overview of the considerations for designing a closed-loop neuromodulation system. To help provide additional context for this design paradigm, we now briefly describe a few historical examples from cardiac rhythm management devices that illustrate the key concepts from a complete system perspective.

8.3 Predicate Therapy Devices Using Smart-Sensors in a Dynamic Control Framework: Lessons Derived from Closed-Loop Cardiac Pacemakers

An early example for the application of the dynamic control framework can be found in cardiac devices used to treat bradycardia. The goal of this system can be thought of as setting a patient-specific set point of hemodynamic output to consistently support daily activities. A block diagram of the closed-loop bradycardia pacemaker is shown in Figure 8.4.

Initial pacemakers were open-loop designs that would provide roughly one pacing pulse per second independent of the patient's intrinsic heart rate or activity level. These open-loop devices operated on a simple principle of supplying a guaranteed minimal pacing frequency, enabled with the "clock" counter increasing until reaching a terminal count (time), which triggered a pacing pulse and reset of the counter. The counter-based algorithm provided a set rate of stimulation which could be adjusted manually through the level of the terminal count variable. Note that this fixed-rate/amplitude pacing is quite similar in operation to state-of-the-art neuromodulation devices today. The pacemaker strategy was initially acceptable when the technology was introduced, due to the lack of better technology options. But it arguably resulted in non-optimal performance due to lack of response to changes in hemodynamic demand,

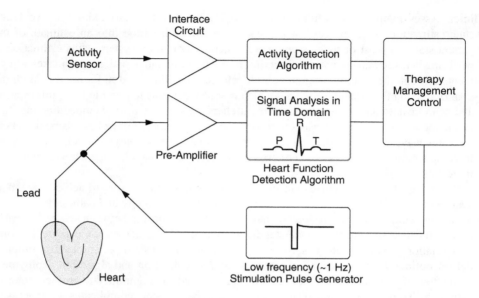

Figure 8.4 Model for bradycardia pacemaker closed-loop system. Note that there is a direct pathway of sensing physiology through an amplifier and an indirect pathway through an accelerometer. Reprinted with permission from John Wiley & Sons [1]

and unnecessary power usage when artificial pacing was not needed, resulting in accelerated battery depletion.

Making the system "smarter" with sensing and algorithms: To overcome these shortcomings, biophysical sensors and algorithms were combined and embedded within the pacemaker to apply smart sensing techniques for dynamic pacing. As shown in Figure 8.4, these sensors can be used for direct and indirect measurements of hemodynamic variables. The direct version of the closed-loop system measures a distinct variable of the hemodynamic function by sensing patient intrinsic heartbeats. As illustrated in Figure 8.5, when an intrinsic heartbeat is present, the counter resets its state. This potentially allows for significant energy

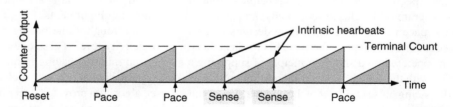

Figure 8.5 Inhibitory responsive pacing. A counter increases until hitting a terminal count that triggers a heart pace and resets the counter. The terminal count setting determines the minimum pacing rate. Any detected intrinsic heartbeats reset the counter without a pacing event, allowing for the heart to take over whenever possible and extending battery life. Reprinted with permission from John Wiley & Sons [1]

savings in the device, assuming that the intrinsic heart rate is greater than the minimal level determined by the terminal variable. The pacemaker fires only when the measured rate goes below a minimal level required for hemodynamic support.

Making the systems even "smarter": While suitable for episodic bradycardia, a shortcoming of the previously mentioned system is that it does not compensate for the variable hemodynamic demands of the patient (e.g., periods of exercise or extended rest). To address this need, a sensor that correlates with hemodynamic demands was developed. While several direct physiological signals were explored for this measurement (oxygenation measurements, sympathetic drive, etc.), designers ultimately developed an activity sensor based on a piezoelectric crystal mounted within the device. As illustrated conceptually in Figure 8.6, as the patient's *activity* varies, the terminal counter is dynamically adjusted to fine-tune the pacing rate. The degree to which the terminal counter is sensitive to activity is a clinician-selected value. Once set up and calibrated in clinic, the device works autonomously for the patient.

Considerations for "direct" and "indirect" measurements of physiological signals: This brief overview of pacemaker systems provides two key observations for engineers to consider when designing closed-loop devices: first, adaptive stimulation titration has the potential to provide more desirable actuation of the body's processes, and second, "indirect" measurements that strongly correlate with physiological variables can sometimes be sufficient for improving efficacy. The later point is especially important within the context of robust manufacturing and high reliability, where more sophisticated sensor systems might be challenged. This observation motivates one point-of-view for analyzing smart sensor systems – considering whether the measurement is an "indirect" measure of underlying physiology based strongly on correlation, or a "direct" measurement of physiological activity. In reality physiological measurements are not always easily partitioned, and the definition is obviously subject to interpretation, but it does highlight variations in strategy that might be useful to the system designer, as the next few sections will show.

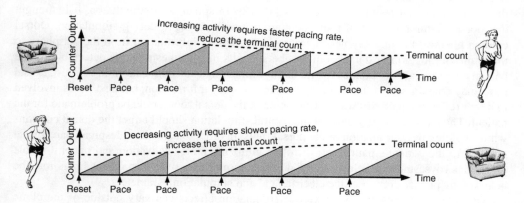

Figure 8.6   Rate-responsive pacing. The terminal count setting still determines the minimum pacing rate, but is now set dynamically by monitoring the activity of the patient. This allows for the pacing system to provide more variable hemodynamic settings that are appropriate to the immediate needs of the patient. Reprinted with permission from John Wiley & Sons [1]

8.4 The Application of "Indirect" Smart Sensing Methods: A Case Study of Posture Responsive Spinal Cord Stimulation for Chronic Pain

8.4.1 Overview of the Posture Responsive Control System

Conventional spinal cord stimulation (SCS) systems provide patients with a programmer to manually adjust stimulation during the course of the day to maintain desired effect, often described as paresthesia, which is believed to be central to the mechanism by which SCS systems alleviate pain in current systems [19]. Analogous to the pacemaker example above, recent work demonstrates that a component of the dynamics of pain and paresthesia can be variable in some patients with changes in posture and activity [19].

Based on these observations, position-adaptive stimulation is a new therapy modality that augments manual control with the capability to adjust stimulation intensity automatically. This feature is based on sensed body position and activity using an accelerometer, coupled with an algorithm for titrating stimulation based on inertial state. The goal is to maintain effective and consistent therapy based on what is deemed as most beneficial for a given patient's needs. In one approach, this method is implemented as a feed-forward algorithm that maps stimulation parameters to detected inertial state. The key elements to consider in the design of this system can be developed within the framework of a dynamic control system for spinal cord neural modulation, applying smart sensor architectures. In particular, we now describe several of the key technical blocks described previously in Figure 8.3, and their application for this problem.

8.4.2 The Design Challenge: Defining the Desired Patient State

As discussed earlier, a key design input for constructing a closed-loop system is defining the driving target for the system. One of the goals of neurostimulation is to obtain maximal selective activation or silencing of neural structures that will lead to clinical benefit while avoiding those structures that result in side effects [11, 12]. In spinal cord stimulation, it is thought that dorsal column and dorsal root fibers are primarily engaged during stimulation. Dorsal column fibers contain mostly nerves transmitting sensory information from the periphery to the brain. Stimulation of dorsal column fibers results in paresthesia that is detectable by the patient over several dermatomes rostral to the stimulating cathode location. In contrast, dorsal roots may contain nerves that not only transmit sensory information, but also fibers involved in motor reflex and pain pathways; stimulation of the dorsal roots could be problematic for the patient. Thus, it is generally thought that neural stimulation should target the dorsal columns while avoiding overstimulation of the dorsal roots, in order to achieve widespread paresthesia targeted to the patient's pain area(s) while minimizing motor activation and uncomfortable sensations (dysesthesia). Selectivity of dorsal column fibers over dorsal root fibers may be achieved by precise control of the electric field and stimulation parameters.

Pain patterns and intensities as well as stimulation effects can vary outside of the clinic on a daily or hourly basis because of changes in posture and other variables. Specifically, the amplitude and/or energy for stimulation when the patient is lying down is significantly lower compared with standing or sitting with a mean difference in the range of 11–35% [16–19]. The spinal cord moves within the subarachnoid space with changes in posture which result in variations in the distance between the stimulating electrodes (placed outside the

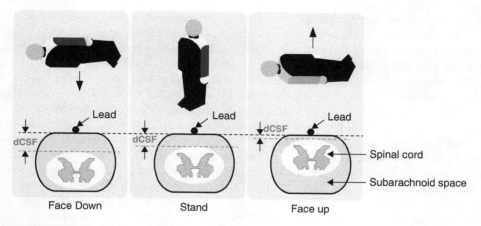

Figure 8.7 Abstracted model for dCSF variations in lead-spinal cord positions due to posture. Reprinted with permission from John Wiley & Sons [1]

subarachnoid space) and the target neurons as shown conceptually in Figure 8.7 [20]. Largely attributed to the effects of gravity, the spinal cord moves closer to the electrodes in a supine position and moves farther away from the electrodes in the prone position. The distance between the electrodes and the spinal cord is primarily determined by the thickness of the dorsal cerebrospinal fluid (dCSF). The variations in dCSF with vertebral level as well as with position result in variations in the amplitudes required for effective stimulation [16, 21–25]. Intuitively, larger amplitudes will be required to activate fibers as dCSF increases. Individual differences in anatomy or spine dynamics can also affect these parameters and may vary considerably from patient to patient.

This observation can be understood from a first-principles analysis of the bioelectrical system. In spinal cord stimulation, the amplitude range for a patient is generally titrated between the perception threshold and discomfort threshold. The perception threshold is the lowest amplitude at which a paresthesia sensation may be detected by the patient. As described above, a confounding issue in SCS is that the titrated therapy varies as the patient's posture and activity vary. The principle understanding of this variability comes from modeling the threshold of neuronal processes as a function of electrode position. The current required for a neuron to reach depolarization threshold (I_{th}) is proportional to the square of the distance of the neuron from the electrode (r) and can be described by the current-distance relationship:

$$I_{th} = I_0 + kr^2 \tag{8.1}$$

where I_0 is the offset and k is the slope, and a symmetrical cylinder is assumed for the axon [13, 14]. The value of k for the activation of central nervous system neurons can vary between $100\,\mu A/mm^2$ to $4000\,\mu A/mm^2$ using 0.2 ms pulses [13]. The clinical outcome of this process is that as the amplitude increases past the perception threshold, the patient perceives an increased intensity of the paresthesia and an expansion of the areas of stimulation-induced paresthesia. Eventually, the amplitude will be great enough to reach the discomfort threshold, which is the amplitude that results in intolerable side effects, such as abnormal movements, pain, or unpleasant sensations (e.g., shocking or jolting). The range between the perception threshold and the discomfort threshold is known as the usage range or therapeutic range.

A typical ratio between the discomfort and perception thresholds varies between 1.4 and 1.7, but may be as high as 2.8 when using a transverse tripole configuration [10, 15]. A larger difference between the discomfort and perception thresholds is preferred in order to ensure flexibility for amplitude titration [11].

Computer models can also be employed to achieve a clearer understanding of how dCSF changes impact the volume of tissue activation that is desired for effective therapy. These models have shown that given a fixed location of stimulation electrodes, the intensity of the electric field at the level of the dorsal columns decreases as dCSF increases [25]. As posited in Equation 8.1, an increase in dCSF results in a decrease in the activating of a dorsal column fiber and an increase in dorsal root activation. Thus, constant amplitude stimulation will generate variations in the activation of dorsal column and dorsal root fibers as patient position is varied (Figure 8.8a). The net outcome is a patient-perceived paresthesia intensity shift. To compensate for this, the stimulus amplitude must vary as the dCSF varies. The model would suggest that the amplitude decrease needed between the supine and standing dCSF values was 47%, corresponding to the similar trend in clinical data as reported in the published literature [16, 18, 19].

A key question is whether impedance fluctuations are the primary driver of variable activation patterns. If so, this would imply that we can compensate for posture fluctuations by simply employing a constant current stimulation source. Returning to the computer model, simulations with constant current stimulation predict that changes in the electrode impedance would not vary significantly with dCSF. The computer model predicts that electrode impedance changed less than 0.3% as dCSF was varied from 3.6 mm to 5.8 mm. Thus cord movement within the subarachnoid space has a very modest effect on electrode impedance. This is because most of the impedance is determined by: the electrical conductivity and thickness of the encapsulation layer surrounding an electrode, and the electrical conductivity of the tissue medium near the electrodes, including the dura and fat [22, 27, 28]. The lack of variation in impedance with varying dCSF values can also be attributed to the highly conductive CSF, which results in 80–90% of the stimulation current flowing through the CSF. Therefore, spinal cord movement within this highly conductive medium has little impact on impedance [22, 25, 29]. The non-significant differences in impedance as a function of patient position have also been confirmed in clinical studies [18, 19]. Therefore, constant current is not thought to be a solution for this posture related problem.

The relationship between patient posture and spinal cord position does motivate another approach towards an automated solution, using principles that build upon concepts described earlier. By adding a posture-responsive feedback system we can provide dynamic titration to potentially compensate for variations in the dCSF variable. One such algorithm that adapts the stimulation amplitude, in response to changes in patient position, has already been shown in clinic [19]. The required enabling technologies for realizing this system *in-vivo* include a sensor as the embedded position detector and a control algorithm for therapy titration within the device, which, when combined, fulfill the desired embodiment of a "smart" physiological sensor. Devices with the capability to automatically adapt stimulation amplitudes with patient position may provide more consistent volumes of tissue activation (Figure 8.8a).

8.4.3 The Physical Sensor: Three Axis Accelerometer

The key design input for SCS adjusted to position is to measure the appropriate variable for dynamic therapy titration. As discussed in the modeling of neural stimulation, the

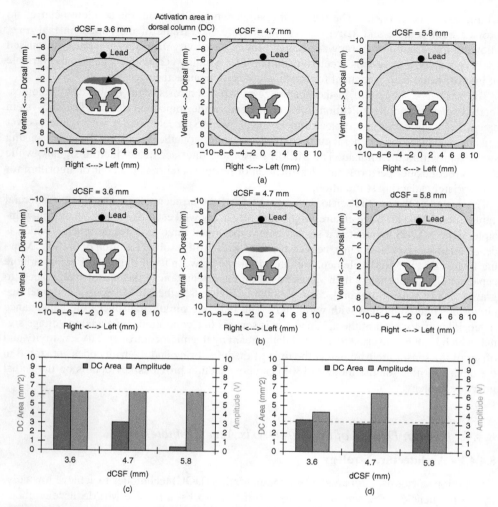

Figure 8.8 Neural activation patterns generated by SCS at varying dCSF values (3.6, 4.7, and 5.8 mm) using modeling. (a) Illustration of the activation area in the dorsal column with constant amplitude stimulation. (b) Illustration of the activation area in the dorsal column with varying amplitude stimulation. (c) Column chart for constant amplitude stimulation with varying dCSF. (d) Column chart for varying amplitude stimulation with varying dCSF. Both column charts present the dorsal column recruitment area (DC area in mm², left) and stimulus amplitude (volts, right) as a function of dCSF. Reprinted with permission from John Wiley & Sons [1]

measurement of impedance is not an ideal information source for countering postural effects. What is needed is a measurement tool that captures the dynamic variation in the distance between the electrode and the spinal cord. While direct biophysical measurements using techniques such as ultrasound or light scattering off of the spinal cord might suffice, they are not currently practical due to technical limitations like power dissipation and component integration on the lead. An indirect but highly correlative measurement can be obtained with a three-axis accelerometer. The inertial measurements yielded from the accelerometer could provide both posture and activity levels in the reference frame of the patient in real time, allowing for titration of therapy as needed. Similar to activity measures previously discussed for pacemakers, a three-axis accelerometer provides a highly correlative signal to the variable of interest while being practical for implantation and manufacturing.

The hardware subsystem in the posture-responsive control system enables physiological inertial sensing and classification to be embedded in the implanted device. The subsystem consists of a three-axis micro-electro-mechanical (MEMS) accelerometer sensor, micropower sensing interface, and a general purpose microcontroller; together they form the smart sensing subsystem within the implant. The MEMS sensor and interface detect inertial signals, while the microcontroller performs the classification of postures and runs the control algorithm for appropriately adapting stimulation.

The design of the inertial sensor leverages state-of-the-art technologies used in commercial applications. This leverage ensures high reliability using well established manufacturing principles. The MEMS element is a surface micro machined three-axis accelerometer, which is designed to survive shocks in excess of 10,000G while still resolving motion of 1/100 g. Moving fingers on the MEMS element are inter-digitated between fixed fingers to form a variable capacitor. The capacitance then varies based on the orientation of the sensor with respect to gravity and/or patient motion [35]. As shown in Figure 8.9, the MEMS sensor and interface circuit are interconnected with wirebonds. The interface circuit transduces the capacitance independently for each of the x, y, and z axis sensors to a continuous time analog voltage signal, which is then converted to a digital data stream by the microcontroller. The sensitivity and offset of the sensor are trimmed at the time of manufacturing and the trim codes are stored in the device in non-volatile memory [30]. The next section provides key details on the actual interface strategy.

8.4.4   Design Details of the Three-Axis Accelerometer

8.4.4.1   Architecture Strategy

The motion sensor uses dynamic offset compensation (DOC) techniques to achieve low noise at low frequencies. The sensor contains two major blocks: a passive MEMS accelerometer and an interface application-specific integrated circuit (ASIC). The MEMS sensor is a passive

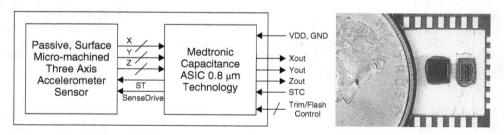

Figure 8.9   Inertial sensing detail: Acceleration sensor with MEMS sensor on the left, and custom integrated circuit for sensing on the right. Physical connections are shown in the photograph on the right. The final sensor is encapsulated for manufacturing. Reprinted with permission from IEEE (Ref: 2987780851019)

surface micro-machined accelerometer, with a sensitivity of 1 fF/g, differential. Given the parasitics shunting the system and typical clocking voltages, the output voltage from the sensor is 1 mV/g, and a resolution requirement of < 5 mg requires 5 µV of front-end stability in a 0.1 Hz to 10 Hz bandwidth. This level of stability could be achievable with large area transistors, but our space constraints limited us to smaller geometry transistors that had excessive 1/f noise. Since the capacitive MEMS sensor requires AC excitation for readout, correlated double sampling (CDS) techniques were implemented to suppress the interface $1/f$ noise.

8.4.4.2   Interface Circuit Overview

A 0.8 um complementary metal-oxide semiconductor (CMOS) ASIC was designed to provide the self-contained interface necessary for reliable precision sensing [12]. The ASIC contains both the sense interface circuit for capacitive pick-off, as well as the supporting circuitry such as references, non-volatile memory for offset and sensitivity trims, and clock state machines. To suppress the offset and $1/f$ noise imperfections in the interface circuit, correlated double-sampling (CDS) techniques were applied to the front end [32]. The choice of CDS over chopper stabilization (CHS) was driven by the need to low-pass filter the motion signal on the chip, and CDS architectures are more amenable to sampled-data circuit architectures [33].

The sensor interface chain is designed to transduce small capacitive deflections into an analog output signal with sufficient noise margin for posture and activity applications. Referring to Figure 8.10 for the circuit and clocking scheme, the sensor interface node is reset during a reduced interval of the master clock, phase Φ_1', while the preamp output demodulation capacitor, C_S, is tied to the reference (ground). Opening Φ_1' switches early enough in the Φ_1 cycle allows for the charge injection error, the kT/C noise from the sensor interface, DC offset and $1/f$ noise from the amplifier to be sampled onto the demodulation capacitor during the last half of Φ_1 [34]. Without rejection of the input kT/C noise, this noise source could easily limit overall system performance at the 500 Hz system clock, with no margin for sensor contributions. During Φ_2, the sensor is excited and the amplifier drives the difference voltage arising from motion across the sampling capacitor into the output sample and hold integrator. The bandwidth of the preamplifier and the series resistor on C_S were scaled to limit the aliasing of noise into the signal band due to the sampling of the CDS architecture. A feedback capacitor, C_{fb}, provides a counter charge that sets the gain on the output of the sample and hold, as well as biasing the output node nominal setpoint at a user-supplied reference "$V_{ref}/2$." Analyzing the complete chain, the net gain through the signal chain is set by

$$\frac{V_{out}}{g} = 2V_{ref}\frac{\Delta C}{gC_{tot}}A_o\frac{C_S}{C_{fb}}, \tag{8.2}$$

where V_{ref} is the supply voltage for ratiometric operation, C is the sensor capacitance, C_{tot} is the total shunt capacitance seen at amplifier input, A_o is 50, $\Delta C/g/C_{tot}$ is $0.5 \times 10^{-3}/g$, and C_S/C_{fb} is 1.5. The net gain is trimmed to 100 mV/g at 1.8 V by adjustment of C_S.

8.4.4.3   Results Summary for the CDS-Based Accelerometer

The complete three-axis accelerometer was prototyped and eventually qualified for human use; typical design performance results are summarized in Table 8.1. The sensitivity, noise, and

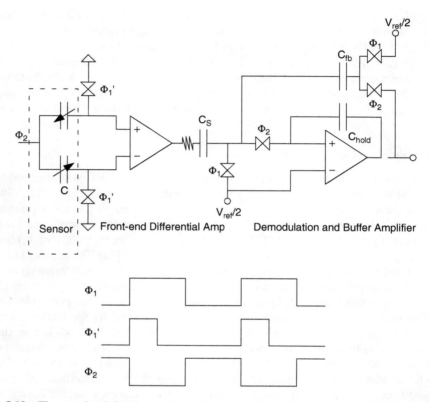

Figure 8.10 The correlated double sampler with staged reset on the sense node. By allowing for the reset charge injection and noise to be sampled on the sampling capacitor, kT/C noise and injection drift are suppressed from the measurement. Reprinted with permission from IEEE (Ref: 2987780851019)

Table 8.1 Key CDS accelerometer results

Specification	Value	Units/Comments
Supply Voltage	1.7 to 2.2 V	
Supply Current	1.05 µA	
Sensitivity(untrimmed)	125 mV/g	
Noise (X, Y channel)	3.5 mg rms	0.1 Hz to 10 Hz
Noise (Z channel)	5 mg rms	0.1 Hz to 10 Hz
Nonlinearity	< 1%	Harmonic Distortion
Operating Temperature Range	20°C to 45°C	Offset < 0.25 g
		Power < 2 µW
Functional Range	−20°C to 105°C	Part Responsive
Shock Survival	> 10000 g	

non-linearity met the expectations from SpectreRF simulation and theoretical calculations. The accelerometer noise performance for a given power usage is the primary design metric. Referencing [35], a noise efficiency figure-of-merit was defined for a generalized capacitive sensor to judge the relative merit of this design. Using this metric, the CDS architecture described here represents the state-of-the-art for low voltage micropower capacitive sensing and meets the needs for a chronic physiological motion sensor.

8.4.5 Making the Sensor "Smart" with State Estimation: The Position Detection Algorithm and Titration Algorithm

The next step in the design process is to translate raw acceleration measurements into a meaningful posture and activity estimation to enable titration of stimulation. This translation is achieved in partitioned stages of classification and adaptive stimulation control. First, the microcontroller samples the analog outputs from the accelerometer. The firmware embedded in the microcontroller then performs additional low-pass filtering and data processing to determine posture orientation and activity level. Based on orientation and rules for adaptation, the algorithm sends adjustments to the host therapy controller. This partitioned strategy isolates the sensing and stimulation algorithms to ensure the novel posture detection algorithm does not impact the delivery of therapy established in predicate devices.

The sensed posture and/or activity states are defined with regions of interest that can map to discrete therapy programs available in the host therapy processor. Initial estimates of patient position are determined empirically. The patient assumes each of the supine, lying on right side, lying on left side, prone, and upright positions and the mean tri-axial accelerations of the patient in the each of these positions, denoted as vectors V_S, V_R, V_L, and V_{UP} respectively, are measured. These initial posture estimates are referred to as orientation vectors. Examples are provided here for clarity.

(a) **Detecting upright** The upright posture geometry, Figure 8.11(a), resembles a cone that is centered on the upright orientation vector (V_{UP}) with a subtend angle of Θ_{UP}. If the angle between the real-time accelerometer vector, denoted posture trend vector, and the angle of V_{UP} is smaller than Θ_{UP} then the patient is classified in the upright position. The parameter Θ_{UP} can be tailored to the individual.

(b) **Detecting lying down (supine, on right side, on left side, or prone)** The geometry of lying positions is illustrated in Figure 8.11(b). If the angle between the posture trend vector and the angle of V_{UP} is greater than Θ_{LD} then the patient is classified as in a lying posture. The specific lying down posture (supine, on right side, on left side, or prone) is determined by the position with the minimum angle between the real-time accelerometer vector and one of V_S, V_R, V_L, and V_P. V_P is the mean acceleration vector in the prone position determined analytically by $V_P = -V_S$. The parameter Θ_{LD} can also be tailored to the individual.

(c) **Transition zone hysteresis** The transition zone, Figure 8.11(c), is the region between the upright cone and the lying region. This region provides hysteresis to prevent the detected positions from dithering between upright and lying when the patient is slightly reclined or rocking on the edge of the upright cone. If the patient is reclining or rocking on the edge of the upright cone, then the upright position is detected. In this case, lying is detected if and only if the posture trend vector passes beyond the transition zone into the lying region.

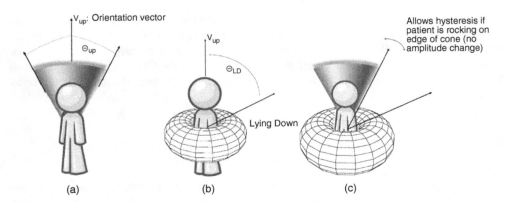

Figure 8.11 Determination of posture. (a). Determination of the upright cone for classification. (b). Determination of the lying cone for classification. (c). Consideration for hysteresis during transitions. Reprinted with permission from John Wiley & Sons [1]

(d) **Detecting activity** The posture detection algorithm also detects when a patient is upright + mobile (active), based on the upright cone geometry Figure 8.11(a). Upright + mobile detection incorporates intensity levels and time components. A patient must first be upright prior to being classified as upright + mobile. After being detected as upright, the patient will be classified as upright + mobile if they are above a programmable intensity level for more than approximately 30 seconds.

(e) **Objective patient data** Since patient position and activity can be detected, this technology enables the collection of objective patient physical activity data, including the time spent lying down, upright, or mobile as well as statistics on the types, frequency, and duration of position changes. The technology also enables the collection of the number, duration, and intensity of activity episodes. Clinicians can then objectively track how their patient's behavior evolves over time as part of standard chronic pain management, and subjectively assess the amount of patient relief in these positions to assess efficacy of the algorithm.

8.4.6   "Closing the Loop": Mapping Inertial-Information to Stimulation Parameters for Posture-Based Adaptive Therapy

The design of the closed-loop therapy system is achieved by mapping inertial signals of posture and activity to the appropriate stimulation parameter, with the goal of automatically maintaining a constant volume of tissue activation as the patient performs activities of daily living. The high-level system flow diagram is illustrated in Figure 8.12. A position detection algorithm detects the supine, lying on right side, lying on left side, prone, and upright (sit/stand) positions as well as when the patient is mobile (e.g., walking). Each therapy program can be assigned to a unique therapy amplitude per position or mobility category and per clinician guidance.

The position adaptive stimulation system also has a learning feedback loop triggered manually by the patient controller. Manual adjustments associated with a specific position or mobility category will update the therapeutic amplitude in the memory of the position adaptive controller. Thus, the system can associate the desired therapy pattern to maintain

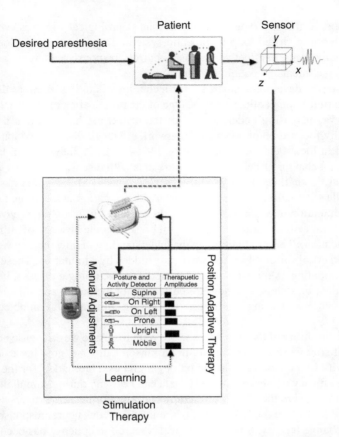

Figure 8.12 Spinal cord stimulation feedback control for a spinal cord stimulator. Reprinted with permission from John Wiley & Sons [1]

paresthesia based on patient feedback, and provides a level of automation in on-going stimulation titration.

This architecture is embodied in the RestoreSensor® system, which is the first responsive neuromodulation device CE-marked and FDA-approved for treatment of a neurological disorder. The design of this system illustrates the key principles of designing an "indirect" smart sensor and integrating it into a neurological implant, using correlates of inertial sensing to map to desired stimulation parameters. The next section discusses how more direct sensing methods might also be used in the future to expand on these techniques.

8.5 Direct Sensing of Neural States: A Case Study in Smart Sensors for Measurement of Neural States and Enablement of Closed-Loop Neural Systems

As understanding of the mechanisms underlying neurological diseases continues to improve, there is increasing interest for investigating and dynamically modulating nervous system activity to improve therapy outcomes for various neurological diseases, with the goal of eventually realizing a closed-loop therapeutic system. The objectives in closed-loop neural modulation include more effective disease control, responsive therapy adjustment, and minimizing both clinical and patient burden.

A key next step in developing smart sensor technology is to realize more direct sensing of neural states to fulfill unmet clinical needs. One of the principal methods to achieve this direct observation is by increasing understanding of the neural mechanisms that underlie disease. We are generally focused on observing the network effects of diseases, which are believed to be represented in local field potentials (LFPs) [36–44]. LFPs may contain key information about functional behavior of networks that correlates with disease symptoms, as a biomarker. Sensing distinct fluctuations of LFPs in the spectral content of the signal has typically been performed while stimulation is off (passive system identification). Observing the system in the presence of stimulation may not only increase the available information *in-vivo*, but may also reveal novel neural activity patterns that are not present in the absence of stimulation (active system identification). This could improve the understanding of how therapy works or uncover specific biomarkers of various diseases previously hidden by stimulation. These data could also be useful for validating animal and computer models of neurologic disease, in which chronic physiological data in a natural setting are currently not available, and for improving neural stimulation paradigms by using neural response during stimulation as an objective biomarker for therapy.

To maximize the utility of the smart sensor, we want to ensure we can sense neural activity in the presence of stimulation. For example, in Parkinson's disease, growing evidence suggests a role for activity in the beta band in the basal ganglia as a biomarker for the severity of the disease and the efficacy of therapy [36–41]. Beta band activity during stimulation could potentially be used to measure the effects of therapy and monitor disease progression. Ultimately, it may be possible to close the loop by using the instantaneous neural response to stimulation as a signal to change therapy. Similarly, clinical research in epilepsy has been limited by the inability to continuously measure seizure-related neural activity due to stimulation [42–44]. Concurrent stimulation and neural sensing may not only enable accurate seizure count information, but also help minimize temporal delay between seizure detection and adaptation of the stimulation. Recent clinical trial data indicate that this approach reduced seizure frequency [74], however, it is unclear if "closed loop" stimulation therapy is more effective then open loop stimulation [75]. Maintaining sensing during stimulation to illuminate the dynamics of neural networks, rather than simply blanking the signal chain during stimulation and eliminating the data, could also be useful in closed-loop neural systems. Finally, concurrent sensing and stimulation may provide an avenue for improving the performance of brain-machine interface (BMI) technologies by enabling sensory feedback directly to neural structures. One example is by providing neural stimulation that emulates tactile feedback might enhance the chronic performance of these systems [50].

The primary challenge of concurrent sensing and stimulation is the fact that the stimulation amplitude is typically 100 dB to 120 dB (five to six orders of magnitude) larger than the relevant underlying neural activity, thereby making it difficult to isolate neural signals due to amplifier saturation, non-linearities, downmodulation, and aliasing. Furthermore, current neuromodulation therapies often deliver stimulation continuously over a wide frequency range, from less than one hundred to several hundred Hertz or greater, making signal discrimination and channel blanking difficult. Several groups have been investigating simultaneous sensing and stimulation, and precise temporal and spatial resolution down to the scale of single action potentials has been demonstrated [45–49]. Since our work is aimed at chronic implantation, our sensing architecture was optimized for identification of emerging neurologic disease biomarkers without consuming power that would compromise implant longevity.

This section of the chapter describes the design and validation of a bidirectional brain-machine-interface (BMI) with the capability for concurrent sensing and stimulation. The design provides a case study of smart sensing techniques applied to direct observations of the nervous system through electrical measurement, including modeling of a complete closed-loop dynamic system *in-vivo*. [76] provides a detailed overview of these methods. At the time of this writing, the technology is still investigational in nature, and not commercially available.

8.5.1 *Implantable Bidirectional Brain-Machine-Interface System Design*

8.5.1.1   Overall Electrical System Architecture

The overall system architecture of the bi-directional BMI is depicted in Figure 8.13. The prototype is built on an existing neurostimulator architecture to leverage proven technology that is viable for chronic implantation. To extract information from the brain, a custom designed Brain Activity Sensing Interface IC (BASIC) is added for sensing neural activity. Connections from the sense and stimulation electronics to electrodes are made through a set of switch matrices and isolation-protection circuitry at the header block of the device; electrode combinations are then attached at this block for flexible BMI architectures. In addition, the custom three-axis accelerometer from the prior case study is included to provide sensing for posture and activity. Sensed signals are passed to a microprocessor for performing control and algorithms. Interactions between the original neurostimulator and the algorithm microprocessor are established by an interrupt vector and I^2C port similar to the indirect inertial system described earlier. A static random access memory (SRAM) module is included for recording events and general data logging. The telemetry subsystem allows for new algorithms to be downloaded into the device and data to be uploaded to an external data logger. The rest of this section highlights the major design features of the BMI architecture.

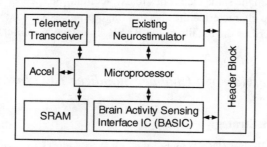

Figure 8.13   Electrical system block diagram for an implantable BMI. Reprinted with permission from IEEE (Ref: 2987780190113)

8.5.1.2 Sensing Strategy and BMI Sensing Architecture

The choice of neural recording strategy is a balance between information content and technical feasibility. While single-cell recordings and EEG data are viable for many applications, a good balance of trade-offs for our application is provided by the recording and analysis of local field potentials (LFPs). LFPs generally represent the ensemble activity of an *in vivo* neural population around the electrode and are believed to be more chronically robust [51]. In addition, LFPs encode highly meaningful data for neurological disease [36], and they are emerging as a viable candidate for BMI applications [51]. Current theories of neuroscience are proposing that LFPs encode network activity from ensembles of neural networks. Network rhythms are believed to encode the binding of brain areas as they perform computations; building on this theme, neurological disease is believed to arise in part from perturbations of these spectral fluctuations, and provide a "spectral fingerprint" of disease states [70]. In our opinion, LFPs represent the best balance between current technological limitations of electrode systems and meaningful biomarkers correlated with pathological neural activity, especially when restricting the discussion on electrodes available for current neuromodulation devices [52]. In addition, the concept of a spectral fingerprint for disease motivates our approach to designing the smart sensor.

High signal resolution and low system power consumption, which are essential for an implantable BMI, are difficult to achieve even for LFPs with moderate frequency content. However, the band power fluctuations in LFPs are generally at least an order of magnitude slower than the frequencies at which they are encoded. This motivates a BASIC architecture that directly extracts energy in key neuronal bands and tracks the relatively slow power fluctuations prior to digitization and algorithmic analysis, similar to the spectral processing paradigm of AM demodulation to extract the audio signal from a high-frequency carrier signal prior to complex processing [52, 66].

The BASIC analog preprocessing block extracts bandpower at key physiological frequencies from LFPs with an architecture that is flexible, low-noise, and power efficient. As described in [52], the signal chain of the BASIC implements a short-time Fourier transform (STFT) by using a modified chopper-amplification scheme. This architecture provides both gain and spectral estimation with power efficient processing. The BMI sensing interface includes four sense channels that are implemented on the BASIC, which can be configured as power sensing channels over a broad range of spectral bands from DC to 500 Hz. Two of the four channels can be configured for recording 200/400/800 Hz-sampled time domain waveforms. The power channels can sense from the same electrodes to extract multiple spectral bands simultaneously at the same site to help characterize the spectral fingerprints associated with brain states. The following section describes the details of the BASIC interface amplifier.

8.5.2 Design Overview of a Chopper Stabilized EEG Instrumentation Amplifier

8.5.2.1 Architecture Strategy

Dynamic offset cancellation (DOC) techniques exist to address excess low frequency noise, but are challenging in low power design [32, 33]. When choosing between Chopping Stabilization (CHS) and correlated double sampling (CDS) for a neural amplifier, CHS is generally more attractive since it does not cause significant aliasing of noise and thereby has the lowest theoretical noise. Given the potential for low noise, CHS has been explored for a variety of biomedical applications [53–55]. At low power, however, finite bandwidth in the chopper signal chain can create problems. In particular, the amplifier limited settling time creates even harmonics that lead to distortion and sensitivity errors. Techniques to eliminate distortion are available, but often result in excess current and signal path complexity [53].

The chopper architecture presented here circumvents issues of distortion using feedback; details of the design were previously published in [55]. The goal of this CHS architecture is to overcome chopper amplifier dynamic limitations by combining AC feedback and chopping at only low-impedance nodes within the signal path. Figure 8.14 illustrates the time domain behavior of the concept. After an input step, the up-modulated error signal passes through the chopper stabilized transconductor and is integrated as a baseband signal. The integrator's output is then up-modulated and fed back through a scaled shunt path until the error signal is eliminated. By enforcing chopping at only low-impedance nodes and using baseband integration, the settling constraints are relaxed and harmonic distortion is suppressed.

The use of AC feedback also yields benefits for sensitivity accuracy. Since the signal flow is now AC, gain can be set through ratios of on-chip capacitors, providing excellent sensitivity tolerance. The net sensitivity error is then set by differences in the settling time-constants in the input and feedback paths. Scaling of the chop frequency and/or balancing of the settling time suppresses second harmonics to relatively insignificant levels.

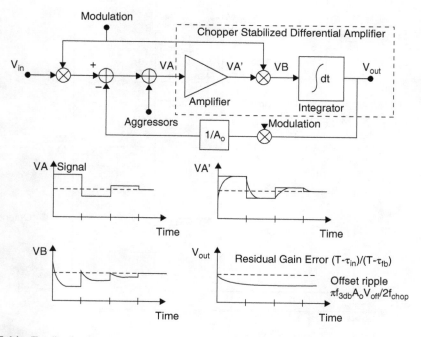

Figure 8.14 Feedback of up-modulated signal significantly suppresses distortion and increases headroom. The use of AC feedback allows for signals to be scaled by on-chip capacitors, and the signal chain implements switching at low impedance nodes. Reprinted with permission from IEEE (Ref: 2987780427485)

A final architectural benefit of AC feedback is that it allows for larger front-end gain by pre-filtering the up-modulated offset. At the output, the residual AC offset signal is $\pi f_{3\,db} A_o V_{off} / (2 f_{chop})$, where $f_{3\,db}$ is the low-pass corner frequency of the feedback loop, A_o is the net gain, and f_{chop} is the chop frequency. Offset filtering allows more gain to be placed in the front-end amplifier, suppressing sensitivity to second-stage imperfections. The next section provides a detailed overview of the CHS amplifier prototype.

8.5.2.2   Micropower Mixer-Amplifier

The core element of the CHS instrumentation amplifier is the "mixer-amplifier." As described in the previous section, the constraint of chopping (modulating) at low-impedance nodes can be achieved by modifying a folded-cascode amplifier [55], which also allows the currents to be partitioned to minimize noise. Referencing Figure 8.15, the folded-cascode architecture requires only two additional sets of switches: the first at the source of Bias N2 demodulates the desired AC signal and upmodulates front-end offsets, while the second is embedded within the self-biased cascode to up-modulate the errors from M8/M9. Source degeneration of M6/M7 and Bias N2 attenuates offsets and excess noise. The output of the transconductance stage is at baseband, allowing the integrator to both compensate the feedback loop and filter upmodulated

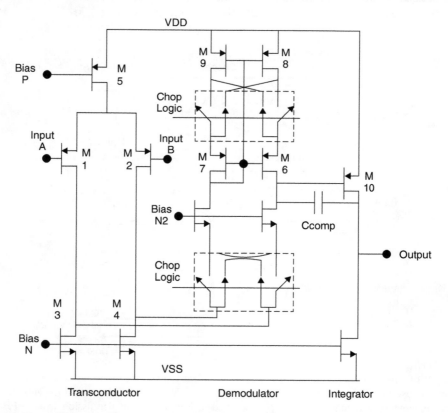

Figure 8.15   Adding modulation switches to a classical folded-cascode amplifier provides the core of the chopper-stabilized amplifier. Reprinted with permission from IEEE (Ref: 2987780427485)

offsets and noise. Input and feedback structures are added to this mixer-amplifier to achieve the desired signal processing.

8.5.2.3 Time-Domain Amplification: Feedback Strategy for Sensitivity and High-Pass Filtering

The amplifier design can perform in two modes of operation: time-domain amplification and spectral processing mode. In the time-domain mode of operation, the gain and filtering characteristics of the instrumentation amplifier are set by adding continuous-time switched capacitor networks around the mixer-amplifier. Referencing Figure 8.16, the input differential voltage is up-modulated by cross-coupled switches to the input capacitors, C_i, providing a differential input to the mixer-amplifier. This approach yields high common-mode rejection in the measurement band and allows for rail-to-rail input swing. To provide both AC modulation and a set-point for the single-ended output, the voltage to C_{fb} is switched between the mixer output and a user-supplied reference potential, V_{ref}. A second shunt feedback loop sets the high-pass characteristic for the amplifier. This high-pass was initially implemented monolithically for both high-accuracy and ease of filter corner adjustment through digital control. A key attribute of the high-pass integrator is that it samples the signal after amplification and low-pass filtering, thereby minimizing the net aliasing from this sampled-data feedback architecture. Alternative schemes for high-pass filtering include passive front-end filtering, which can be beneficial for some modes of operation as we will see.

8.5.2.4 Spectral Estimation: Hardware Strategy for Approximating the Short-Time Fourier Transform

Neural Signal Processing: Spectral Analysis and Chopping Strategy
The other mode of BASIC operation is spectral extraction of biomarkers directly from the LFP. Analog preprocessing is used to extract key biomarker information from the neural field

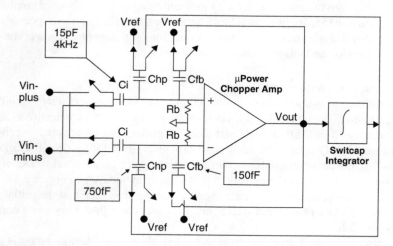

Figure 8.16 Simplified circuit implementation of an instrumentation amplifier suitable for implantable EEG recordings. Reprinted with permission from IEEE (Ref: 2987780427485)

potential prior to digitization to minimize power usage. To achieve this function, the signal chain must extract the energy from a specified band defined by the band-center, δ, with a bandwidth about δZ defined by f_{BW}. Because the science of field potentials is rapidly evolving, we would like to maintain the most flexibility in setting both δ and f_{BW}.

The following mathematical treatment for spectral analysis helps to support our circuit architecture. The spectral density of the desired signal is derived from the conjugate product of the Fourier transform, which includes a windowing function $w(t)$ that embeds the bandwidth, f_{BW}.

$$\phi(f) = \frac{X(f)^* X(f)}{2\pi},$$

$$\text{where } X(f) = \int_{-\infty}^{\infty} x(t)w(t)e^{-j2\pi ft}dt. \tag{8.3}$$

Expanding the spectral power $\phi(f)$ using Euler's identity, we see that the net signal energy can be represented by the superposition of two orthogonal signal sources, *in-phase* and *quadrature* (90° out-of-phase),

$$\phi(f) = \left| \int_{-\infty}^{\infty} x(t)\, w(t)[\cos(2\pi ft)]\; dt \right|^2$$

$$+ \left| \int_{-\infty}^{\infty} x(t)\, w(t)[\sin(2\pi ft)]\; dt \right|^2 \tag{8.4}$$

Both terms ought to be considered since the phase relationship between the *in-vivo* neural circuit and the interface IC are not correlated. Note that this is the same phase ambiguity encountered and dealt with by an incoherent AM communication system.

Equations 8.3 and 8.4 motivate the design of our analog signal chain for flexible spectral analysis. Along with significantly amplifying the signals, one needs to multiply the input neural signal by a sine and cosine term at the band-center ($\delta = 2\pi f$) window or equivalently set the effective bandwidth, square the signals, and then add them together with a final low-pass filter prior to digitization. Arguably, the most challenging portion is a pure multiplication of the neural signal $x(t)$ with the tone at δ. By a minor manipulation of the chopper amplifier design previously discussed [55, 56], we can achieve both robust amplification and spectral extraction that is both highly flexible and robust to process variations with acceptable power consumption, modest noise penalty, and minimal addition of silicon area.

Spectral Extraction Architecture Design

Chopper stabilization is a well-known noise-eliminating and power efficient architecture for amplifying low-frequency neural signals in micropower biomedical applications [55–57]. As previously discussed in [58], chopper stabilized amplifiers can be adapted to provide wide dynamic range, high-Q filters. The key design difference from [56] is the offset of clock frequencies within the chopper amplifier to re-center a targeted signal band to DC in a manner similar to superheterodyne AM receivers [59]. As shown at node VA in Figure 8.17, we use an equivalent modulation strategy to classic chopper stabilization such that the initial F_{clk} modulation frequency places the signal well above the excess low frequency noise corners ($1/f$, popcorn noise) [32].

After amplification, as a diversion from the classical chopper, demodulation is performed with a second clock of frequency, $F_{clk2} = F_{clk} + \delta$ that is shifted from the first clock by the

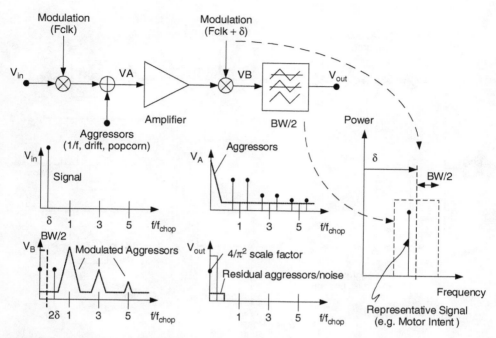

Figure 8.17 Concept of merging heterodyning and chopper stabilization for flexible pass-band selection. Reprinted with permission from IEEE (Ref: 2987780715658)

desired band-center. The frequency convolution of the modulated signal with the second clock re-centers the neural signal initially at δ to dc and 2δ at the node VB. Since the biomarkers are encoded as low frequency fluctuations of the spectral power, we filter out the 2δ component with an on-chip low-pass filter with a bandwidth defined as $f_{BW}/2$; signals on either side of δ are folded into the passband at V_{OUT}. The heterodyning chopper suppresses harmonics as the square of the harmonic order, to yield a net transfer function of

$$V_{out}(f) = \frac{4}{\pi^2} \sum_{n\_odd} \frac{1}{n^2} V_{in}(f + \delta n) \cos(\varphi), \tag{8.5}$$

where n denotes the harmonic order, and ϕ is the phase between the δ clock and the field potential input.

To first order, the proposed chopper heterodynes the frequency content of the signal by the clock separation, δ, with a scale factor of $4/\pi^2$. The robustness of this design comes from the same features that make heterodyning attractive for AM radio applications – the center frequency is set by a programmable clock difference, which is relatively simple to synthesize on-chip, while the bandwidth and effective Q are set independently by a programmable low-pass filter. Additional odd harmonics do fold in, but fall off proportionally with order.

The operation shown in Equation 8.3 is implemented on the BASIC, as shown in Figure 8.18, with parallel in-phase (I) and quadrature (Q) paths, using clocks derived from a master at twice the chopper frequency. As shown in the block diagram of the signal chain in Figure 8.18, the final power extraction is achieved with the superposition of the squared in-phase and

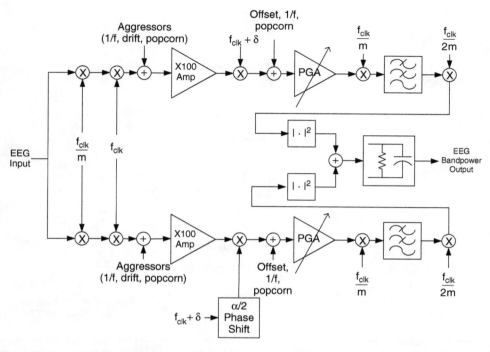

Figure 8.18   Tunable heterodyning neural receiver extracts signal power within the physiologically relevant band. The dual-nested chopper architecture uses two different chopper frequencies to improve the power-bandwidth tradeoff while eliminating offsets and low frequency noise. Reprinted with permission from IEEE (Ref: 2987780715658)

quadrature signals, yielding a transfer function for the entire signal chain of

$$V_{EEG\_Power}(f) \propto \left[\frac{4}{\pi^2} \sum_{n\_odd} \frac{1}{n^2} V_{in} \left(f + \delta \cdot n \right) \right]^2, \tag{8.6}$$

The residual sensitivity to odd harmonics is not a major concern in our application because the signal power in the cortical circuits generally falls off with a $1/f$ law [60]. This means that the net measured power of the neuronal signals at the third harmonic is effectively down by 48 dB, hence maintaining an acceptable selectivity to the pertinent band. For applications that require better harmonic attenuation, a notched clock strategy to suppress higher-order harmonic content [61] can be implemented.

Overall Chopper Modulation Strategy
Several chopper modulation techniques are actually required to achieve microvolt signal resolution with the NPIC's spectral analysis strategy. The total signal chain with modulation is detailed in Figure 8.18. As discussed in the previous section, the "core" chopper modulation, with two clocks separated by δ, provides the mechanism for selecting the band of interest. Although this does achieve the necessary frequency heterodyning, two practical issues remain.

The first issue is that the residual offsets in the core chopper can be on the order of several microvolts. The problem with this residual offset is that it is superimposed on the signal of interest, which causes significant signal perturbations in the output signal as the phase of the biomarker beats against the δ clock. To fix this problem, we implemented a "nested" chopper switch set [62, 63] before the first chopper amplifier, and after the programmable gain amplifier (PGA). The small residual offsets are then up-modulated and filtered out using the $f_{BW}/2$ selection filter. The nested loop runs nominally at $F_{clk}/64$, 128 Hz, to minimize residual charge injection offset, but fast enough to minimize perturbations to low-frequency dynamics [62, 63]. Note that since the PGA is also embedded in the loop, its residual $1/f$ noise and offset is also suppressed at the lower rate. The use of the passive low-pass architecture in the $f_{BW}/2$-selection block minimizes additional contributions of offset to the signal chain after the nested chopper.

The second issue is that residual offsets in the output multiplier blocks create an intermodulation product that also creates significant distortion when trying to resolve microvolt signals. The use of a low-frequency chopper prior to multiplication corrects that issue; note that since the multiplier squares the signal, we do not require a subsequent explicit down-modulation block, the offset just creates a stable offset that can be trimmed out. Recent work has used similar techniques to stabilize a broad range of multiplication circuits and mixers [64].

8.5.2.5   Results Summary for the CHS BMI Sensing Amplifier

In summary, the CHS BASIC amplifier architecture developed here has many key advantages for sensing neural activity. The use of continuous-time modulation of input and feedback signals provides high gain accuracy and linearity through use of on-chip capacitors. In addition, the switching of the input between two capacitors provides high common-mode rejection ratio (CMRR) and rail-to-rail common-mode input swing which is important for stimulation systems. The use of continuous time techniques throughout the design yields low noise due to minimal aliasing of signals or kT/C sources. Finally, the use of heterodyning switching techniques in the chopper amplifier allows for estimation of the short-time Fourier transform in a power efficient way.

The CHS BASIC amplifier design was prototyped in a 0.8 μm process. The prototype results met design expectations from theoretical calculations and SpectreRF simulations; the results are summarized in Table 8.2. The primary goal of this design was achieving low noise in the LFP signal band between 0.5 Hz and 100 Hz. To gauge the relative merit of the design compared to others in the recent literature, we used the noise efficiency factor described in [65]. The value of 3.6 for the diagnostic LFP design is one of the lowest published to date and is achieved by the elimination of excess $1/f$ noise at low frequencies. The use of spectral

Table 8.2   System performance summary

Minimal Detectable Signal Power	$< (0.5\,\mu\text{Vrms})^2$
Noise Spectral Density (Time Domain)	$150\,\text{nV}/\sqrt{\text{Hz}}$
Bandpower Center Frequency (δ)	dc to 500 Hz, programmable
Bandwidth of Spectral Estimate	1 Hz to 20 Hz, programmable
BASIC + Classifier Algorithm Power	25 μW (typical)
Real-time Wireless Uplink Capability	11.7 kbps @ 175 kHz (ISM)

processing does modestly impact the signal-to-noise floor of the system, but this is deemed an acceptable trade-off for benefits in system-level power consumption, as will be discussed in the use conditions.

Recent work has built on the core principles explored in this brain interface IC. The application space to include cardiac sensing and detection has been extended [71], and [72] has improved the input impedance using positive feedback loops. Additional prototyping technology work is focused on developing this front-end interface into a complete epilepsy seizure detection system [73].

8.5.2.6   Making the BMI Interface Smart: Signal Processing and Algorithm Architecture

While the BASIC processes and amplifies neural signals, it requires the addition of classifiers to identify key brain states in order to truly make it a smart sensor. The challenge of processing signals is balancing power consumption, flexibility and performance. Since biomarkers of interest have already had their spectral power extracted by the BASIC STFT and the spectral power changes vary slowly compared to the LFP frequencies that encode the biomarkers, sampling and processing can be done at sampling rates on the order of Hz. This allows for a system partition, as illustrated in Figure 8.19, of analog pre-processing to extract key information and reduce dynamic range, while running complex digital algorithms at slow clock rates. This partition generally results in an acceptable power budget for a chronically implantable device; similar "neuromorphic" principles are discussed in [66] for a range of applications, including cochlear implants and retinal prosthesis.

Given the unknown nature of an "ideal" neurological feedback strategy, it is imperative that the system is highly configurable. In this case study, the microprocessor controls the BASIC chip via control registers, enabling adjustments to gain, STFT parameters for spectral estimation, and electrode connectivity through telemetry and algorithm control. To maximize flexibility, algorithms can always be adjusted via telemeterized firmware updates.

The algorithm running in the processor is used to appropriately classify the signal and estimate patient state. This allows the BMI to actuate stimulation therapy appropriately and/or

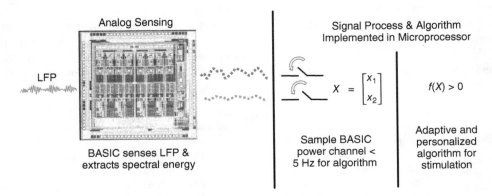

Figure 8.19   Partitioning for signal processing and algorithms. Reprinted with permission from IEEE (Ref: 2987780190113)

measure key diagnostics. Recent research is suggesting that patient-specific algorithms can be useful for improving the sensitivity and specificity of this classification. Clinician-supervised machine learning is a good way to accomplish a patient-personalized algorithm, and this system is designed to enable this patient-specific definition in a power-efficient way [67]. As the merging of the BASIC and detection algorithms is critical to the design of a "smart sensor" for brain state detection, the concept will be developed more fully in the closed-loop validation example exploring thalamo-cortical circuits in the ovine brain.

8.5.3 Exploration of Neural Smart Sensing in the Brain: Prototype Testing in an Animal Models

8.5.3.1 Bi-Directional BMI System Features

Referring to the electrical system architecture in Figure 8.13, several other key blocks are included in the device for system operation. For actuating the neural network, an existing Medtronic neurostimulator is employed in the system for stimulation therapy, which also serves as a platform for the overall bi-directional BMI system. The device communicates through a wireless telemetry link for system configuration, algorithm programming, and data uplink. A 1 MB SRAM is included on the platform for recording algorithm-defined or externally-generated events, time-domain waveforms, and general data logging. The data can be downloaded through the wireless link for analysis and investigation at 11.7 kbps using the 175 kHz ISM band.

8.5.3.2 System Integration

A complete implantable BMI system was prototyped using established, state-of-the-art, medical device technology. The prototype system is shown with a cutaway window in Figure 8.20. The BASIC is fabricated using a 0.8 μm CMOS process and stacked on the SRAM to provide a module with small form-factor. The electrode-interconnect and algorithm processor are in

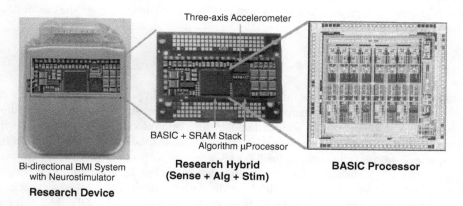

Three-axis Accelerometer

BASIC + SRAM Stack
Algorithm μProcessor

Bi-directional BMI System
with Neurostimulator
Research Device

**Research Hybrid
(Sense + Alg + Stim)**

BASIC Processor

Figure 8.20 Prototype implantable bi-directional BMI system with key elements shown in device. Reprinted with permission from IEEE (Ref: 2987780190113) (see Plate 3)

close proximity to maintain signal integrity. The right side of Figure 8.20 is a close-up of the side of the hybrid board containing the new sensing and algorithm electronics. The other side (not pictured) contains the stimulation electronics. This device has complete bi-directional BMI functionality and is suitable for chronic preclinical research.

8.5.3.3   General System Characterization

The prototype system including implantable circuits, electrodes and telemetry was tested in a saline tank with recorded patient data. The BASIC was verified to consume $10\,\mu A$ from a 2 V supply, achieving a signal resolution of $< 1\,\mu V_{RMS}$ for a 5 Hz power spectral estimation of two channels of operation (one/hemisphere). The linear support vector classification algorithm drew an additional $5\,\mu W$ to classify signals in real-time with 1 s estimation updates. In addition to demonstrating basic BMI functionality, the system was also verified to withstand ESD, electrocautery and defibrillation, which is critical for a robust and practical BMI system. The performance is summarized in Table 8.3, which gives a representative snapshot of a state-of-the-art, bidirectional BMI with smart sensing capability.

8.5.3.4   Practical Considerations for Smart Sensing: Controlling Stimulation-Sense Interactions

As highlighted in the introduction, a significant challenge in combining sensing and stimulation in a bi-directional BMI is dealing with signal contamination. The signals we are interested in sensing are in the order of microvolts, while the signals we are introducing (the stimulation) are in the order of volts. The extraction of a biomarker that is six orders of magnitude lower than therapeutic stimulation is a significant challenge.

Several methods are employed in the prototype to allow for simultaneous sensing and stimulation. One method is simply to have separate leads for each function; but this comes at the cost of increased surgical complexity, and we often want to measure activity in the vicinity of our stimulation target. For simultaneous sensing and stimulation from the same lead, careful placement of the leads and sense-stimulation configuration can take advantage of the reciprocity theorem of electromagnetism. Stated mathematically, we attempt to design the electrode and anatomical approach such that

$$\phi_A - \phi_B = \frac{\vec{E}_{AB} \bullet \vec{I d}}{I_{AB}} \to 0 \tag{8.7}$$

Intuitively, we can think of this mathematical relationship imposing a symmetry constraint on the sense-stimulation configuration. Figure 8.21 shows an example where the sensing dipole $(A \leftrightarrow B)$ is placed symmetrically about a unipolar stimulation electrode $(C \leftrightarrow D)$ with far-field return. Figure 8.22 shows how the electrodes are connected to the device. When the dipole from therapy stimulation is orthogonal to the biomarker sensing vector, our chances for extracting a signal are greatly increased as the problem is now one of classical common-mode rejection.

Another key method employed in all electrode configurations is to take advantage of the spectral filtering properties of the BASIC. In particular, the architecture of the BASIC is capable of rejecting signals that are out of its tuned band. Saturation is avoided by filtering the signal before significant gain is applied as part of the STFT processing. This allows for the

Table 8.3 Summary of key performance specifications related to the sensing blocks, embedded algorithms, stimulation, and telemetry for the bi-directional BMI. Note that the stimulation sub-system has capabilities consistent with existing commercial-based designs

LFP/ECoG Sensing		Inertial Sensor (Extensible)	
Operating Power Dissipation (Time Domain)	100 μW/channel	Operating Power (3-axis measurement)	2 μW
Operating Power Dissipation (Spectral Mode)	5 μW/channel	Inertial Algorithm Power Dissipation	25 μW
Function mode	Time domain/ Bandpower	Sensitivity	125 mV/g (.01 g/LSB)
MUX, channels available PC Dual Lead Implant System Assumed	Input mux allows $12->4$ DOF down selection of best channels for upload	Dynamic Range	$+/-5$ g (Falls, footsteps, high impact activity)
Minimal Detectable Signal	<1 μV$_{RMS}$	Noise (X,Y axis)	3.5 mg$_{RMS}$ (0.1 Hz to 10 Hz)
Spot Noise Spectral Density	150 nV/$\sqrt{Hz}$	Noise (Z axis)	5 mg$_{RMS}$ (0.1 Hz to10 Hz)
Bandpower Center Frequency	dc to 500 Hz	Nonlinearity	$<1\%$
Bandwidth of Spectral Estimate	1 Hz to 20 Hz	Shock Survival	$>10,000$ g
CMRR/PSRR	>80 dB	**Telemetry**	
High Pass Corners	0.5 Hz to 8 Hz	Physical Layer	Established 175 kHz (ISM)
		Data Capacity	4 DOF/preprocessed
		Training Mode	2 DOF/ raw high data rate
Input Range (Stim compliance)	$>+/-10$ V	**Memory Buffer (Monitoring Diagnostics)**	
Embedded Algorithm Characteristics		SRAM	8 Mb
Algorithm Power	5 μW/channel (typical)	**Stimulation Capability**	
Algorithm Type (Embedded)	Support Vector Machine (Linear kernel, 4DOF)	Stimulation Channels	8 for bilateral (4/lead) (unipolar/bipolar)
Algorithm Upgrade Capability	In-vivo through telemetry and embedded bootloader	Stimulation Parametrics	Predicate Approved (Activa PC)

possibility of delivering stimulation therapy in one spectral band and sensing in another at the same time through the same lead but not the same electrode. This constraint is compatible with many deep brain stimulation (DBS) systems which have biomarkers well separated spectrally from therapy stimulation and sensing dipoles bounding a unipolar-driven stimulation target. These techniques show promise for simultaneous stimulation and sensing of the same neural circuit, establishing feasibility for real-time adaptive therapy titration.

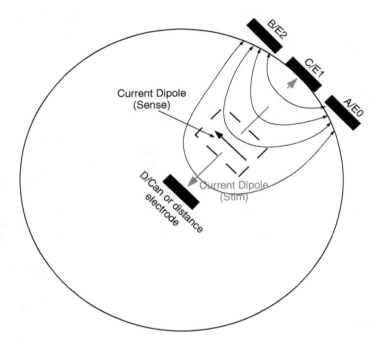

Figure 8.21 Diagram of lead placement and stimulation configuration exploiting the reciprocity relationship. Reprinted with permission from IEEE (Ref: 2987780190113)

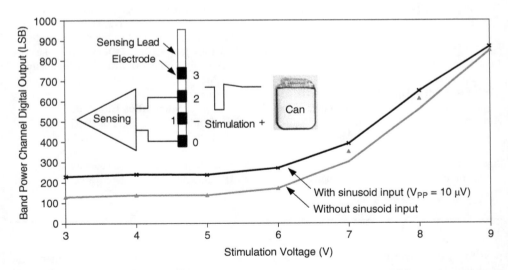

Figure 8.22 Sensing capability in the presence of stimulation. Reprinted with permission from IEEE (Ref: 2987780190113)

The concepts described here for enabling the system to perform sensing during the delivery of stimulation were tested in a saline tank model. Figure 8.22 shows the results from a test where 145 Hz stimulation was delivered between contact 1 and the neurostimulator. A 24-Hz signal, representing typical β band biomarkers, with $10\,\mu V_{PP}$ amplitude was injected into the tank and sensed across contacts 0 and 2 using the BASIC. This was compared to results obtained using the same stimulation but no test signal. The separation of the two curves indicates a promising ability to sense during delivery of stimulation, especially since significant clinical therapy is delivered using 5 V of amplitude or less.

8.5.3.5 Defining Smart Sensing for the Brain: Applying Subject-Specific Algorithms and Classification for Estimating Brain States

The intent of building this prototype bi-directional BMI is to provide a platform for research that is adaptable to a number of neurological disorders. As mentioned earlier when motivating the design of the BASIC and the smart sensing strategy, the common thread for these disorders is that biomarkers are believed to be encoded in LFPs as distinct fluctuations in the frequency spectrum [36, 52]. With appropriate lead placement and BASIC tuning, it might be possible to differentiate neural states and use this information for diagnostics and/or therapy titration.

For example, Figure 8.23 shows a spectrogram of LFP data collected from a Parkinson's patient's DBS leads; note the high correlation between energy in the beta band and the pathological symptoms associated with the disorder [36; data courtesy of Abosch and Ince, University of Minnesota]. In the "off state," the patient is showing symptoms of not being on medication (L-dopa) and has issues with initiating and executing motion, and there is significant energy in the beta band. When the L-dopa takes effect around the ten-minute mark, the beta band energy subsides and the patient's symptoms are greatly diminished.

The generation of a smart sensor involves detection of these signals with the BASIC amplifier and subsequently classifying them appropriately into an estimated brain state. Figure 8.24 shows how these spectral data have the potential to classify the patient's clinical state in

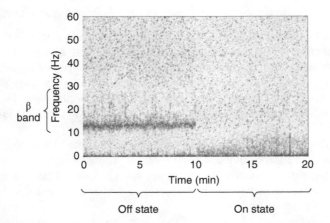

Figure 8.23 Spectrogram of LFP data collected from Parkinson's patient. Reprinted with permission from IEEE (Ref: 2987780190113) (see Plate 4)

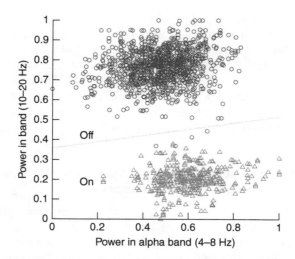

Figure 8.24   Classification of Parkinson's patient state based on a supervised, patient-specific machine learning algorithm that is downloaded into the BMI. The classifier for this patient during this epoch has > 95% sensitivity and specificity. Reprinted with permission from IEEE (Ref: 2987780190113)

real-time with high sensitivity and specificity, applying support vector classification on the prototype controller "trained" with a supervised learning process similar to that described in [67]. The classified states are then fed to the therapy/diagnostic prototype controller to explore stimulator therapy settings using algorithms or to provide a clinician feedback based on quantitative diagnostics. The next section demonstrates this concept *in-vivo* in an animal model for exploring thalamo-cortical circuits relevant to epilepsy.

8.5.4   Demonstrating the Concepts of Smart Sensing in the Brain: Real-Time Brain-State Estimation and Stimulation Titration

The research tool described thus far was tested in a chronically implanted ovine model for exploring neural circuits relevant to seizures. In this experiment, seizure-like neural activities are induced by using proper stimulation parameters in the normal ovine model. The goal of the system validation protocol was to demonstrate that the smart sensor design principles described in this chapter are effective in a representative intended use application. The validation was performed in a series of protocols with increasing complexity and with the intention to:

- demonstrate the ability to measure disease-related neural data in a biological environment in the presence and absence of therapeutic levels of stimulation;
- demonstrate the ability to automatically detect a desired neural state in the presence and absence of stimulation using a multi-dimensional embedded algorithm embodied in the "user-defined event detector"; and
- for extensibility purposes, demonstrate the ability to change stimulation parameters in real-time, based on neural state in real-time and based on the detection state of the embedded algorithm.

Together, this protocol demonstrates all the key principles of a smart sensor for estimating brain states using a complex feedback loop with the nervous system.

8.5.4.1 Methods: Chronic Ovine Model to Collect Data

To demonstrate validation, the research system was chronically implanted in a large ovine seizure model. The study was conducted under an IACUC-approved protocol. Following anesthesia, 1.5T MRI (magnetic resonance imaging) results were collected and transferred to a surgical planning station. Trajectories for a unilateral anterior nucleus (AN) of thalamus DBS lead (Medtronic model 3389) and unilateral hippocampus (HC) lead (Medtronic model 3387) were planned, and the leads were implanted using a frameless stereotactic system (NexFrame from Medtronic, Inc.). Once lead placement was confirmed based upon electrophysiological measures, Medtronic model 37083 extensions were connected to the DBS leads, tunneled to a post-scapular pocket, and connected to the chronically implantable device. Figure 8.25 illustrates the overall system placement and setup. Following closure of all incisions, anesthesia was discontinued, and the animal was transferred to surgical recovery.

Following a two-week recovery period, stimulation and recording sessions were conducted on a weekly basis for approximately two-hour periods with the animal resting in a sling. To reflect more relevant neurological dynamics, the sheep was not anesthetized during these sessions. The coincident response to stimulation in the hippocampus and thalamus was characterized by sensing during a sequence of no stimulation, and then slowly increasing amplitudes to 3.5 V in either target. The electrode contacts for concurrent sensing and stimulation were selected to minimize stimulation distortion. The frequency of stimulation was fixed at 120 Hz based on empirical measurements to minimize aliasing in the beta band. The pulse width was set to 90 μs. While it did not necessarily cause a seizure, approximately half of stimulations above 2.5 V correlated with seizure-like activity, enabling this model to generate a useful validation signal for the system. The maximal amplitude was set by the onset of behavioral side effects at this frequency. During most of the data collection, the stimulation duration was 30 seconds. For the closed-loop data collection, stimulation was dependent on classifier output. After completion of the session, data was downloaded and digital spectral analysis

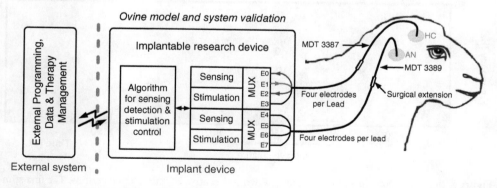

Figure 8.25 Placement and setup of neuromodulation system implanted in an ovine model. Reprinted with permission from IEEE (Ref: 2987780619986)

was performed off-line to confirm the results. Between sessions, data were collected at timed intervals using the embedded loop recorder with the animal freely moving in its own living environment and downloaded at the following monitor session.

8.5.4.2 *Results*: Validating the Device Maintains Meaningful Sensing during Stimulation

One of the key challenges for the system was ensuring that observations do not represent stimulation artifacts. The presence of after-discharges, seizure-like activity after stimulation, on approximately half of the stimulations above 2.5 V in the HC served as a useful validation signal. Since after-discharges had a magnitude of approximately $20\,\mu V_{rms}$ to $30\,\mu V_{rms}$, the underlying seizure activity that preceded them was obscured during stimulation periods in the time domain. However, evidence of seizure-like activity could be observed using frequency domain analysis. Figure 8.26 shows evidence of the putative seizure beginning approximately 7 seconds after the onset of HC stimulation. This activity is not an interference of stimulation, since it does not appear on all stimulation periods. The second stimulation pattern did not induce a seizure, and it serves as a control for the performance sensing channel characteristics in the presence of large stimulation inputs.

The seizure-like activity in the HC was centered at approximately 20 Hz with multiple stimulation interferences outside the signal band. Note that those stimulation inferences are high-Q with fixed frequencies in the spectrogram; therefore, they can be readily distinguished from

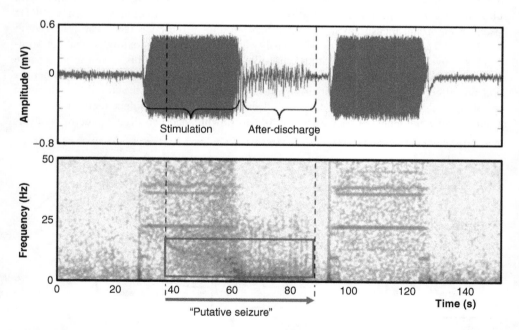

Figure 8.26   Time-domain and spectrogram of two consecutive HC stimulation periods. The first stimulation block resulted in an after-discharge whereas the second stimulation block did not. Reprinted with permission from IEEE (Ref: 2987780619986) (see Plate 5)

real neural activity. These frequency bands were then used to perform real time detection through "training" a support vector machine to differentiate states, including a measure of the stimulation interference as an input.

8.5.4.3 Validating Algorithm Detection During Stimulation

Seizure activity was detected using power bands corresponding to the seizure (beta band) and stimulation (~80 Hz) to train the linear discriminant classifier embedded in the device. This detector, on the back-end microprocessor, processed estimations at 5 Hz after extraction of the spectral bands in the BASIC. Note that simply tuning to the frequency of stimulation would saturate the channel, so 80 Hz was selected to take the stimulation artifact into account without saturation. Linear discriminant coefficients were determined in a four-step process using supervised machine "learning" and support vector machines.

- In the first step, time-domain information was collected from all electrode combinations to determine the pair that was most correlated with seizure-like activity. As shown in Figure 8.26, activity in the 5 Hz to 20 Hz band corresponded to the putative seizure, whereas activity in the 40 Hz band was dominated by stimulation.
- In the second step, the device was programmed to measure these bands of interest to serve as the raw data for algorithm development using the heterodyning mode of the sensing channel. This was a necessary step due to the scale differences in power estimation between the power channel extraction using the device and off-line digital signal processing of the time-domain signal.
- In the third step, coefficients for a linear discriminant function were calculated with an off-line support vector machine algorithm using the power channel as the independent variables, and true seizures determined from time domain as the dependent variable. True seizures were defined as a period of high beta band activity.
- In the last step, the linear discriminant coefficients were uploaded onto the device for real-time seizure classification.

When viewing spectrograms derived from time-domain data containing stimulation artifact, horizontal lines are related to the simulation and/or sampling the signal in the presence of stimulation. When the stimulation engine turns on or off, a brief broadband dose of energy is sent through the sensing circuitry. This causes a vertical line to appear in the spectrogram. True physiological biomarkers generally do not have these vertical or horizontal properties and can therefore be distinguished from these artifacts.

Figure 8.27 shows how the extracted power data were divided into putative seizure and non-seizure classes. The x-axis represents sensed stimulation power and the y-axis represents beta power. A 2D classifier is needed because using the seizure band alone does not adequately separate the seizure/no-stimulation condition from a no-seizure/stimulation condition based on spectral leakage of stimulation into the beta band.

The detection algorithm was tested chronically *in-vivo*. The linear discriminant coefficients were kept constant for five weeks after being uploaded to the device without any appreciable loss of seizure detection performance. Figure 8.28 shows the results of the seizure detection at a representative monitor. The seizure is detected during the stimulation period and persists several seconds after stimulation has ended. The algorithm performs well against time-domain

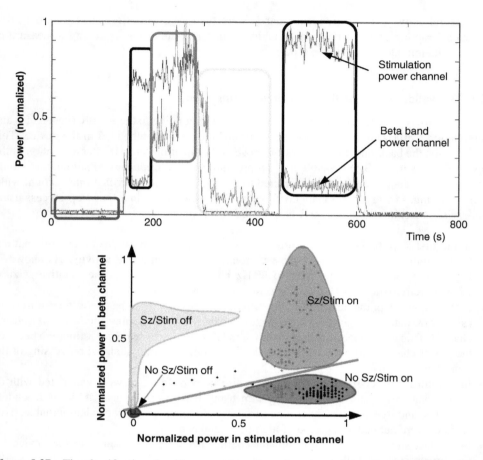

Figure 8.27 The classification algorithm separates seizure from non-seizure data in the presence and absence of concurrent stimulation. The test data shown here are separate from training data used to develop the algorithm. Reprinted with permission from IEEE (Ref: 2987780619986) (see Plate 6)

output when the stimulation is off. During stimulation, the embedded algorithm still maintains the capability to detect seizure onset, which cannot be readily observed in time-domain. The physiological response to stimulation is still present and measureable in the prototype system, providing critical observations to the effect of stimulation. This response is often blanked out by existing implantable research systems, which do not have the capability to sense in the presence of stimulation.

8.5.4.4 Validating Closed-Loop Operation: Real-Time Stimulation Titration Conditional on Classifier Detection

To explore the extensibility of the device for future closed-loop use cases, our last validation step was to use the sensing and classification processes to change stimulation in real time

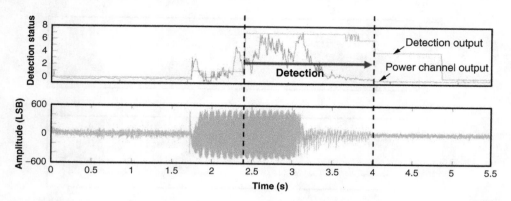

Figure 8.28   Detection of seizure with and without stimulation. The top panel shows the status of the classification algorithm, where the upper waveform represents detection output and the lower waveform represents power channel output. Detection is marked with a black arrow. The bottom panel shows the raw time-domain data. Reprinted with permission from IEEE (Ref: 2987780619986)

based on the measured state of the neural network. This was accomplished by coupling the output of the linear discriminant to the stimulation control of the device. The device was programmed to shut off stimulation when the linear discriminant triggered detection indicative of seizure induction. This experiment is an investigational prototype of an algorithm concept that might eventually be applied as a closed-loop hippocampal stimulator for the treatment of epilepsy, where stimulation putatively suppresses seizures and the system would mitigate the risk of unnecessary stimulation by monitoring for induced seizures and taking appropriate action.

The linear discriminant graph of Figure 8.29 shows the onset of a seizure. Seizure detection occurs during a stimulation period and cannot be identified in the time domain graph. Though the onset of seizure cannot be visualized in the time-domain graph, it is seen in frequency domain as activity in 5 Hz to 20 Hz. Figure 8.29 also shows that the onset of seizure activation was detected by the linear discriminant leading to an immediate shutoff of the stimulation during the second stimulation period. This contrasts with the first stimulation period, acting as a self-control, in which there was no seizure, therefore, no detection and the 30 second stimulation timed out rather than being terminated by the linear discriminant. This demonstration illustrates all key components of the investigational system design working to enable dynamic closed-loop operation concurrent with stimulation: sensing in the presence of stimulation, embedded detector operation, and the ability to titrate the stimulator.

Although tempting to conclude that immediate cessation of stimulation helped to suppress the after discharge, compared to Figure 8.26, more data collection is required to build up the statistical basis for this initial observation.

8.5.4.5   Discussion: The Potential of Smart Sensing for Advancing Therapeutic Neural Systems

The ability to sense neural activity and stimulate neural tissue is a fundamental need in the study and treatment of neurological disease. In this case study, we have validated the smart

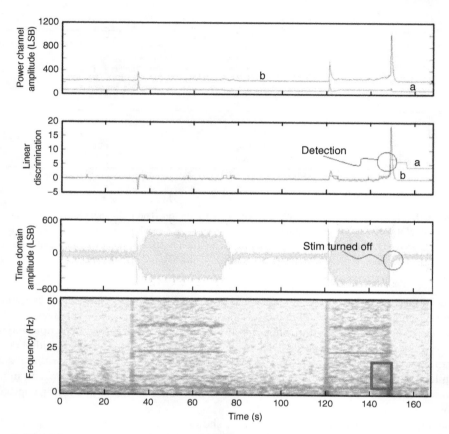

Figure 8.29   Closed-loop seizure detection and stimulation termination. The top graph shows the power in the two bands of interest. Trace (a) in the second graph shows the output of the discriminant. The detection is circled. Trace (b) is a metric of distance to the detection boundary. The third and fourth traces are similar to Figure 8.26 and show the time and frequency domain traces. Reprinted with permission from IEEE (Ref: 2987780619986) (see Plate 7)

sensing prototype ability to perform concurrent neural stimulation and sensing, embedded event detection using user-defined parameters, and the extensibility for closed loop operation. Why might this be useful?

A major challenge in applying a smart neural sensing system is the ability to concurrently sense, properly interpret the sensing signal and apply therapeutic stimulation of the nervous system. This ability is important in order to increase the amount of useful neural data, minimize time delays in responding to biomarkers, and investigate biomarkers and neural circuit dynamics that may only be present in neural signals during stimulation. This is particularly important in therapies where stimulation is on for the majority of the time, such as in movement disorder therapies, and a closed-loop feedback control strategy might require monitoring of brain states and accurate algorithm control in the presence of stimulation. The challenge of simultaneous sensing and stimulation primarily arises from the feedforward effect of stimulation, which is up to six orders of magnitude greater than the desired neural signal. Though some of this feedforward effect can be minimized using hardware design, residual stimulation interference remains due to the constraints of practical implantable systems. These include imbalances in the biological interface and power limitations.

In the validation work we have described in this case study, we have taken a systematic approach to mitigate the impact of stimulation artifacts on sensing and algorithm performance. The solution is not a targeted single-element resolution method, but instead relies on managing sense-stimulation interactions throughout the signal chain. These methods include symmetric sensing across the stimulation electrode, careful selection of stimulation parameters, and use of well-designed algorithms to account for stimulation effects.

The methods do enforce modest constraints on the system, such as symmetry of sensing around stimulation electrodes and frequency spacing between physiological biomarkers and stimulation. In addition, our design focuses on spectral LFP processing and does not transfer easily to time-domain, spike-based processing. However, these constraints are generally acceptable in several applications found in movement disorders, epilepsy, and brain-machine interfacing for prosthesis, where potentially useful biomarkers appear to be represented as fluctuations in frequency spectrum.

To ensure that the methods are potentially translational and not just theoretical, the prototype system was validated in a chronic implant in an ovine over a period of more than one year. During this time, the system was used to measure seizure activity from the HC, automatically detect seizure activity using a classification algorithm, and change stimulation based on the classification algorithm output. In this application, the desired signal magnitude was in the order of $10\,\mu V_{rms}$. Biomarkers in other neurological disease states, such as Parkinson's disease, may require signal resolution in the order of $1\,\mu V_{rms}$. Validation of these low-level signals still depends primarily on bench measurements.

While our benchtop tests have demonstrated signal resolution below $1\,\mu V_{rms}$, further work will be needed to validate this performance *in-vivo*. Furthermore, some disease state biomarkers, such as tremors associated with Parkinson's disease, may not be reflected in the neural data available, motivating the use of the system accelerometer sensor in the future to capture important aspects of the patient assessment.

The core features of the prototype smart-sensing-enabled device validated in this work form a rich basis for optimization of stimulation therapies based on observing the physiology of neural systems. Ultimately, through a deeper understanding of neural processing, physiology, and disease mechanisms, tools enabled with smart sensing have the potential to help better optimize neuromodulation therapies.

8.6 Future Trends and Opportunities for Smart Sensing in the Nervous System

This chapter posited the use of smart sensors in biomedical systems as a key enabler of closed-loop systems. While smart sensors have a strong historical foundation in cardiac systems, the ability to chronically perform sensing and patient state estimation using neural and non-neural biomarkers is relatively novel. Smart sensing and algorithm systems allow the study of the nervous system through a dynamic control framework. This framework facilitates research into neurological diseases as well as more sophisticated therapies. Ultimately, it may be possible to perform continuous patient state monitoring, estimation of state, and stimulation titration with the potential for shorter time delays; improved patient outcomes; and reduced patient and clinician burden. As devices with more sensing features come on-line [31, 66, 67] the opportunities might grow even larger.

The pathway to achieving closed-loop neuromodulation, however, will require a new look at the way devices are designed. Specifically, the design will not necessarily be a simple mapping of cardiac pacemakers to the neuromodulation space. As illustrated in Figure 8.30, architectural similarities do exist between the two systems. These include smart sensing strategies for direct monitoring of biophysical state through biopotential measurements and indirect measurements through sensors like the accelerometer. These sensors may measure and feedback parameters like position and movement or more direct symptom output, such as tremor amplitude or distribution in the body. These signals and algorithms can then be used to provide therapy through a control management algorithm.

The differences between the cardiac and nervous systems do require nuanced changes in feedback approach. Examples of differences include the varied signal characteristics from both the biopotential and inertial sensors; the signals are orders of magnitude different in amplitude (millivolts in cardiac potentials, μvolts in neural potentials) and/or different in signal coding (time-domain in pacemakers, time-frequency in neural stimulators). Stimulation rates are also significantly different; while cardiac signals are easily resolved from pacing artifacts with channel blanking, neural signals and stimulation often overlap temporally and require alternative filtering approaches like spectral analyses. In addition, the risk of false positives and false negatives are quite different between these therapies in terms of patient outcome.

We explored two case studies that help draw these principles out. In particular, we introduced the notion of both indirect and direct smart sensors of signals relevant for neurological treatments. While not entirely distinct, indirect sensors rely primarily on correlations of a measurement to an underlying process, while direct sensors attempt to make more of an immediate measure of key state variables. Each sensor type has a potential role in emerging treatment systems. Taken together, they will help enable us to initiate the era of closed-loop systems designed for the treatment of neurological disease.

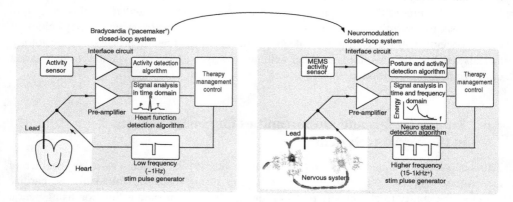

Figure 8.30 Mapping pacemaker architectures to neuromodulation systems. Designers must be mindful of the similarities and differences between the cardiac and nervous system to avoid oversimplifying the mapping of existing architectures to future therapies

Disclosure

The authors are employees and stockholders in Medtronic, Inc.

References

[1] T. Denison and B. Litt. "Advancing Neuromodulation through Control Systems: A General Framework and Case Study in Posture-Responsive Stimulation," *Neuromodulation*, 2014, in press.

[2] M. Rosa, S. Marceglia, D. Servello, *et al.*, "Time dependent subthalamic local field potential changes after DBS surgery in Parkinson's disease," *Experimental Neurology*, vol. 222, no. 2, pp. 184–90, April 2010.

[3] K. Paralikar, P. Cong, W. Santa, D. Dinsmoor, B. Hocken, G. Munns, J. Giftakis, and T. Denison, "An Implantable Optical Stimulation Delivery System for Actuating an Excitable Biosubstrate," *Solid-State Circuits, IEEE Journal of*, vol. 46, no. 1, pp. 321–332, January 2011.

[4] I. Diester, M. Kaufman, M. Mogri, R. Pashaie, *et al.*, "An optogenetic toolbox designed for primates," *Nature Neuroscience*, vol. 14, pp. 387–397, January 2011.

[5] I. Osorio, M.G. Frei, and S.B. Wilkinson, "Real-time automated detection and quantitative analysis of seizures and short-term prediction of clinical onset," *Epilepsia*, vol. 39, no. 6, pp. 615–627, June 1998.

[6] C.R. Craddock, P.E. Holtzheimer, X.P. Hu, H.S. Mayberg, "Disease state prediction from resting state functional connectivity," *Magnetic Resonance in Medicine*, vol. 62, no. 6, pp. 1619–162, December 2009.

[7] P. Tass, D. Smirnov, A. Karavaev, *et al.*, "The causal relationship between subcortical local field potential oscillations and Parkinsonian resting tremor," *Journal of Neural Engineering*, vol. 7, pp. 1–16, 2010.

[8] G.C. Miyazawa, R. Stone, and G.F. Molnar, "Next generation deep brain stimulation therapy: modeling field steering in the brain with segmented electrodes," *Society for Neuroscience*, vol. 693, no. 21, San Diego, CA, 2007.

[9] P.J. Hahn and C.C. McIntyre, "Modeling shifts in the rate and pattern of subthalamopallidal network activity during deep brain stimulation," *Journal of Computational Neuroscience*, vol. 28, no. 3, pp. 425-441, 2010.

[10] J. Holsheimer, B. Nuttin, G.W. King, W.A. Wesselink, J.M. Gybels, "Clinical evaluation of paresthesia steering with a new system for spinal cord stimulation," *Neurosurgery*, vol. 42, no. 3, pp. 541–547, 1998.

[11] G. Barolat, S. Zeme, and B. Ketcik, "Multifactorial analysis of epidural spinal cord stimulation," *Stereotactic and Functional Neurosurgery*, vol. 56, no. 2, pp. 77–103, 1991.

[12] J. Holsheimer and W.A. Wesselink, "Optimum electrode geometry for spinal cord stimulation: the narrow bipole and tripole," *Medical and Biological Engineering & Computing*, vol. 35, no. 5, pp. 493–497, 1997.

[13] E.J. Tehovnik, "Electrical stimulation of neural tissue to evoke behavioral responses," *Journal of Neuroscientific Methods*, vol. 65, no. 1, pp. 1–17, 1996.

[14] S.D. Stoney, W.D. Thompson, and H. Asanuma, "Excitation of pyramidal tract cells by intracortical microstimulation: effective extent of stimulating current," *Journal of Neurophysiology*, vol. 31, no. 5, pp. 659–669, 1968.

[15] R.B. North, D.H. Kidd, J.C. Olin, and J.M. Sieracki, "Spinal cord stimulation electrode design: prospective, randomized, controlled trial comparing percutaneous and laminectomy electrodes-part I: technical outcomes," *Neurosurgery*, vol. 51, no. 2, pp. 381–389, 2002.

[16] T. Cameron and K.M. Alo, "Effects of posture on stimulation parameters in spinal cord stimulation," *Neuromodulation*, vol. 1, no. 3, pp. 177–183, 1998.

[17] J.C. Olin, D.H. Kidd, and R.B. North, "Postural changes in spinal cord stimulation perceptual thresholds," *Neuromodulation*, vol. 1, no. 4, pp. 171–175, 1998.

[18] D. Abejon and C.A. Feler, "Is impedance a parameter to be taken into account in spinal cord stimulation?" *Pain Physician*, vol. 10, no. 4, pp. 533–540, 2007.

[19] C.M. Schade, D. Schultz, N. Tamayo, *et al.*, "Automatic adaptation of spinal cord stimulation intensity in response to posture changes," *North American Neuromodulation Society*, Las Vegas, December 2–6, 2009.

[20] J. Holsheimer, J.A. den Boer, J.J. Struijk, and A.R. Rozeboom, "MR assessment of the normal position of the spinal cord in the spinal canal," *American Journal of Neuroradiology*, vol. 15, no. 5, pp. 951–959, 1994.

[21] J.J. Struijk, J. Holsheimer, G. Barolat, J. He, and H.B. Boom, "Paresthesia thresholds in spinal cord stimulation: a comparison of theoretical results with clinical data," *IEEE Transactions on Rehabilitation Engineering*, vol. 1, no. 2, pp. 101–108, 1993.

[22] J. He, G. Barolat, J. Holsheimer, and J.J. Struijk, "Perception threshold and electrode position for spinal cord stimulation," *Pain*, vol. 59, no. 1, pp. 55–63, 1994.

[23] G. Barolat, "Epidural spinal cord stimulation: Anatomical and electrical properties of the intraspinal structures relevant to spinal cord stimulation and clinical correlations," *Neuromodulation*, vol. 1, no. 2, pp. 63–71, 1998.

[24] J. Holsheimer and G. Barolat, "Spinal geometry and paresthesia coverage in spinal cord stimulation," *Neuromodulation*, vol. 1, no. 3, pp. 129–131, 1998.

[25] J. Holsheimer, G. Barolat, J.J. Struijk, and J. He, "Significance of the spinal cord position in spinal cord stimulation," *Acta Neurochiropodists Supplement*, vol. 64, pp. 119–124, 1995.

[26] G.C. Molnar, E. Panken, and K. Kelley, "*Effects of spinal cord movement and position changes on neural activation patterns during spinal cord stimulation*," *American Academy of Pain Medicine*, San Antonio, Texas, 2010.

[27] C.R. Butson, C.B. Maks, and C.C. McIntyre, "Sources and effects of electrode impedance during deep brain stimulation," *Clinical Neurophysiology*, vol. 117, no. 2, pp. 447–454, 2006.

[28] K. Alo, C. Varga, E. Krames, J. Prager, J. Holsheimer, L. Manola, *et al.*, "Factors affecting impedance of percutaneous leads in spinal cord stimulation," *Neuromodulation*, vol. 9, no. 2, pp. 128–135, 2006.

[29] W.A. Wesselink, J. Holsheimer, and H.B. Boom, "Analysis of current density and related parameters in spinal cord stimulation," *IEEE Transactions on Rehabilitation Engineering*, vol. 6, no. 2, pp. 200–207, 1998.

[30] T. Denison, W. Santa, G. Molnar, and K. Miesel, "*Micropower Sensors for Neuroprosthetics*," *IEEE Sensors Conference*, Atlanta, GA, October 2007.

[31] A.G, Rouse, S. Stanslaski, P. Cong, R. Jensen, P. Afshar, D. Ullestad, R. Gupta, G. Molnar, D. Moran, and T. Denison, "A chronic generalized bi-directional brain–machine interface," *Journal of Neural Engineering*, vol. 8, no. 3, 2011.

[32] C.C. Enz and G.C. Temes, "Circuit techniques for reducing the effects of op-amp imperfections: autozeroing, correlated double sampling, and chopper stabilization," *Proceedings of the IEEE*, vol. 84, no. 11, pp. 1584–1614, 1996.

[33] K. Makinwa, "Dynamic Offset Cancellation Techniques in CMOS," ISSCC 2007 tutorial.

[34] M. Lemkin and B.E. Boser, "A three-axis micromachined accelerometer with a CMOS position-sense interface and digital offset-trim electronics," *IEEE Journal of Solid-State Circuits*, vol. 34, no. 4, pp. 456–468, 1999.

[35] T. Denison, K. Consoer, W. Santa, M. Hutt, and K. Mieser, "A 2μW Three-Axis Accelerometer," *IEEE Instrumentation and Measurement Conference*, Warsaw, 2007.

[36] A. Kühn, F. Kempf, C. Brücke, *et al.*, "High-frequency stimulation of the subthalamic nucleus suppresses oscillatory β activity in patients with Parkinson's disease in parallel with improvement in motor performance," *The Journal of Neuroscience*, vol. 28, no. 24, pp. 6165–6173, June 11, 2008.

[37] L. Rossi, S. Marceglia, G. Foffani, *et al.*, "Subthalamic local field potential oscillations during ongoing deep brain stimulation in Parkinson's disease," *Brain Research Bulletin*, vol. 76, no. 5, pp. 512–521, 2008.

[38] N.F. Ince, A. Gupte, T. Wichmann, J. Ashe, T. Henry, M. Bebler, L. Eberly, and A. Abosch, "Selection of optimal programming contacts based on local field potential recordings from subthalamic nucleus in patients with parkinson's disease," *Neurosurgery*, vol. 67, no. 2, pp. 390–397, 2010.

[39] F. Yoshida, I. Martinez-Torres, A. Pogosyan, *et al.*, "Value of subthalamic nucleus local field potentials recordings in predicting stimulation parameters for deep brain stimulation in Parkinson's disease," *Journal of Neurology, Neurosurgery & Psychiatry*, vol. 81, pp. 885–889, 2010.

[40] M. Weinberger, N. Mahant, W.D. Hutchison, A.M. Lozano, E. Moro, M. Hodaie, A.E. Lang, and J.O. Dostrovsky, "Beta oscillatory activity in the subthalamic nucleus and its relation to dopaminergic response in Parkinson's disease," *Journal Neurophysics*, vol. 96, pp. 3248–3256, 2006.

[41] B. Wingeier, T. Tcheng, M. Koop, B.C. Hill, G. Heit, and H.M. Bronte-Stewart, "Intra-operative STN DBS attenuates the prominent beta rhythm in the STN in Parkinson's disease," *Experimental Neurology*, vol. 197, pp. 244–251, 2006.

[42] W.C. Stacey and B. Litt, "Technology insight: neuroengineering and epilepsy- designing devices for seizure control," *Nature Clinical Practice Neurology*, vol 4, no. 4, pp. 190–201, 2008.

[43] C.H. Halpern, U. Samadani, B. Litt, J.L. Jaggi, and G.H. Baltuch, "Deep brain stimulation for epilepsy," *Neurotherapeutics*, vol. 5, no. 1, pp. 59–67, 2008.

[44] B. Litt and A. Krieger, "Of seizure prediction, statistics, and dogs: a cautionary tail," *Neurology*, vol. 68, no. 4, pp. 250–251, 2007.

[45] S. Venkatraman, K. Elkabany, J.D. Long, Y. Yao, and J. Carmena. "A system for neural recording and closed-loop intracortical microstimulation in awake rodents," *IEEE Transactions on Biomedical Engineering*, vol. 56, no. 1, pp. 15–21, 2009.

[46] A. Jackson, J. Mavoori, and E. Fetz. "Long-term motor cortex plasticity induced by electronic neural implant," *Nature*, vol. 444, no. 2, pp. 56–60, 2006.

[47] J. Mavoori, A. Jackson, C. Diorio and E. Fetz. "An autonomous implantable computer for neural recording and stimulation in unrestrained primates," *Journal of Neuroscientific Methods*, vol. 148, pp. 71–77, 2005.

[48] M. Azin, D.J. Guggenmos, S. Barbay, R.J. Nudo, and P. Mohseni, "A battery-powered activity-dependent intracortical microstimulation IC for brain-machine-brain interface," *IEEE Journal of Solid-State Circuits*, vol. 46, no. 4, pp. 731–745, 2011.

[49] M. Azin, D.J. Guggenmos, S. Barbay, R.J. Nudo and P. Mohseni, "A miniature system for spike-triggered intracortical microstimunlation in an ambulatory rat," *IEEE Transactions on Biomedical Engineering*, vol. 58, no. 9, pp. 2589–2597, 2011.

[50] S.S. Hsiao, M. Fettiplace, and B. Darbandi, "Sensory feedback for upper limb prostheses," *Progress in Brain Research*, vol. 192, pp. 69–81, 2011.

[51] A. B. Schwartz, X. T. Cui, D. J. Weber, and D. W. Moran. "Brain-controlled interfaces: movement restoration with neural prosthetics," *Neuron*, vol. 52, pp. 205–220, 2006.

[52] A. Avestruz, W. Santa, D. Carlson, R. Jensen, S. Stanslaski, H. Helfenstine, and T. Denison, "5µW/Channel spectral analysis IC for chronic bidirectional brain-machine interfaces," *IEEE Journal of Solid-State Circuits*, vol. 43, no. 12, pp. 3006–3024, December 2008.

[53] P. Yazicioglu, R. Merken, R.F. Puers, and C. Van Hoof, "A 60uW 60 nV/rtHz Readout Front-End for Portable Biopotential Acquisition Systems," *ISSCC Digest of Technical Papers*, San Francisco, CA, February 2006.

[54] R. Burt and J. Zhang. "A Micropower Chopper-Stabilized Operational Amplifier using a SC with Synchronous Integration inside the Continuous-Time Signal Path," *ISSCC Digest of Technical Papers*, San Francisco, CA, February 2006.

[55] T. Denison, K. Consoer, K. Kelly, A. Hachenburg, and W. Santa, "A 2µW, 94 nV/rtHz, chopper-stabilized instrumentation amplifier for chronic implantable EEG detection," *ISSCC Digest of Technical Papers*, pp. 162–594, San Francisco, CA, February 2007.

[56] T. Denison, K. Consoer, W. Santa, A.T. Avestruz, J. Cooley, and A. Kelly. "A 2 µW 100 nV/rtHz Chopper-Stabilized Instrumentation Amplifier for Chronic Measurement of Neural Field Potentials," *IEEE Journal of Solid-State Circuits*, vol. 42, pp. 2934–2945, 2007.

[57] R.F. Yazicioglu, P. Merken, R. Puers, and C. Van Hoof, "A 200mW Eight-Channel Acquisition ASIC for Ambulatory EEG Systems," *ISSCC Digest of Technical Papers*, pp. 164–603, San Francisco, CA, February 2008.

[58] T. Denison, W. Santa, R. Jensen, D. Carlson, G. Molnar, and A.-T. Avestruz. "An 8 µW Heterodyning Chopper Amplifier for Direct Extraction of 2 µVrms Neuronal Biomarkers," *ISSCC Digest of Technical Papers*, pp. 164–165, San Francisco, CA, February 2008.

[59] W. Siebert, *Circuits, Signals and Systems*, MIT Press, 1986.

[60] W.J. Freeman, L.J. Rogers, M.D. Holmes, and D.L. Silbergeld, "Spatial spectral analysis of human electrocorticograms including the alpha and gamma bands," *Journal of Neuroscientific Methods*, vol. 95, pp. 111–121, February 15, 2000.

[61] S.R. Shaw, D.K. Jackson, T.A. Denison, and S.B. Leeb, "Computer-aided design and application of sinusoidal switching patterns," in 6th Workshop on Computers in Power Electronics, pp. 185–191, 1998.

[62] A. Bakker, K. Thiele, and J. Huijsing. "A CMOS nested chopper instrumentation amplifier with 100 nV offset," *ISSCC Digest of Technical Papers*, pp. 156–157, San Francisco, CA, February 2000.

[63] A. Bakker, K. Thiele, and J.H. Huijsing, "A CMOS nested-chopper instrumentation amplifier with 100-nV offset," *IEEE Journal of Solid-State Circuits*, vol. 35, pp. 1877–1883, 2000.

[64] A. Hadiashar and J.L. Dawson. "A Chopper Stabilized CMOS Analog Multiplier with Ultra Low DC Offsets," in *Proceedings of the 32nd European Solid-State Circuits Conference*, pp. 364–367, 2006.

[65] R.R. Harrison and C. Charles, "A Low-power Low-noise CMOS Amplifier for Neural Recording Applications," *IEEE Journal of Solid-State Circuits*, vol. 38, no. 6, pp. 958–965, 2003.

[66] R. Sarpeshkar, "Brain power: Borrowing from biology makes for low–power computing," *IEEE Spectrum*, pp. 24–29, May 2006.

[67] A. Shoeb, D. Carlson, E. Panken and T. Denison, "A micropower support vector machine based seizure detection architecture for embedded medical devices," *EMBC 2009*, pp. 4202–4205, 2009.

[68] F.T. Sun, M.J. Morrell and R.E. Wharen, "Responsive cortical stimulation for the treatment of epilepsy," *Neurotherapeutics*, vol.5, pp. 68–74, 2008.

[69] H.W. Moses and J.C. Mullin, "Rate-Modulated Pacing," in *A Practical Guide to Cardiac Pacing*, Lippincott Williams & Wilkins, 2007.

[70] M. Siegel, T.H. Donner and A.K. Engel, "Spectral Fingerprints of Large Scale Neuronal Interactions," *Nature Neuroscience Review*, vol. 13, no. 2, pp. 121–34, 2012.

[71] R.F. Yazicioglu, P. Merken, R. Puers, and C. Van Hoof. "A low power ECG signal processor for ambulatory arrhythmia monitoring system," *IEEE VLSI Symposium*, pp. 19–20, 2010.

[72] Q. Fan, F. Sebastiano, H. Huijsing, and K. Makinwa. "A 1.8 µW1µV-offset capacitively-coupled chopper instrumentation amplifier in 65 nm CMOS," *Proceedings of the European Solid-State Circuits Conference*, pp.170–173, September 2010.

[73] J. Yoo, L. Yan, D. El-Damak, M. Bin Altaf, A. Shoeb, and A. Chandrakasan, "An 8-channel scalable EEG acquisition SoC with fully integrated patient-specific seizure classification and recording processor," *ISSCC Digest of Technical Papers*, pp. 292–294, San Francisco, CA, February 2012.

[74] R. Fisher, V. Salanova, T. Witt, R. Worth, T. Henry, R. Gross, and K. Oommen, "Electrical stimulation of the anterior nucleus of thalamus for treatment of refractory epilepsy," *Epilepsia*, vol. 51, no. 5, pp. 899–908, May 2010.

[75] M.J. Morrell, "Responsive cortical stimulation for the treatment of medically intractable partial epilepsy," *Neurology*, vol. 77, no. 13, pp. 1295–1304, September 2011.

[76] S. Stanslaski, P. Afshar, P. Cong , J. Giftakis, P. Stypulkowski, D. Carlson, D. Linde, D. Ullestad, A. Avestruz, T. Denison, "Design and Validation of a Fully Implantable, Chronic, Closed-Loop Neuromodulation Device With Concurrent Sensing and Stimulation," *IEEE Transactions on Neural Systems and Rehabilitation Engineering*, vol. 20, no. 4, pp, 410–421, 2012.

9

Micropower Generation: Principles and Applications

Ruud Vullers, Ziyang Wang, Michael Renaud, Hubregt Visser,
Jos Oudenhoven and Valer Pop
imec/Holst Centre, Eindhoven, The Netherlands

9.1 Introduction

The continuously decreasing power consumption of silicon-based electronics has enabled a broad range of battery-powered handheld, wearable and even implantable devices. Spanning six orders of magnitude, the typical power consumptions of a variety of electronic devices are listed in Table 9.1, together with their corresponding energy autonomy.

All these devices need a compact, low-cost and lightweight energy source, which enables the desired portability and energy autonomy. Nowadays, batteries represent the dominant energy source for the devices listed in Table 9.1 and alike. Despite the fact that the energy density of batteries has increased by a factor of 3 during the past 15 years, their presence in many cases has a large impact on, or even dominates, the overall dimension and operational cost of devices. For this reason, alternative energy sources have become the focus of worldwide research and development. One possibility is to replace batteries with energy-storage systems featuring higher energy density, e.g., miniaturized fuel cells [1]. A second approach consists of supplying the required energy to the device in a wireless mode. This solution, already used for radio-frequency identification (RFID) tags, can be extended to more power-hungry devices but requires dedicated transmission infrastructures. A third method is to harvest energy from the ambient by converting, for example, waste heat, vibration/motion energy, or RF radiation into electricity.

Among the different approaches, the energy-harvesting technology is widely considered as the enabling factor for the development of wireless-sensor networks. Wireless-sensor networks are generally made up of a large number of small, low-cost sensor nodes working in

Smart Sensor Systems: Emerging Technologies and Applications, First Edition.
Edited by Gerard Meijer, Michiel Pertijs and Kofi Makinwa.
© 2014 John Wiley & Sons, Ltd. Published 2014 by John Wiley & Sons, Ltd.

Table 9.1 Power consumption and energy autonomy of battery-operated electronic devices

Device Type	Power Consumption	Energy Autonomy
Smartphone	1 W	8 hours
MP3 player	50 mW	15 hours
Hearing Aid	1 mW	5 days
Wireless Sensor Node*	100 μW	Lifetime
Cardiac Pacemaker	50 μW	7 years
Quartz watch	5 μW	5 years

*when powered by energy harvester and energy storage device.
Reproduced by permission of Elsevier

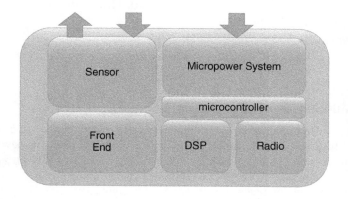

Figure 9.1 Schematic representation of the components in a typical WSN

collaboration to collect data and transmit them to a base station via a wireless link. They are poised to bring about a huge impact in sectors like health care, machinery, transportation and energy, and so on.

As schematically illustrated in Figure 9.1, a typical wireless-sensor node (WSN) consists of micropower module, sensor/actuator, front-end processing unit, digital-signal processor and radio. The average power consumption of a WSN has been estimated to vary between 1 μW and 20 μW in recent works [2, 3]. The power consumption strongly depends on the complexity of the measured physical effect, the application algorithm and the transmission frequency. Experimental results show that 90 μW is sufficient to drive a pulse oximeter sensor, to process data and to transmit them at an interval of 15 seconds [4]. In another example, 10 μW turns out to be sufficient for the measurement and transmission of temperature every 5 seconds [5]. The value of 100 μW reported in Table 9.1 is therefore representative of those relatively complex sensor nodes, deployed in systems operating at a relatively high data rate. Additionally, in the foreseeable future, the power consumption will further decrease thanks to the incessant advancement in low-power and ultra-low-power circuit design.

In many cases, wireless-sensor networks are intended to operate for a period of years. Because of the large number of sensor nodes and their small dimension, replacing depleted batteries is either unpractical or simply not feasible. Installing an enlarged battery to ensure energy autonomy leads to increased system volume and higher cost. The combination of an energy harvester with a small-sized rechargeable battery (or an energy-storage system, like a supercapacitor) is the best approach to deliver energy autonomy of the network throughout its entire lifetime. For instance, if the power consumption of a WSN is approximately $100\,\mu W$, the lifetime of a primary battery is expected to be only a few months [5]. In contrast, the combination of a rechargeable battery and an energy harvester with an output power of $100\,\mu W$ is able to realize energy autonomy for the whole lifetime.

Nevertheless, abolishing the energy-storage system altogether is not an option in most cases. As shown in Figure 9.2, the peak current flowing through the wireless transceiver during transmit and receive operation exceeds what can be achieved by using the energy harvester alone. Furthermore, buffering is also desired to guarantee continuous operation during times with no power generated. Depending on the specific application, the energy-storage system can be either a battery or a capacitor.

Table 9.2 summarizes the source power and harvested power by using different energy harvesting technologies. It suggests that energy harvesters can deliver an output power between $10\,\mu W$ and $1\,mW$. Such an output power is already sufficient to drive the low-power and ultra-low-power devices, particularly when duty cycling is switched on.

With regard to successful implementation of harvesting devices, three determining factors have been identified:

- The price of the harvesting device has to be reduced relative to the price of a complete WSN.
- The output power delivered should be sufficient to sustain the desired functionality of the WSN.
- Power consumption of the WSN has to be minimized via the use of ultra-low-power optimization techniques and technology breakthroughs.

The cost consideration is clearly related to the targeted application. For example, in the case of infrastructure monitoring for predictive maintenance, the price of an individual WSN

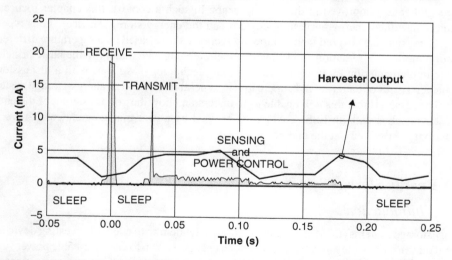

Figure 9.2 A typical scenario of the power consumption of a WSN. Since the power consumption does not fully match the output power of the energy harvester, an energy buffer and power management IC in between are necessary. Reproduced by permission of Elsevier

Table 9.2 Characteristics of various energy harvesting technologies [6]

Source	Source power	Harvested power
Ambient Light		
Indoor	$0.1 \, mW/cm^2$	$10 \, \mu W/cm2$
Outdoor	$100 \, mW/cm^2$	$10 \, mW/cm^2$
*Vibration/motion**	$0.5 \, m@1 \, Hz$	$4 \, \mu W/cm^2$
Human	$1 \, m/s^2 @50 \, Hz$	
*Vibration/motion**	$1 \, m@5 \, Hz$	$100 \, \mu W/cm^2$
Industrial	$10 \, m/s^2 @1 \, kHz$	
Thermal Energy		
Human	$20 \, mW/cm^2$	$30 \, \mu W/cm^2$
Industrial	$100 \, mW/cm^2$	$1 \, mW/cm^2$ to
		$10 \, mW/cm^2$
RF Cell phone	$0.3 \, \mu W/cm^2$	$0.1 \, \mu W/cm^2$

*Depending on source, either expressed in amplitude or acceleration at a certain frequency.
Reproduced by permission of Elsevier

can be relatively high, because the accumulated cost reduction is far greater than the initial investment. Condition monitoring of infrastructures is also exactly the field where the first energy harvesters have appeared on the market. On the other hand, for most other applications, current-harvesting technologies are still far too expensive. A possible route to cheaper harvesting devices is to use micromachining technology for manufacturing. The devices can thus be fabricated on a wafer basis in a batch mode, thereby significantly lowering the cost. Nevertheless, reducing the size of a harvesting device affects not only the cost but also the output power.

Thanks to the widespread interest from academia and industry, energy harvesting is expected to be significantly improved in the coming years. In such a context, this chapter focuses on the micromachined energy harvesting devices and the energy storage devices. Four types of energy harvesters are covered in the following sections: those based on temperature difference, vibration/motion RF radiation and PV, with attention being paid mainly to the basic principles and the implementation by using micromachining technology. We start with a discussion of the characteristics of batteries and supercaps (in general referred to as energy storage systems, ESS). The reason is that this will enable us to understand how the voltage output of the energy harvester matches with an ESS. This gap is bridged by power management, which we will discuss only very briefly at the end of each section.

9.2 Energy Storage Systems

9.2.1 Introduction

Energy storage systems (ESS) are frequently used in tandem with energy harvesting devices for several reasons. First, they may be required as an energy backup to ensure a stable power supply at moments when the energy harvester delivers a less than desired output power. A second function is to serve as energy buffer: many wireless sensors consume relatively high peak currents during transmit and receive operations while the continuous background consumption is much lower (Figure 9.2). These sensors benefit from the incorporation of an ESS, which is recharged continuously by an energy harvester. This ESS subsequently delivers the high peak currents. A third field of application for ESS is in sensor devices where the ESS itself serves as the main energy source.

All these functions impose different requirements on the ESS. For an energy backup, a rechargeable ESS needs to have a relatively high capacity to deliver long-term autonomy. Its capacity is usually higher than that of an energy buffer but still lower in comparison to an energy source. Furthermore, it should have a relatively low self-discharge rate, because the charge often has to be stored for a long period of time. For an ESS employed as energy buffer, a high peak current is frequently drawn while the energy storage time can be relatively short. When an ESS is used as the main energy source, its capacity needs to be high and its self-discharge rate low. However, the ESS is not necessarily a rechargeable system in this case.

There are three main types of micro scale ESS: supercapacitors, micro-batteries (typically Li-ion) and solid-state thin-film batteries. Their typical characteristics are compared in Table 9.3. For a specific application, the choice of an ESS is determined by the matching between the total requirements of the sensor system and the capabilities of the ESS. In the next section, the basic principles of supercapacitors, micro-batteries and solid-state batteries are described in the context of smart sensor systems.

9.2.2 Supercapacitors

Supercapacitors, often referred to as electrochemical capacitors or electrochemical double-layer capacitors (EDLC), consist of two electrodes, an electrolyte and a separator (Figure 9.3(a)) [7]. When the electrodes become charged due to an externally applied voltage, the charge carriers in the electrolyte (ions) will compensate for this by accumulating at the interface of the electrolyte with the electrodes. This effect is the formation of an electrochemical double-layer, which is the basic charge storage mechanism for supercapacitors. The number of ions that can accumulate at the surface of the electrodes is highly dependent on the surface area of these electrodes. The capacity of a supercapacitor, and thus the energy stored in it, are therefore directly related to the active surface area of its electrodes. To increase the capacity of the supercapacitors, high surface area materials, like activated carbon, are

Table 9.3 Typical characteristics of three types of ESS [6]

	Supercapacitor	Battery	
		Li-ion	Thin film
Operating voltage (V)	1.25	3–3.7	3.7
Energy density (Wh/L)	6	435	<50
Specific energy (Wh/kg)	1.5	211	<1
Self-discharge rate (%/month) at 20°C	100	0.1–1	0.1–1
Cycle life (cycles)	>10,000	2000	>1000
Temperature range (°C)	−40 to 65	−20 to 50	−20 to 70

commonly used as the electrodes. One should, however, realize that a porous electrode with pores smaller than the ions in the electrolyte is not useful to increase the capacity.

The voltage of a supercapacitor is linearly related to its state of charge. The maximum voltage below which a supercapacitor remains stable is determined by the electrochemical stability of the electrolyte. The upper voltage is limited to around 1.2 V, when an aqueous electrolyte is used. In comparison, supercapacitors with an organic electrolyte are available with a voltage level up to 3 V.

Supercapacitors are especially useful to serve as energy buffer. Because neither electro-chemical reaction nor charge transfer from the electrodes to the electrolyte is involved, the supercapacitor can operate very fast and deliver high current pulses. However, the energy density is relatively low while the self-discharge rate is high. Therefore, supercapacitors are less suitable for application as either backup energy source or primary energy source, where a relatively large amount of energy needs to be stored for a long period of time.

9.2.3   Lithium-Ion Batteries

Similar to supercapacitors, lithium-ion batteries consist of two electrodes, an electrolyte and a separator (Figure 9.3(b)) [7]. The electrolyte is in this case a lithium-containing non-aqueous liquid. The frequently used materials for the positive and negative electrodes are lithium cobalt oxide ($LiCoO_2$) and graphitic carbon, respectively. The charge storage mechanism is, however, slightly more complicated than for supercapacitors. It is based on not only charge separation in the electrolyte, but also on electrochemical reactions taking place at the electrodes. When the battery is charged, lithium is released from $LiCoO_2$, following the chemical reaction below:

$$LiCoO_2 \underset{Discharge}{\overset{Charge}{\rightleftarrows}} Li_{1-x}CoO_2 + xLi^+ + xe^- \quad 0 \leq x \leq 0.5 \tag{9.1}$$

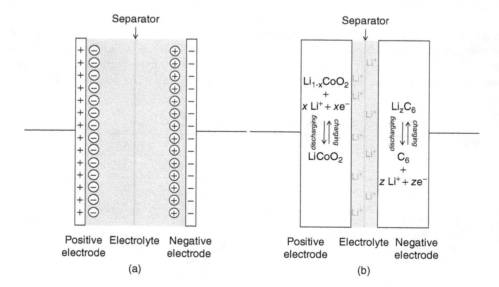

Figure 9.3   Schematic overview of (a) a supercapacitor and (b) a Li-ion battery

For one unit of $LiCoO_2$, 0.5 lithium ions can be released (1 lithium ion per 2 $LiCoO_2$ units). The lithium ions that are made available during this reaction are transported through the electrolyte to the negative electrode, referred to as the anode. The chemical reaction taking place at this anode follows:

$$C_6 + zLi^+ + ze^- \xrightleftharpoons[Discharge]{Charge} Li_zC_6 \qquad\qquad 0 \leq z \leq 1 \qquad\qquad (9.2)$$

Typically, one lithium ion is stored per six carbon atoms. When the battery is discharged, the direction of these reactions is reversed.

Contrary to the case of supercapacitors, it is not the surface area of the electrodes but the amount of material and therefore the volume of the electrodes that determines the capacity of Li-ion batteries. Therefore, much higher volumetric energy densities can be obtained. However, the maximum current that can be drawn from a battery is generally lower than that from a supercapacitor, because the chemical reactions and the charge transfer processes are much slower than double-layer effects.

Unlike supercapacitors, the open-cell voltage of a battery is only loosely dependent on the state of charge. This is mainly because the electrochemical potential of the redox reactions (9.1) and (9.2) is well defined. A typical Li-ion battery based on the reactions (9.1) and (9.2) has an open-cell voltage around 3.7 V. In practice, the voltage under discharging conditions will be somewhat lower, because of the internal resistance of the battery and limitations inside the electrodes. The processes that increase the resistance are associated with, for example, mass transfer and reaction rate kinetics. These influences on the battery voltage are all dependent on the current. Therefore, applying higher currents will result in a lower voltage. Batteries operated at higher currents will as a result also deliver a lower energy. On the other hand, all these effects that decrease the discharging voltage and the energy will result in a higher voltage and energy required for charging, leading to an energetically less efficient battery [8].

A crucial part of battery design is the packaging: lithium is for example very reactive with water, oxygen, carbon dioxide and nitrogen, which are all major constituents of ambient air. Therefore, Li-ion batteries should be well encapsulated. Several designs of packaging are possible. For macro-scale batteries, this often includes metallic cylindrical cells or prismatic cells. The volume of the packaging is in this case very small compared to the volume of the active electrode materials. Therefore, relatively high energy densities can be obtained. For micro-batteries, the electrodes and electrolyte are miniaturized and placed into a button cell or laminated sheets of plastics and metals. The relative volume of the packaging is thus much larger and the energy density of the battery will be significantly lower than for macro-scale batteries. However, the energy density of these micro-batteries is generally still higher than that of supercapacitors. Moreover, the self-discharge rate is much lower than that of supercapacitors, which makes micro-batteries more suitable as energy backup or primary energy source.

Micro-batteries cannot deliver high peak currents because of their relatively large internal impedance. Some authors therefore suggest the use of a hybrid system that combines a battery with a supercapacitor [9, 10]. It was indeed demonstrated that such a hybrid system is more capable of delivering high peak currents. It should, however, also be noted that a price has to be paid for this peak current increase: the volume of the energy storage (battery + supercapacitor) will increase significantly, with a factor varying from 1.5 up to 10. Much effort is therefore invested in the decrease of the impedance of (micro-) batteries.

9.2.4   Thin-Film Lithium-Ion Batteries

Thin-film lithium-ion batteries are a special case of Li-ion batteries. Their operating principle is essentially the same as that of liquid-electrolyte batteries. However, the electrodes and electrolyte are in the form of solid thin films with thicknesses in the order of μm. These films are usually deposited on a substrate of silicon, glass or some polymer material and the stack is encapsulated by a sealing material, as illustrated in Figure 9.4a.

The thickness of the electrolyte is much smaller than that of conventional batteries. A thin-film battery contains a solid electrolyte layer which is typically 1 μm thick, while the separator in a conventional battery normally has a thickness of $\sim 20\,\mu m$. This fact means that the volumetric energy density of a thin-film battery can in theory be higher than that of a conventional battery of the same size. However, because the substrate (packaging) of thin film batteries is relatively thick, this benefit is generally lost. A second advantage of the solid electrolyte system is that the absence of organic liquid electrolytes eliminates the risk of leakage and therefore allows a wider range of applications. The solid electrolyte is also more stable at higher temperatures, widening the range of applications and increasing the flexibility of process steps in the device manufacturing. On the other hand, solid electrolytes have a disadvantage. They generally have a lower ionic conductivity than liquid electrolytes, which induces a larger voltage drop due to the ionic transport within the solid electrolyte.

To solve this problem, several research groups have proposed an approach using 3D thin-film batteries. The aim for these 3D batteries is to increase the surface area of the positive and negative electrode, without increasing the footprint area of the battery. This can be achieved, for example, by using a microporous substrate or a first electrode with a 3D structure. A schematic example of such a 3D battery can be observed in Figure 9.4(b). In this image it can be observed that the size of the battery remains the same while the effective area of the battery stack (positive electrode / solid electrolyte / negative electrode) increases. When the same current (A) is applied to a 3D battery, the current density (A/cm^2) inside the battery stack is lower. Thus, the

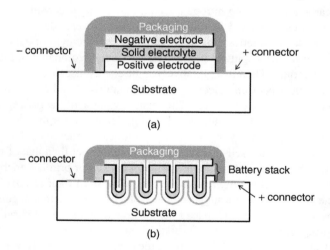

Figure 9.4   Schematic representation of a thin-film battery in a planar (a) and 3D (b) geometry

voltage drop will become lower accordingly. The increased battery voltage will result in an increased energy density (i.e. voltage × current × time/volume) for a 3D battery compared to an equivalent battery with a planar configuration. Moreover, a larger surface area also provides the possibility to have a larger electrode volume. The capacity will then be higher, resulting in an even further increased energy density [11].

9.2.5 Energy Storage Applications

For successful applications of ESS in smart sensor devices, several important issues need to be properly addressed. First, the power requirements, including energy, voltage, autonomy, life time, and so on, need to be considered. These technical requirements have been addressed in the previous sections. However, other requirements also play an important role. One is the shape of the ESS, particularly with regard to the size limitations. For example, button cells are a standard shape of batteries and capacitors. Despite a relatively small size, they may not be suitable for applications where a very small thickness is required. For this purpose, laminated sheet-batteries or thin film batteries are more suitable. These batteries cover a generally larger area than button cells, but are less than 1 mm thick and therefore can be more easily incorporated into smart packaging or smart cards. Additionally, because foil-based batteries can be made flexible, they can be integrated onto bendable surfaces. This fact leads to a less restricted system design. Some examples of the shapes of ESS are given in Figure 9.5.

A very important aspect regarding the use of ESS in smart sensor systems is the end-of-life disposal and recyclability. In the European Union, there are no general rules for supercapacitors. Generally, these devices, together with the device they power, fall under the WEEE (Waste Electrical and Electronic Equipment) directive. This directive rules that the amount of several substances, like lead, mercury and cadmium, must be very low and moreover, manufacturers or importers of the devices are required to ensure proper recycling schemes for these devices [12].

Batteries fall under another directive and their recycling is promoted by the European Union. Manufacturers or importers are required to have a return schedule for batteries. Moreover, batteries should be installed in electronic systems in such a way that they can be readily removed. There are, however, some exceptions to this rule when "for safety, performance, medical or data integrity reasons, continuity of power supply is necessary" and "a permanent connection between the appliance and the battery" is required [13].

Battery related laws in the US are enacted to ensure that certain metals are recovered from electronic waste and to stimulate recycling of single-use and rechargeable batteries [14]. These laws are, however, less stringent than their European counterparts.

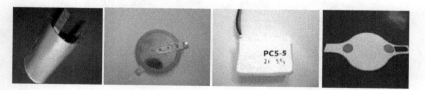

Figure 9.5   Various form factors of ESS. From left to right: laminated foil battery, button cell, prismatic cell and printed thin-film battery

For applications of ESS, one should therefore not only look at the technical requirements but also remain informed about the development of directives and laws in the countries where the device will be used.

9.3 Thermoelectric Energy Harvesting

9.3.1 Introduction

The term *thermoelectric effect* in general encompasses three different but related physical phenomena: the *Seebeck effect*, the *Peltier effect* and the *Thomson effect*. Initially, interest in the thermoelectric effects was driven by the intention to exploit the Seebeck effect for power generation. Nevertheless, only after the 1950s when a handful of high quality semiconductor materials had been characterized, were a variety of modern thermoelectric devices realized successfully. An example is the radioisotope thermoelectric generator (RTG), which converts the heat dissipated from radioactive isotopes into electricity. The RTGs, made of high-temperature resistant silicon germanium (SiGe) alloys, could operate unattended for more than 20 years in the Voyager I spacecraft [15]. Recently, the potential of using the thermoelectric effect for waste heat recovery has been gradually acknowledged by the industry. A German car manufacturer, BMW AG, has deployed commercial thermoelectric generator (TEG) module in between the exhaust gas pipeline and the cooling fluid pipeline [16]. An output power of about 200 W has been recuperated in the benchmarking test drive at 130 km/h.

The Seebeck effect relates to the generation of an electrical potential difference within conducting materials, either metal or semiconductor, which are subject to a temperature difference [15]. It is essentially an electrical effect caused by a temperature difference. With regard to an isolated conducting material, the aforementioned phenomenon is referred to as the absolute Seebeck effect (ASE), as schematically illustrated in Figure 9.6(a). The absolute Seebeck coefficient (α) at a given temperature is thus defined as the sensitivity of the absolute Seebeck voltage V_{SA} to the temperature change. In other words, this coefficient is mathematically expressed as $\alpha = dV_{SA}/dT$. Because the gradient of the potential difference can be the same as or opposite to the existing temperature gradient, the sign of α, normally denoted in $\mu V/K$, can be either positive or negative.

The most frequent way of exploiting the Seebeck effect is to form a thermocouple by electrically connecting two dissimilar conductors at one set of their terminals. When a temperature difference is present between the closed and the open end of the thermocouple, a voltage is

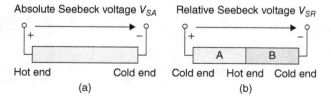

Figure 9.6 Schematic illustration of (a) absolute Seebeck voltage V_{SA} in an isolated conducting material under a temperature difference; (b) relative Seebeck voltage V_{SR} generated in a thermocouple consisting of two dissimilar conducting materials, A and B, under a temperature difference

built up across the unpaired terminals of the open end, as described in Figure 9.6(b). The resulting voltage is dubbed as the relative Seebeck voltage. Similar to α, the relative Seebeck coefficient, α_{AB}, is defined as $\alpha_{AB} = dV_{SR}/dT$. For a thermocouple made up of two dissimilar conducting materials A and B, its α_{AB} is then expressed as $\alpha_{AB} = \alpha_A - \alpha_B$, where α_A and α_B are the absolute Seebeck coefficients for material A and B, respectively. Hence, to obtain a large α_{AB}, the two conducting materials are required to have absolute Seebeck coefficients with the opposite signs. Because both α and α_{AB} denote the Seebeck coefficient of conducting materials, albeit in different contexts, hereafter they are not differentiated explicitly and generally written only as the Seebeck coefficient. So is the case for the Seebeck voltage.

9.3.2 State-of-the-Art

Thermoelectric energy harvesting is based on a number of serially connected thermocouples subject to a common temperature difference. The generated Seebeck voltage can thus drive an electrical current through an external load connected to the open ends of the thermocouple chain. Hence, the heat flow through the thermocouples is converted into electricity. Compared to the other methods for recuperating energy from the ambient, thermoelectric energy harvesting offers a unique set of advantages, such as wide applicability on different objects, high mechanical robustness thanks to lack of moving parts, and uninterrupted operation irrespective of the weather condition.

Along with the progress in MEMS technology, the miniaturization of TEGs has become realistic in the past decade. Among the developed devices, there exists a large variation in terms of device geometry, scale of integration, choice of materials and fabrication method. Nevertheless, according to the orientation of thermocouple relative to the substrate, be it a silicon substrate or a flexible polymer foil, most of the miniaturized TEGs can be divided into two categories: in-plane devices, in which the thermocouple legs are parallel to the substrate, and cross-plane devices, in which the thermocouple legs are perpendicular to the substrate. The schematic configurations are shown in Figure 9.7. Because the cross-plane devices have been fabricated by using both thick film and thin film technologies, this category is then further divided into two subtypes according to the used fabrication method.

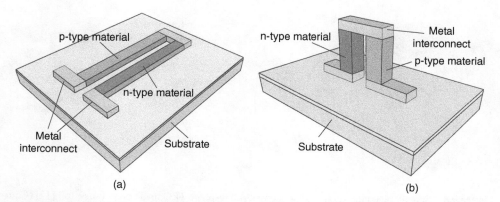

Figure 9.7 Schematic configuration of a thermocouple in (a) in-plane TEG device; and (b) cross-plane TEG device (see Plate 8)

In-plane TEG In-plane TEGs are mostly made of antimony (Sb), bismuth (Bi) or bismuth telluride (BiTe) by electroplating or sputtering on polymer foils, such as Kapton foil [17–19]. Polymer foils are preferred as the substrate for several reasons: first, they typically have a low thermal conductivity, for example, 0.12 W/m/K, which makes the heat loss through the polymer foils less pronounced; second, they have a thermal expansion coefficient, for example, $20 \times 10^{-6} \mathrm{K}^{-1}$, similar to those of the thermoelectric materials; third, they are a low-cost substrate material [18]. One common feature of the in-plane TEGs is the relatively large geometry of thermocouples, for example, tens of μm for width and hundreds of μm or even several mm for length, thus allowing the use of technically adequate but low-cost fabrication methods, such as screen printing, as opposed to the conventional MEMS microfabrication technologies. Thanks to the structural flexibility of polymer foils, the in-plane TEGs can be coiled up into a spiral shape in such a way that the resulted device can be erected in a self-standing manner. Illustrated in Figure 9.8(a), this scheme leads to more thermocouples per unit footprint area [17]. Another advantage of the in-plane TEGs is the high aspect ratio of the thermocouples, because the thermocouple geometry is free from restrictions, like the maximum thickness of thin film deposition and the depth-of-focus in the contact photolithography. Consequently, the thermal resistance of each thermocouple is significantly increased. In the meantime, the contact resistance between thermoelectric materials and metal interconnect is reduced because of the usually large contact area. The main disadvantage is the poor thermal contact between thermocouple junctions and the heat source or heat sink [17]. This issue can be alleviated by applying thermal interface materials (TIMs), like various types of thermal grease. A second drawback is the thermal shunting effect of polymer foils, which are thermally parallel to the thermocouples. Methods to deal with this drawback include the adoption of an even thinner polymer foil or, in a better case, the peeling-off of polymer foil partially or even completely.

Some in-plane TEGs are fabricated on Si substrate instead of polymer foils. Imec/Holst Centre has succeeded in delivering a series of in-plane TEGs, among which one type consists of only free-standing poly-Si or poly-SiGe thermocouples, as schematically shown in Figure 9.8(b) [20]. Although it is a challenging task to remove the complete supporting layers underneath the thin film thermocouples, this has been accomplished by fine tuning the stress in the thin film stack. One potential assembly scheme for boosting the output performance

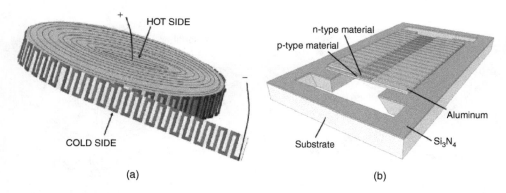

(a)   (b)

Figure 9.8 Schematic configuration of (a) a coiled-up in-plane TEG made on polymer foil [17] (Courtesy of Weber, Reproduced by permission of Elsevier); and (b) a free-standing poly-SiGe in-plane TEG made on Si substrate [20]. With kind permission from Springer Science and Business Media B.V. (see Plate 9)

is to bundle several in-plane thermopile chips on the same substrate by erecting them and connecting them electrically in series. As another example, IMTEK has developed an in-plane TEG device based on n-type poly-Si and aluminum (Al) on a membrane of thin film SiO_2, which behaves as a structural support but a thermal shunting path as well [21].

Table 9.4 compares various in-plane TEG devices covered in this section. Both the output voltage and the output power are normalized to the temperature difference and the area for the ease of comparison. This table shows that the in-plane TEG devices usually have a relatively high output voltage but in the meantime suffer from a low output power. This is mainly attributed to the large internal electrical resistances.

Cross-plane thick film TEG Cross-plane thick film TEGs are characterized by film thickness ranging from tens of μm to hundreds of μm. Thermocouples made of such a thick film are usually fabricated by electroplating in a predefined mould, conventional mechanical machining or even manual manipulation. Therefore, the dimensions of thermocouples are relatively large, particularly when compared to those of cross-plane thin film TEGs described in the next section. For instance, each thermocouple leg in the TEG device developed by Seiko measures 80 μm × 80 μm × 600 μm, as shown in Figure 9.9 [22]. Cross-plane thick film TEGs usually contain only a limited number of thermocouples, mostly not larger than 200. This fact, in combination with the non-optimized thermoelectric properties of BiTe compound, determines the moderate output voltage frequently observed in the devices of this category.

Table 9.5 lists the technical details of the cross-plane thick film TEGs found in literature. Note that those parameters not explicitly given in the references are extracted from graphs or figures if possible. Typical for cross-plane thick film TEGs, the output voltage is usually at mV level per unit temperature difference.

On the other hand, the output power can reach μW level for the same device. This is attributed to the usually low internal electrical resistance, given the relatively large geometries and contact areas with metal interconnect. For instance, the internal electrical resistance of the BiTe TEG device made by Jet Propulsion Laboratory (JPL) is only $12\,\Omega \sim 30\,\Omega$ [23].

Cross-plane thin film TEG Cross-plane thin film TEGs are usually made of poly-Si or poly-SiGe by resorting to thin film MEMS technology. Fabricated in a batch process, thin

Table 9.4 Comparison of various in-plane TEG devices

Institution	D.T.S. [18]	TU Dresden [19]	Holst Centre [20]	IMTEK [21]
Substrate	polyimide	epoxy	Si substrate	Si substrate
Material	BiTe	Sb-Bi	Poly-Si / SiGe	Poly-Si - Al
Fabrication	Sputtering	Electroplating	Thin film deposition	Thin film deposition
Geometry	50 μm wide	40 μm wide	100 μm long	120 μm long
Total number of thermocouples	2250	93	278	7500
Output voltage ($mV/K/cm^2$)	310	N/A	390*	166
Output power ($\mu W/K^2/cm^2$)	0.087	N/A	0.01	1.37×10^{-3}

*With the assumption of chip standing on sidewall

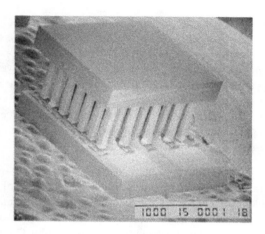

Figure 9.9 SEM photo of the thermocouples in the TEG developed by Seiko (Courtesy of Kishi [22]). Each thermocouple leg measures 80 μm × 80 μm × 600 μm. Reproduced by permission of IEEE

Table 9.5 Comparison of various cross-plane thick film TEG devices

Institution	Seiko [22]	JPL [23]	Micropelt [24]	ETH [25]
Substrate	Si	Si	Si	Si
Material	BiTe	BiTe	BiTe	BiTe
Fabrication	Hot pressing	Electroplating	Sputtering	Electroplating
Geometry*	600 μm in height	60 μm in diameter	20 μm in height	210 μm in diameter
Total number of thermocouples	104	126	540	99
Output voltage	0.52 V/K/cm²	0.2 ~ 0.5 V/K/cm²	0.98 V/K/cm²	N/A
Output power	27 μW/K²/cm²	0.016 ~ 0.1 μW/K²/cm²	114 μW/K²/cm²	0.25 μW/K²/cm²

*Note that not all the structural parameters have been reported in the literature.

film TEGs can potentially be made at a relatively low unit cost. Because of the thin film technology, this type of TEGs also consumes less raw thermoelectric materials, some of which are becoming increasingly expensive. Compared to the other types of TEGs, cross-plane thin film ones are distinctly characterized by small feature size, down to several μm, and meanwhile a large number of thermocouples, from several thousand to tens of thousands, as illustrated in Figure 9.10. For instance, the TEG developed by Infineon has integrated about 15,000 thermocouples within a chip area of only about 6 mm² [26]. A large number of thermocouples, connected electrically in series, help to boost the open-circuit output voltage per unit temperature difference for unit chip area. However, the combination of a small feature size and a long chain brings about the problem of a large internal electrical resistance, which originates from both the serial interconnection and the more pronounced issue of contact resistance. For example, the TEG developed by A ∗ STAR, when scaled to 1 cm², has an internal electrical resistance of 52.8 MΩ, in which 23 MΩ is attributed to the contact between poly-Si and metal [27]. Another knock-on effect is the usually low thermal resistance for TEGs of this category, as indicated by the small fraction of temperature difference falling

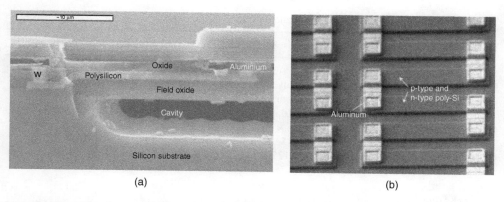

(a)   (b)

Figure 9.10 SEM photos of (a) cross-section of an individual thermocouple in the TEG developed by Infineon (Courtesy of Strasser [26], Reproduced by permission of Elsevier); the Si substrate is undercut to create a cavity underneath the thermocouples for improved thermal isolation, and (b) top view of thermocouples in the TEG developed by A ∗ STAR (Courtesy of Xie [27]); p- and n-type thermocouple legs are arranged in meandering shape

(a)   (b)

Figure 9.11 SEM photos of (a) planar thermocouples [28], Reproduced by permission of Elsevier; and (b) thermocouples fabricated on a 6-μm-high topography (Courtesy of Su [29], © IOP Publishing. Reproduced by permission of IOP Publishing). Both are developed by imec/Holst Centre

along the thermocouples in the characterization of A ∗ STAR TEG [27]. The main reason for this phenomenon is that all the thermocouples are connected thermally in parallel, resulting in a much lower total thermal resistance.

To fully leverage the advantage of thin film TEGs, one needs to maximize the thermal resistance of an individual thermocouple and meanwhile reduce the internal electrical resistance, particularly the contact resistance between semiconductor material and metal. Recently, Imec/Holst Centre has successfully developed a micromachined thermopile consisting of high topography thermocouples, which has a thermal resistance increased by almost a factor of 10 compared to the planar thermocouples of previous generation [28, 29]. The juxtaposition of two generations in Figure 9.11 clearly shows the advancement of technology. Table 9.6 compares various cross-plane thin film TEG devices covered in this section.

Table 9.6   Comparison of various cross-plane thin film TEG devices

Institution	Infineon [26]	A * STAR [27]	Holst Centre [28]	Holst Centre [29]
Substrate	Si substrate	Si substrate	Si substrate	Si substrate
Material	Poly-Si / SiGe	Poly-Si	Poly-SiGe	Poly-Si / SiGe
Fabrication	Thin film	Thin film	Thin film	Thin film
	deposition	deposition	deposition	deposition
Geometry (μm^2)	49×10.9	16×5	30×16	34×10
Total number	15872	31536	2700	1500
Output voltage	2.2	0.125	2.08	10.32
(V/K/cm^2)				
Output power	0.035	7.8×10^{-5}	0.006	0.047
($\mu W/K^2/cm^{2)}$				

9.3.3   Conversion Efficiency

In general, the conversion efficiency of a TEG is determined by two aspects: the available temperature difference and the material properties. Mathematically, the conversion efficiency is expressed as

$$\eta = \frac{T_h - T_c}{T_h} \cdot \frac{M - 1}{M + T_c/T_h} \approx \frac{T_h - T_c}{T_c} \cdot (M - 1), \text{ and } M = \sqrt{1 + ZT}, \qquad (9.3)$$

where T_h and T_c respectively indicate the temperature of hot side and cold side, M is a material factor, and ZT is the material figure-of-merit. Furthermore, ZT is usually defined as

$$ZT = \frac{\alpha^2 T}{\rho \kappa} \qquad (9.4)$$

where α is the Seebeck coefficient, ρ is the electrical resistivity, κ is the thermal conductivity, and T is the absolute temperature.

9.3.4   Power Management

As some thermal harvesters generate very low voltages, associated power management focused on low-voltage startup. The circuit reported in [30], can start working from an input voltage of 0.13 V and is designed to transfer approximately 2 mW. The control power is as high as 0.4 mW, but in view of the milliWatts generated by the harvester, this is acceptable. A power management circuit for two sources of power was presented by [31]. It is able to convert thermally harvested power and RF power. For thermal power management, an integrated boost converter with an external inductor was used. The circuit consumes 70 µW and can transfer approximately 1 mW.

In 2008, a power management circuit for very low power application was demonstrated [32], followed by an improved version later [33] realized in a standard 0.35 µm CMOS technology that can handle up to 1 mW and still has a very low power consumption. The conversion principle is optimized to the characteristics of the TEG. It contains a charge pump with capacitors

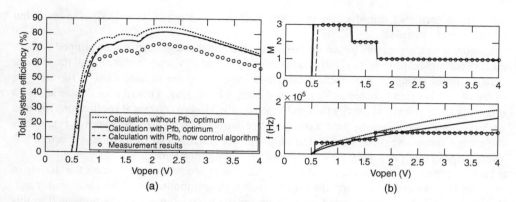

Figure 9.12   Calculation and measurement results for a system connected to a TEG: (a) system efficiency (*Pfb* is the power consumed by the feedback circuit), (b) number of stages M and switching frequency f ([32]. Reproduced by permission of IEEE)

of 2.45 nF and a maximum of eight stages. The total area is 59 mm^2. The measured overall system efficiency, number of stages M and switching frequency f for a TEG with an impedance of 11 kΩ ohm are plotted in Figure 9.12. The circuit starts up at 0.6 V open circuit voltage of the TEG. The measured peak efficiency is 70%.

9.3.5   Conclusion

In order to develop thermoelectric energy harvesting, especially under a limited temperature difference, one would need to deal with two fronts. The first is device design aimed at an increased temperature difference falling along the thermocouples. In the presence of parasitic serial thermal resistors, only part of the overall temperature difference can fall along the thermocouples. The second is to optimize the thermoelectric materials to eventually achieve a high figure-of-merit ZT. As to the cross-plane thin film TEGs with small feature size, one should not ignore the influence of the contact resistance between semiconductor material and metal. With decreasing contact area, the contact resistance tends to play a more important role in the determination of the internal electrical resistance.

9.4   Vibration and Motion Energy Harvesting

9.4.1   Introduction

Mechanical vibrations or motions are existing almost everywhere and therefore are attractive energy sources for generating electrical power. Self-powered electronic wristwatches based on an eccentric mass are well-known commercial examples [34]. With the recent development in low-power portable electronics, a renewal of interest has been observed for miniaturized vibration energy harvesters. These devices convert mechanical energy into electrical energy through electromechanical transducers. The most commonly implemented transduction mechanisms are electromagnetic [35, 36], electrostatic [37–39] and piezoelectric conversion

[40–44]. Alternative transduction methods, such as magnetostriction [45], also exist but remain largely marginal.

Motion-driven generators generally fall into two distinct categories: those equipped with an electromechanical transducer rigidly connected to a source of vibration/motion and those using inertial forces acting on a movable proof mass for exciting the transducer. The former type of harvesters is often denoted as strain-based energy harvesters. This type of device allows a relatively large amount of energy to be extracted easily. Moreover, several applications focused on human body applications have already been commercialized. The best known example is the energy harvesting shoe [46, 47]. The output power generated by these shoes exceeds a few watts, large enough to supply power to many sorts of applications.

However, most of the strain-based energy harvesters are bulky and require the design of resilient transducers. Therefore, they do not fulfill the requirements of wireless-sensor networks, which is the main targeted application for the harvesting principles presented in this chapter. On the other hand, miniaturized inertial energy harvesters can be realized by MEMS technology, which allows large-scale manufacturing at a low cost. The fabricated devices are protected by dedicated packaging to withstand harsh mechanical conditions.

Wireless-sensor networks have a huge potential for both machinery and human body applications. Tire-pressure monitoring systems [48] and patient-health monitoring [49] are respective examples of these applications. The characteristics of the vibrations differ markedly in these two environments. Based on experimental results, Roundy concluded that the vibrations occurring in the vicinity of various machineries, including cars, have dominant components between 60 Hz and 200 Hz with the acceleration amplitude ranging from $10^{-2} ms^{-2}$ to $10^1 ms^{-2}$ [50]. On the human body, von Büren determined that the dominant frequencies of the observed motions are below 5 Hz [51]. Given the large difference between the ranges of frequencies, distinct designs of inertial vibration energy harvesters are required for wireless applications deployed on machines and human body, respectively.

For machine applications, the classical design of inertial harvesters is based on a resonant scheme. Namely, the mechanical element of the transducer consists of a resonator, which is excited in one of its resonant modes to deliver a maximum output power. The low frequencies associated with the human body require different approaches, which are based on non-linear and non-resonant principles, such as up-conversion of the frequency or impacts. The two different approaches are discussed hereafter in separate sections.

9.4.2 Machine Environment: Resonant Systems

Resonant vibration harvesters are by far the most widely investigated in the literature. Fine-machined versions are the earliest emerging commercial devices [52, 53] while micro-machined versions on the other hand are less mature but offer cost effective production methods. Their power levels are to be raised and reliability needs to be improved. Most of miniaturized energy harvesters ($\approx 1 \ cm^3$) provide power in the range of tens to hundreds of microwatts. The resonance frequencies of these systems are typically tens or hundreds of hertz. An overview of the performances of existing inertial harvesters can be found in [54, 55].

In the following, the general principles of inertial energy harvesting are first described. A short discussion about the three main types of electromechanical transducers is then presented. Examples of existing devices are also given. In the end, the optimization of the output power generated with inertial harvesters is elaborated.

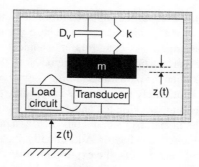

Figure 9.13   Generic lumped model of a resonant inertial vibration energy harvester

General principles The general principle of a resonant inertial vibration harvester can be understood through the lumped model given in Figure 9.13. The mechanical resonator is represented here by a seismic mass m connected to the package by a suspension element of stiffness k. Parasitic dissipations are introduced by the presence of a damper with damping D_v. The package is subject to a vibration $Z(t)$ so that, due to the inertia, the mass undergoes a relative displacement $z(t)$ with respect to its equilibrium position. The mass m is connected to an electromechanical transducer which extracts part of the kinetic energy carried by the mass and transfers it into an electrical load circuit. The latter can be the powered application or an energy storage system.

In the frame of reference attached to the package, the differential equation governing the dynamics of the system can be written as:

$$m\frac{d^2z}{dt^2} + D_v\frac{dz}{dt} + kz + F = -m\frac{d^2Z}{dt^2} \tag{9.5}$$

where F is the force developed by the transducer to counteract the displacement of the seismic mass. The harvested energy can be obtained by computing the work done by F.

In a first approximation, the association between the electromechanical transducer and the load circuit is represented by a viscous damper D_e, inducing a force proportional to the velocity of the seismic mass. In the electrical domain, the viscous damper is equivalent to an electrical resistor. The harvested power P is equal to the power dissipated in D_e. In case of a sinusoidal input vibration $Z(t) = Z_0 \sin(\omega t)$ and assuming steady state operation, P can be expressed in the frequency domain as:

$$P = \frac{1}{2}D_e\omega^2|\bar{z}(\omega)|^2 \tag{9.6}$$

in which the dash superscript indicates the complex transform of the corresponding variable.

Equation (9.6) is maximized when the frequency of the input vibration ω equals the resonance frequency of the mechanical system $\omega_0 = \sqrt{(k/m)}$ and moreover $D_e = D_v$. The corresponding maximum output power is expressed as:

$$P_{max} = \frac{1}{8}m\omega_0^3 Q_m Z_0^2 \tag{9.7}$$

in which the mechanical quality factor of the system Q_m is equal to $m\omega_0/D_v$.

The frequency ω and amplitude Z_0 of the input vibration are the primary parameters under consideration for maximizing the produced power. These parameters are dictated by the vibration source. Next, the seismic mass m and the mechanical quality factor Q_m should be maximized as P_{max} increases proportionally with them. Note that non-physical results are obtained from Equation (9.7) when the parasitic damping is ignored. Namely when Q_m reaches infinity, the produced power is infinitely large. In this case, some physical constraints, such as a maximum allowed displacement, are violated and eventually Equation (9.7) does not hold.

In practice, the electromechanical transducer not only acts as a purely dissipative element but also exhibits reactive and/or inductive behavior. Additionally, it can have non-linear characteristics. Irrespective of the non-linear effects, the expression given in Equation (9.7) represents the theoretical limit for the attainable output power. As shown in the next section, reaching this theoretical limit requires a proper design.

Transduction mechanisms

- ### Electrostatic transduction

  Electrostatic transducers are based on an electrical capacitor with a moveable electrode that can be displaced by the seismic mass. For a linear load circuitry, the motion of the moveable electrode can generate an electrical power only if the capacitor is biased by an external voltage V_0 or by a built-in charge Q_0 (electret based devices).

  MEMS electrostatic harvesters are often made in comb drive structures, such as the one depicted in Figure 9.14 [37]. They allow a large variation of the capacitance per unit displacement of the seismic mass. The device shown in Figure 9.14 is realized with a stack of two wafers. The first wafer contains the electrical polarization source, which is made of an electret, that is, the electrostatic equivalent of a permanent magnet. The second wafer containing the variable capacitor is then bonded onto the first one. Process development is currently ongoing to deal with the remaining technical issues, mainly related to the stability in time of the electret. It is predicted that the output power generated by the fully fledged device can reach several tens of µW when accelerations in the range of $10\,\mathrm{ms}^{-2}$ are considered.

- ### Piezoelectric transduction

  In piezoelectric materials, the barycentres of the positive and negative electrical charges in a crystal cell do not exactly coincide with each other when the cell is deformed. This mismatching results in an electrical polarization of the cell. The deformation-dependent polarization can be used to pump charges from conductors connected to the surface of the piezoelectric material.

  Most piezoelectric energy harvesters are based on bending mechanical elements, that is, beams or membranes, as they allow resonance at a frequency ranging from tens to hundreds of Hz. This frequency range can match the dominant frequencies of ambient vibrations [50]. Lead zirconate titanate (PZT) compounds are the most commonly used piezoelectric material in energy harvesters. Aluminum nitride (AlN) has recently gained more interest because of its standard sputter deposition technique. In terms of output power, energy harvesters based on PZT and AlN have equivalent performance.

  An example of AlN-based MEMS harvester is given in Figure 9.15. It is constructed on a stack of three wafers. The top and bottom wafer act as a primary package while the transducer itself is processed on the middle wafer. These devices generate an output power between 10 and 100 µW/g^2 at their resonance frequencies, which fall into the range between 300 and 1000 Hz.

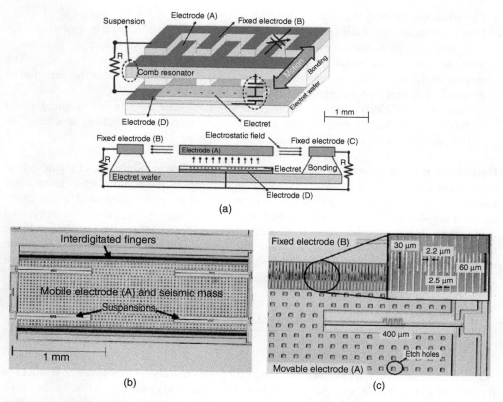

Figure 9.14 (a) Schematic configuration of an electret based harvester; (b) top view of the device fabricated by MEMS process; (c) close-up view on the interdigitated finger area (Courtesy of Sterken [37], Reproduced by permission of IEEE)

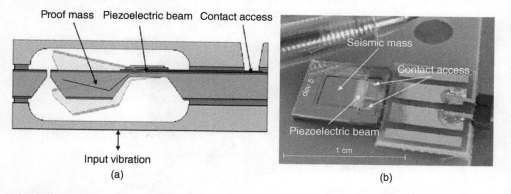

Figure 9.15 (a) Schematic design and (b) photo of a MEMS piezoelectric vibration energy harvester based on aluminum nitride (Courtesy of Elfrink [56], © IOP Publishing. Reproduced by permission of IOP Publishing) (see Plate 10)

- **Electromagnetic transduction**

  Electromagnetic transducers make use of the electromotive force induced by a varying magnetic flux through a conductive coil according to Faraday's law of induction. For energy harvesters, the source of magnetic flux B is often obtained with a permanent magnet. The motion of the seismic mass attached to either the coil or magnet induces the variation of magnetic flux necessary to generate a current in the coil.

  An example of electromagnetic harvester is given in Figure 9.16. In this system, the coil is fixed while an NdFeB magnet is attached to a cantilever-shaped vibrating structure. The device resonates at 52 Hz and generates 46 µW at 0.06 g.

Optimization of the generated power Sterken developed linearized models of the described transducers [58], which are valid for small displacements of the seismic mass. From these models, it is possible to derive a closed form expression of the power dissipated into an external load resistor R. The expression of the harvested power can be written in the same way for the three transduction mechanisms described above. It can furthermore be shown that the maximum P_m of the power is attained when the frequency of the sinusoidal input vibration ω is equal to the resonance frequency of the mechanical resonator ω_0. Moreover, the resistor should match the absolute value of the electrical impedance of the harvester. The expression of P_m is given in Equation (9.8):

$$P_m = \frac{m\omega_0^3 Z_0^2 Q_m}{4} \frac{1}{1 + \sqrt{1 + \frac{1}{K^4 Q_m^4}}} \tag{9.8}$$

where K represents the effective electromechanical coupling factor of the harvester. It relates to the amount of mechanical energy converted into electrical energy during an oscillation cycle. For optimal power generation, the product $K^2 Q_m$ should be maximized to reach the theoretical limit given in Equation (9.7).

As discussed for the simplified model, the frequency ω and amplitude Z_0 of the input vibration and the seismic mass m are the primary parameters under consideration for power

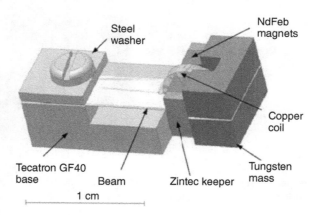

Figure 9.16   Electromagnetic energy harvester based on NdFeB magnets (Courtesy of Beeby [57], © IOP Publishing. Reproduced by permission of IOP Publishing)

generation. Q_m should also be increased by limiting the parasitic dissipations. Lowering the undesired damping due to pressure can be realized by using vacuum packaging [59]. On the other hand, minimizing the parasitic dissipations due to structural and anchor losses remains a complex engineering task.

In addition, the effective electromechanical coupling factor K should also be increased to deliver an optimum output power. K can be expressed in terms of the material properties and the system geometries. Based on a set of slightly different notations, this has been carried out respectively for electrostatic and electromagnetic harvesters in [58]. For electrostatic harvesters, K is optimized by maximizing the polarization voltage and the variation of capacitance per unit displacement of the seismic mass (comb drives design are used for this reason). In case of harvesters based on electromagnetic transduction, coils with low self-inductance are essential and the variation of magnetic flux through the coil per unit displacement of the seismic mass should be maximized. For piezoelectric composite bending structures, K is optimized by using materials having large piezoelectric constants and by implementing a specific ratio between the thickness of the support material and that of the piezoelectric material [60]. Shown in [61], K can range from 0.05 for AlN based MEMS harvesters to 0.3 for ceramic PZT based benders. This fact suggests that the ceramic PZT based benders are better suited for energy harvesting. This is however not the case, as the quality factor of these devices are much smaller than those of AlN based MEMS harvesters. Quality factors in the range of several hundred till several thousand can be obtained with MEMS fabricated piezoelectric harvesters [59] while Q_m is typically below 100 for ceramic PZT based benders.

In terms of power generation, the three transduction mechanisms are equivalent, if the same set of K and Q_m is assumed. However, this conclusion does not hold when the manufacturing technology is taken into consideration. For the relatively large devices fabricated by conventional machining, the electromagnetic transduction is generally preferred as the corresponding manufacturing process is cheap and well established. This is not the case for miniaturized devices manufactured by MEMS technologies. Magnetic materials are problematic from the perspective of integration into a process flow compatible with silicon based technologies. The design of micromachined coils is not straightforward, either. In this aspect, electrostatic and piezoelectric systems (based on IC processing compatible materials, such as aluminum nitride) are more convenient. Typically, the mass m of such MEMS harvesters is in the range of tens and hundreds of mg and their resonance angular frequency ω_0 between 500 and 10000 rad.s^{-1}. Commonly encountered values of K and Q_m have already been mentioned above.

9.4.3   Human Environment: Non-Resonant Systems

The aforementioned resonant harvesters can be adapted to the vibration at a relatively high frequency, for example, in a machine environment. In the case of the human body, the observed motions are characterized by low frequencies and high amplitude. It may not be obvious to design small-sized devices resonating at such low frequencies. Furthermore, the amplitude of the external motion is typically larger than the allowed displacement of the proof mass. Therefore, alternative design making use of the non-linear principles should be developed.

Rotational rather than linear internal motion of the seismic mass is adapted for the low frequency input motion. The most notable example is the self-winding automatic wristwatch. An analysis of the possible operating modes and power limits of rotating mass generators is presented in [62].

Targeting at low frequencies, Miao proposed the system based on electrostatic transduction depicted in Figure 9.17(a) [63]. The harvester structure is very similar to that of the resonant devices previously discussed. Namely, it is made up of a proof mass, which also constitutes the movable plate of a variable capacitor, connected to a package by elastic suspensions. The operating principle is nevertheless different. In Figure 9.17, a charge or voltage source is used to charge the fixed electrode, to which the proof mass is initially stuck as the result. The proof mass remains in this position until the inertial force induced by the external acceleration overcomes the electrostatic attraction. As soon as the equilibrium is lost, the proof mass is released. Due to the displacement, the movable electrode performs work against the electrostatic force, leading to an accumulation of electrical charges. These charges are extracted when the proof mass reaches contact pads, which are located on the opposite side of the device. The first prototype of this configuration with a volume of $0.6\,cm^3$ delivered about $4\,\mu W$, when excited by a sinusoidal acceleration of 1 g at the frequency of 30 Hz. With this concept, the output power can be further increased and moreover even lower frequencies can be reached.

Several "up-conversion" principles of low input frequencies have also been proposed. As exhibited in Figure 9.17(b), Kulah described a device made of a large ferromagnetic mass

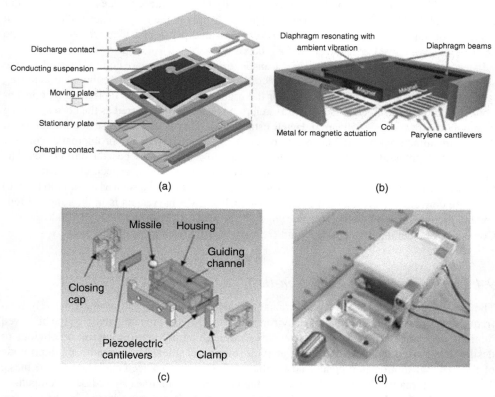

Figure 9.17 Examples of inertial harvesters for human body applications: (a) Courtesy of Mitcheson [63], with kind permission from Springer Science and Business Media B.V; (b) Courtesy of Kulah [64], Reproduced by permission of IEEE; (c) and (d) from [66], © IOP Publishing. Reproduced by permission of IOP Publishing

attached to a package by soft polymer suspensions [64]. This allows a system design that can resonate with low frequency environmental vibrations. In an effort to boost the harvested energy, a non-linear energy extraction mechanism is implemented. High resonance frequency cantilevers, which support a coil and a small metallic mass, are devised close to the large ferromagnetic mass. When the large mass oscillates, it alternatively captures and releases the cantilevers which afterwards undergo free vibrations. As the coils on top also undergo a varying magnetic field, electrical currents are thus induced. A MEMS prototype was recently manufactured but no complete characterization of output power has been presented [65]. As indicated in the preliminary measurement results, the proposed concept outperforms the conventional approaches in terms of power density. Another type of harvesters that can be adapted for human body applications is the impact based system. An example is given in Figure 9.17(c) and (d) [66]. It is made of a frame containing a guiding channel for a free sliding metallic missile. Two piezoelectric cantilevers are attached onto the periphery of the frame so that when the frame is shaken, the missile occasionally impinges on the piezoelectric benders. A prototype of this device with a volume of about $14\,cm^3$ delivered $50\,\mu W$ when rotating at an angular speed of over $180°$ per second. The output power is increased by a factor 12 when the system was forcefully shaken (approx. $9.7\,Hz$ and $10\,cm$ amplitude).

9.4.4   Power Management

When harvesting vibration energy, an AC voltage is generated. Therefore, the input voltage of the power management system can also be negative. As most types of load cannot handle negative voltages, the circuit has to perform rectification and also an adjustment of the DC-level of the voltage. In the seminal work by Shenck and Paradiso [67] a complete power management system was presented. Rectification is performed by a regular diode bridge which was suitable given the high voltages generated. A linear regulator was used for voltage regulation which leads to a low efficiency for the DC–DC-converter. The control circuit consumes only $15\,\mu A$. The power output can be further optimized by doing a joint mechanical and electrical system optimization.

For a piezoelectrical system, The synchronous, switched harvesting with inductor (SSHI) technique reported by Guyomar *et al.* in [68] introduced the key advance of using a switched inductor to flip the charge on the capacitor twice per cycle. Since the extra charge does not have to be drawn from an external supply, the losses can be minimal – limited primarily by the finite Q of the path containing the inductor. They actively modify the voltage on the piezoelectric capacitance, meaning the charge from the current source is forced into a higher voltage, corresponding to increased work being done and correspondingly, an increase in the electrical damping and output power. Gains have been reported between 2 and 10. A more elaborate discussion on power management for piezo electric harvesting can be found here [69].

9.4.5   Summary

Mechanical vibrations are present in many environments where applications for wireless-sensors networks can be found. They constitute then an interesting source of energy that can be converted into electricity for powering the nodes of the networks. The electromechanical conversion can be implemented through piezoelectric, electromagnetic and electrostatic transducers. When machine environments are considered, the typical frequencies of the existing vibrations are in the range of tens to hundreds of Hz. Harvesters based on an inertial design are adapted to these types of situations. MEMS fabricated inertial harvesters with a total size of a few cm^3 allows generating tens and hundreds of microwatts in machine environment, which is sufficient to power simple sensors. The inertial designs are, however, not adapted to the human body environment, where the typical frequencies of the encountered vibrations are below 10 Hz. In this case, alternative designs based on non-linear principles are necessary.

9.5 Far-Field RF Energy Harvesting

9.5.1 Introduction

In an environment where temperature gradient, vibration or ambient light cannot be used for energy harvesting, we can employ microwave power transmission (MPT). MPT can be used for directly powering a sensor node or for charging a battery or capacitor at a distance that, in turn, powers the WSN. The intercepted power of a wirelessly transmitted signal is proportional to the size of the collecting aperture. A substantial class of miniature, autonomous sensors is characterized by extremely low power consumption and/or low duty cycles, for example, temperature sensors, presence detectors. Therefore, the realization of miniature radio frequency (RF) energy harvesting devices – having a relatively small collecting aperture – has become feasible.

The history of far-field wireless power transmission by radio waves (excluding near-contact inductive or magnetic resonance power transfer [70]) dates back to the experiments of Heinrich Hertz in the 1880s. Hertz conducted his experiments to prove Maxwell's theory of electromagnetics [70]. The modern history of wireless power transmission started with the experiments performed by Brown in the 1960s, resulting in a microwave-powered model helicopter [70, 71]. These experiments paved the way for Glaser to propose the Solar Power Satellite (SPS) concept [72, 73]. According to the SPS concept, the solar energy in space is collected and subsequently converted into RF energy, which is then beamed to the Earth and eventually converted into electrical energy [70, 72–77]. The SPS concept offers an alternative energy source for the future.

Due to the termination of SPS efforts in the United States, the 1980s and 1990s showed scarce activity in the field of free-space power transmission [76]. Since 2000, interest in the field is growing again, as can be observed in Figure 9.18.

This interest is partly initiated by the introduction of Short-Range Devices, focusing on the available Industry, Science and Medical (ISM) frequency bands around 0.9 GHz, 2.4 GHz, 5.8 GHz and higher. For these frequencies, the wavelengths become sufficiently short for the realization of miniature wireless products, occupying typical volumes of one to a few cm^3. An RF power supply for such a system consists of an antenna (having dimensions in the order of a quarter to half a wavelength) coupled to a high-frequency rectifying circuit. The combination of a *rect*ifying circuit and an ant*enna* is commonly denoted as a *rectenna*.

9.5.2 General principle

A general RF-harvesting system – including the source – is schematically depicted in Figure 9.19.

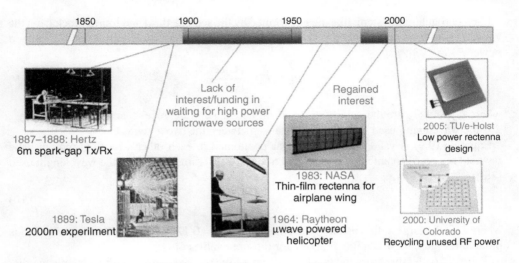

Figure 9.18   Brief history of wireless energy transfer

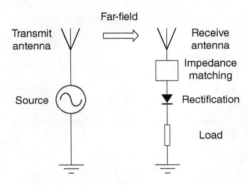

Figure 9.19   General RF-harvesting system

From the left to the right, we encounter a microwave source connected to a transmit antenna followed by the free space being bridged by electromagnetic waves. Then we continue with the RF harvester that consists of a receive antenna, an impedance and filtering network, a rectification circuit and a load. In some cases, a low-pass filter is inserted between the recti-fication circuit and the load. In the following, we will discuss the elements of the system and the components of the RF harvester.

Friis Transmission Equation The reason for including the microwave source, transmit antenna and free space into the system shown in Figure 9.19 is to illustrate the effect of far-field free space transmission and, more specifically, the power spreading with distance from the source. The far-field of an antenna is defined as the region at a sufficiently large distance away from the antenna such that the electromagnetic field behaves locally as a Transverse Electro Magnetic (TEM) wave. The power distribution in the far-field is a function of direction only, not of distance; the power amplitude of course is.

For a two-antenna system, like the one formed by the second, third and fourth block from the left in Figure 9.19, the received power P_R may be expressed as a function of the transmitted power P_T by [78]:

$$P_R = P_T \frac{G_T G_R \lambda^2}{(4\pi)^2 r^2} \qquad (9.9)$$

where G_T and G_R are the gain of the transmit antenna and the receive antenna respectively[1], λ is the wavelength used[2] and r is the distance between the two antennas.

Equation (9.9) is valid only for antennas positioned in each other's far-field regions. The far-field region of an antenna r_{ff} is related to its physical dimensions and the wavelength used [78]:

$$r_{ff} \geq \frac{2D^2}{\lambda} \qquad (9.10)$$

where D is the largest dimension of the antenna. For $\lambda = 0.125$ m ($f = 2.40$ GHz), (P_R/P_T) is plotted in Figure 9.20 as a function of r for different values of G.

Figure 9.20 clearly shows the quadratic decay of the received power with an increasing distance and the partial compensation for this decay by choosing a higher antenna gain. The figure shows that in practical situations only a small amount of power will be available. It is thus evident that this RF power should be converted into usable DC power as efficiently as possible.

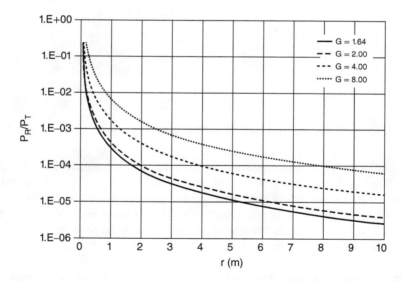

Figure 9.20 Received power, normalized to transmitted power as a function of distance for $\lambda = 0.125$ m. The curves start at distances equal to the far-field condition (Equation 9.10)

---

[1] Gain is a figure-of-merit characterizing how much power is radiated or received at maximum compared to a hypothetical uniform radiating or receiving antenna.

[2] Wavelength λ and frequency f are related through the speed of light: $c = \lambda f$.

Impedance and Filtering Network To maximize the RF-to-DC power transfer, the receive antenna needs to be impedance matched to the rectifying circuit. Next to the power transfer maximization, the impedance transformation network in Figure 9.19 also serves the purpose of filtering out higher harmonic frequency components. The rectifying circuit consists of one or more non-linear elements, for example, Schottky diodes, which will generate signals at multiples of the operational frequency. The impedance matching and filtering network prevents these signals from being reradiated by the receive antenna of the RF harvester. If we do not use standard antennas (i.e. having a $50\,\Omega$ input impedance) but have the possibility to design dedicated antennas, conjugately matched to the rectifying circuit, we can omit the impedance transformation and the filtering network.

Conjugate Matching Due to the conjugate matching, the RF-to-DC power transformation is maximized. Higher harmonic frequency components are mismatched to the antenna and thus will not be radiated. Omitting the impedance transformation and the filtering network has shown to increase the power transfer efficiency from 40% to 52% at a 0 dBm RF input power level[3] [79]. Models for designing antennas having a complex input impedance can be found in [80].

9.5.3 Analysis and Design

The design procedure for a conjugately matched RF harvester is as following:

- Decide on the frequency to be used. Necessary considerations include the use of ISM frequencies, the maximum allowed transmit power[4] and the size of the receive antenna that is directly related to the wavelength used.
- Determine the RF input impedance of the rectifying circuit.
- Design an antenna having an input impedance that is equal to the complex conjugate of the value found in the previous step. Use for example for the design of the antenna.

For designing the RF harvester, we need to be able to analyze the components of the harvester. The analysis tools for the antenna can be found in [80]. Here, we will briefly discuss the analysis of the rectifying circuit. The analysis is complicated by the fact that we are dealing with non-linear elements subject to both RF and DC analysis.

RF Analysis The most commonly used and also least complicated rectifier circuit consists of a single Schottky diode. A Schottky diode is capable of switching in the GHz frequency regime. The equivalent circuit of a loaded rectifier, employing a single packaged Schottky diode, is shown in Figure 9.21.

In this circuit, V_g is the high-frequency source, having an internal resistance R_g. The diode is modeled as an ideal diode d having a series resistance R_s and a voltage-dependent junction capacitance C_j. C_p and L_p are the parasitic capacitance and inductance due to the packaging. All these values can be found in the diode's data sheet. The diode is loaded with a parallel circuit consisting of a capacitor C_L and a resistor R_L.

---

[3] dBm denotes power relative to 1 mW expressed in dB. So, 0 dBm equals 1 mW, −10 dBm equals 0.1 mW, etc.

[4] International and national regulations in general do not specify P_T but the product of P_T and G_T which is known as the Effective Isotropic Radiated Power (EIRP = $P_T G_T$).

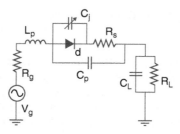

Figure 9.21   The equivalent circuit of a loaded rectifier, employing a single packaged Schottky diode

The circuit shown in Figure 9.21 can be analyzed in the time domain by using an adaptive step-size Runge-Kutta method [81]. Next, the results are Fourier-transformed to the frequency domain to obtain the input impedance and the DC output voltage. However, due to the different time constants of the system (period of the source and time constant of the load), the solving method may become unstable.[5]

To overcome this problem, we can separate RF and DC analysis by enlarging the value of the user-supplied capacitor C_L. The effect is that for RF signals the capacitor will act as a short circuit and the Runge-Kutta time-marching algorithm now becomes a very efficient way to analyze the equivalent circuit.

DC Analysis With the RF and DC analysis separated by virtue of the load capacitor C_L, we can now obtain the DC output voltage through an analysis by applying the Ritz-Galérkin averaging method [82]. This results in the following relation between the output voltage V_0 and the available incident power P_{inc}:

$$I_0 \left(\frac{q}{nkT} \sqrt{8 R_g P_{inc}} \right) = \left(1 + \frac{V_0}{R_L I_s} \right) e^{\left(1 + \frac{R_g + R_s}{R_L} \right) \frac{q}{nkT} V_0} \qquad (9.11)$$

where $I_0(x)$ is the zero-order modified Bessel function of the first kind having argument x, n is the diode's ideality factor, q is the electron charge, k is Boltzmann's constant, T is the temperature denoted in Kelvin and I_s is the diode's saturation current.[6]

When attached to an antenna, P_{inc} is equal to P_R, where P_R can be calculated using Equation (9.9) and where the source resistance is equal to the real part of the antenna input impedance.

9.5.4   Application

Instead of a single diode, a dual diode in voltage doubler configuration can be employed to (approximately) double the output voltage. The doubling circuit is schematically shown in Figure 9.22.

---

[5] The differential equations describing the system become 'stiff'.

[6] The output voltage in Eq. (9.11) is frequency-independent, which is true in practice for a load capacitor $C_L > 0.1\,\mu F$ in the frequency range $0.1\,GHz < f < 2.5\,GHz$, acting as a short circuit.

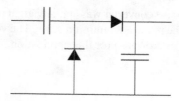

Figure 9.22   Two diodes in a voltage doubling configuration

Figure 9.23   Wirelessly-powered wall clock employing eight rectenna elements with voltage doubler rectifying circuits [79, 80]. Reproduced by permission of Wiley

For RF signals, the capacitors behave as short circuits and the equivalent circuit is that of an anti-parallel diode pair. The input impedance of the voltage doubler is thus half of that of the single diode rectifier. For DC signals, the capacitors behave as open circuits and the equivalent circuit now consists of two DC sources, that is, the diodes, in series connection. The output voltage is thus twice that of the single diode. In reality, the halving of the RF input impedance is correct but the doubling of the DC output voltage is too rough an approximation. The real output voltage will be a bit lower, especially for higher input powers.

A rectenna element has been designed and realized wherein a dedicated microstrip patch antenna is conjugately matched to the voltage doubling circuit. Eight of these elements have been arranged in a series connection for wirelessly powering a 1.5 V electric wall clock at a distance up to 6 m from a 2.45 GHz transmitter, transmitting an EIRP of approximately 3 W [79, 80]. The prototype clock is shown in Figure 9.23.

Since no DC to DC voltage boost converter was applied, eight rectenna elements were used in a series connection to deliver the required minimum 1.2 V. The voltage requirement dominated the power requirement (a few microwatts) for this application.

9.6   Photovoltaic

Photovoltaic cells convert incoming photons into electricity. Outdoor they are an obvious energy source for self-powered systems. Efficiencies range from 5% to 30%, depending on the material used. Indoor the illumination levels are much lower ($10\,\mu W/cm^2$ to $100\,\mu W/cm^2$) and photovoltaic cells generate a surface power density similar or slightly larger than that of the harvesters described above. As photovoltaic technology is well developed and many reviews have been published (e.g. [83]) it will not be discussed here. Indoor use requires a fine-tuning of the cell design to the different spectral composition of the light and the lower level of illumination [84].

Depending on the indoor illumination level and the installed location as well as the orientation of the solar cell, the output power of an amorphous silicon solar cell can vary from less than a few $\mu W/cm^2$ up to hundreds of $\mu W/cm^2$ [85]. To manage the variation in both the illumination and panel size, a power-management circuit that is able to efficiently handle a wide range of input power is necessary. A solar cell can be modeled as a light-controlled current source in parallel with a diode. Its output current is determined by the output voltage in an exponential relation. At one point, the solar cell reaches its maximum power point (MPP). This MPP can be tracked by tuning either the output voltage or the load impedance [86].

In [87] an IC is reported consisting of an inductive boost converter targeting solar cells at indoor conditions. The converter bridges a solar cell array and an energy storage system (ESS), which can be either a Li-ion battery or a super-capacitor. The efficiency of the converter without MPPT was measured under various conditions, as shown in Figure 9.24. The converter is able to convert input power from $5\,\mu W$ up to $10\,mW$. The measured peak efficiency is around 87%. The whole control circuit consumes a static current of $0.65\,\mu A$ from the battery when not converting power. While converting minimal power ($5\,\mu W$) and maximum power ($10\,mW$), the current consumption is $0.8\,\mu A$ and $2.1\,\mu A$, respectively.

9.7   Summary and Future Trends

9.7.1   Summary

This chapter is dedicated to the state-of-the-art in the miniaturized energy harvesters and energy storage devices, the combination of which are widely considered as a feasible solution to address the rapidly growing need of autonomy for wireless-sensor network. Made up of an array of wireless connected WSNs, wireless-sensor networks are poised to bring about a huge impact in a variety of sectors. Its development trend towards further miniaturization, larger scale and longer autonomy has made the batteries, the traditional power supply, obsolete. Moreover, with the observed continuous technological advancement, the power consumption of electronic components has been and will be, in the foreseeable future, steadily decreasing to such a level that the energy harvesters can suffice.

Within such a context, various types of energy harvesting and storage devices have attracted intensive interest from both academia and industry. The eventual market acceptance of energy

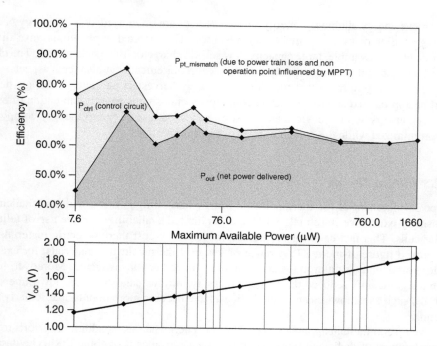

Figure 9.24 Experimental end-to-end efficiency using a discrete equivalent circuit as a solar cell simulator (top, battery voltage = 3 V). Also shown (bottom) is the corresponding open-circuit voltage [87]. Reproduced by permission of IEEE

harvesting technology hinges on further cost reduction, which can be approached by using MEMS technology. In this chapter, the general principle and the state-of-the-art are elaborated with regard to three types of energy harvesters, namely those based on temperature difference, vibration/motion and RF transfer. Given the necessity of using energy storage systems, an overview is subsequently presented with respect to the supercapacitors, the micro-batteries and the solid-state thin-film batteries.

Based on the Seebeck effect, the thermoelectric generators are mostly made of semiconductor materials, like SiGe or BiTe, on Si substrate or polymer foils. Several frequently used device configurations are presented with their respective MEMS embodiments. The output performance is determined by both the temperature difference and the overall material properties.

Motion-driven energy harvesters generally fall into two categories: resonant and non-resonant system. Used mostly in the machinery environment, the devices in the former category can be realized by using electrostatic, piezoelectric or electromagnetic transduction mechanisms. Devices in the latter category are frequently employed to deal with low frequency and large amplitude found in human body applications.

RF energy harvesters with rectennas provide an interesting alternative, particularly in an environment which other energy harvesting methods cannot properly address. A generic model for RF energy harvester is established and used to analyze the relevant technical aspects, for example, Friis transmission equation and conjugate matching. The design procedure for a conjugately matched harvester is proposed.

To achieve energy autonomy, energy storage devices are often needed as an energy backup, an energy buffer or even the main energy source. The targeted applications have distinct requirements, for example, large capacity, low self-discharge or high peak current. The choice of ESS should be relevant to not only the technical requirements but also legal aspects.

One should keep in mind that the choice of energy harvesters and energy storage devices should be approached at a system level. Namely, by choosing a smart combination of energy harvesters, energy storage devices and sensor devices, the use of energy-hungry components can be minimized while the overall efficiency can be optimized.

9.7.2   Future Trends

Further development of the thermoelectric energy harvesting is faced with the challenge of increasingly expensive BiTe material, due to its limited availability and the use of tellurium in solar cells. This fact calls for the development of low-cost thermoelectric materials with an improved figure-of-merit ZT. A range of low-cost thermoelectric materials, for example, Heusler compounds, skutterudites and clathrates, have been investigated. Moreover, having shown a significantly reduced thermal conductivity, various nanostructured low-dimensional materials, such as nanowires and superlattices, have become more popular particularly in the academia.

To develop commercial applications of vibration energy harvester, additional efforts towards the development of dedicated power conditioning electronics is ongoing [88]. The design of new concepts for low frequency or broadband vibrations has also become an active subject. The reliability of MEMS based devices needs to be investigated. For piezoelectric MEMS devices, the development of high performances thin film materials is a hot topic as well.

Given the proved feasibility of realizing miniature rectennas for sensors, current research is aimed at integrating rectennas into wireless autonomous sensors. The rectennas will be used for charging batteries or capacitors. Using the same antenna both for energy harvesting and for data communication is necessary to keep the sensor small. Additionally, the integration of energy harvesting antenna, communication antenna and battery is currently being evaluated.

Energy storage systems still have a large space for further improvement and exploration. The development and application of new electrode materials can help to increase the energy density. Next, an optimized packaging can also enhance the energy density of supercapacitors and batteries. A large part of the volume of an energy storage device is nowadays occupied by the inactive parts, like substrate, barrier, packaging and other materials that do not directly fulfill an active role in energy storage.

References

[1] Kamarudin, S., Daud, W., Ho, S. and Hasran, U. (2007). Overview on the challenges and developments of micro-direct methanol fuel cells (DMFC). *Journal Power Sources*, 163, 743–754.

[2] Cook, B.W., Lanzisera, S. and Pister, K.S.J. (2006). SoC Issues for RF Smart Dust. *Proceedings of the IEEE*, 94, 1177–1196.

[3] Mitcheson, P.D., Yeatman, E.M., Rao, G.K., Holmes, A.S. and Green, T.C. (2008). Energy harvesting from human and machine motion for wireless electronic devices. *Proceedings of the IEEE*, 96, 1457–1486.

[4] Torfs, T., Leonov, V., van Hoof, C. and Gyselinckx, B. (2006). Body-heat powered autonomous pulse oximeter. In the *Proceedings 5th IEEE Conference on Sensors*, Daegu, South Korea, pp. 427–430.

[5] Pop, V., Penders, J., van Schaijk, R. and Vullers, R. (2009). The limits and challenges for power optimization and system integration in state-of-the-art Wireless Autonomous Transducer Solutions. In the *Proceedings of 3rd Smart Systems Integration*, Brussels, Belgium, pp. 544–547.

[6] Vullers, R.J.M., van Schaijk, R., Doms, I., van Hoof, C. and Mertens, R. (2009). Micropower energy harvesting. *Solid-State Electronics*, 53, 684–693.

[7] Winter, M. and Brodd, R.J. (2004). What are batteries, fuel cells and supercapacitors? *Chemical Reviews*, 104, 4245–4269.

[8] Bergveld, H.J., Danilov, D., Pop, V., Regtien, P.P.L. and Notten, P.H.L. (2009). Adaptive state-of-charge determination. In *Encyclopedia of Electrochemical Power Sources*, Elsevier, The Netherlands, 1, 459–477.

[9] Penella, M.T. and Gasulla, M. (2007). Runtime extension of autonomous sensors using battery-capacitor storage. International Conference on Sensor Technologies and Applications, Valencia, Spain, pp. 325–330.

[10] Holland, C.E., Weidner, J.W., Dougal, R.A. and White, R.E. (2002). Experimental characterization of hybrid power systems under pulse current loads. *Journal of Power Sources*, 109, 32–37.

[11] Oudenhoven, J.F.M., Baggetto, L. and Notten, P.H.L. (2011). All-solid-state lithium-ion microbatteries: a review of various three-dimensional concepts. *Advanced Energy Materials*, 1(1), 10–33.

[12] Directive 2002/96/Ec of the European Parliament and of The Council of 27 January 2003 on waste electrical and electronic equipment (WEEE).

[13] Directive 2006/66/Ec of The European Parliament and of The Council of 6 September 2006 on batteries and accumulators and waste batteries and accumulators.

[14] Rechargeable Battery Recycling Act, United States Public Law 104–142 – MAY 13, 1996.

[15] Rowe, D.M. (1994). *CRC Handbook of Thermoelectrics*, CRC Press, Boca Raton, US.

[16] Eder, D. (2009). Thermoelectric power generation–the next step to future CO_2 reductions? In the *Proceedings of Thermoelectrics Applications Workshop*, San Diego, US.

[17] Weber, J., Potje-Kamloth, K., Hasse, F., Detemple, P., Völklein, F. and Doll, T. (2006). Coin-size coiled-up polymer foil thermoelectric power generator for wearable electronics. *Sensors and Actuators A*, 132, 325–330.

[18] Stark, I. and Stordeur, M. (1999). New micro thermoelectric devices based on Bismuth Telluride-type thin solid films. In the *Proceedings of the 18th International Conference on Thermoelectrics*, Baltimore, US, pp. 465–472.

[19] Qu, W., Plötner, M. and Fischer, W.J. (2001). Microfabrication of thermoelectric generators on flexible foil substrates as a power source for autonomous microsystems. *Journal of Micromechanics and Microengineering*, 11, 146–152.

[20] Wang, Z., van Andel, Y., Jambunathan, M., Leonov, V., Elfrink, R. and Vullers, R.J.M. (2010). Characterization of a bulk-micromachined membrane-less in-plane thermopile. *Journal of Electronic Materials*, in print.

[21] Kockmann, N., Huesgen, T. and Woias, P. (2007). Microstructured in-plane thermoelectric generators with optimized heat path. In the *Proceedings of the 14th International Conference on Solid-State Sensors, Actuators and Transducers*, Lyon, France, pp. 133–136.

[22] Kishi, M., Nemoto, H., Hamao, T., Yamamoto, M., Sudou, S., Mandai, M. and Yamamoto, S. (1999). Micro-thermoelectric modules and their application to wristwatches as an energy source. In the *Proceedings of the 18th International Conference on Thermoelectrics*, Baltimore, USA, pp. 301–307.

[23] Snyder, G.J., Lim, J.R., Huang, C.K. and Fluerial, J.P. (2003). Thermoelectric microdevice fabricated by a MEMS-like electrochemical process. *Nature Materials*, 2, 528–531.

[24] Böttner, H., Nurnus, J. and Volkert, F. (2007). New high density micro structured thermogenerators for standalone sensor systems. In the *Proceedings of the 26th International Conference on Thermoelectrics*, Jeju, Korea, pp. 306–309.

[25] Glatz, W., Schwyter, E., Durrer, L. and Hierold, C. (2009). Bi_2Te_3-based flexible micro thermo-electric generator with optimized design. *IEEE Journal of Microelectromechanical Systems*, 18(3), 763–772.

[26] Strasser, M., Aigner, R., Lauterbach, C., Sturm, T.F., Franosch, M. and Wachutka, G. (2004). Micro-machined CMOS thermoelectric generator as on-chip power supply. *Sensors and Actuators A*, 114, 362–370.

[27] Xie, J., Lee, C. and Feng, H. (2010). Design, fabrication, and characterization of CMOS MEMS-based thermoelectric power generators. *IEEE Journal of Microelectromechanical Systems*, 19(2), 317–324.

[28] Wang, Z., Leonov, V., Fiorini, P. and van Hoof, C. (2009). Realization of a wearable miniaturized thermoelectric generator for human body applications. *Sensors and Actuators A*, 156(1), 95–102.

[29] Su, J., Goedbloed, M., van Andel, Y., de Nooijer, M.C., Elfrink, R., Leonov, V., Wang, Z. and Vullers, R.J.M. (2010). Batch process micromachined thermoelectric energy harvesters: fabrication and characterization. *Journal of Micromechanics and Microengineering*, 20, 104005.

[30] Mateu. L, Pollak, M. and Spies, P. (2007), Power management for energy harvesting applications. Presented at PowerMEMS.

[31] Lhermet, H., Condemine, C., Plissonnier, M., Salot, R., Audebert, P. and Rosset, M. (2008). Efficient power management circuit: from thermal energy harvesting to above-IC microbattery energy storage. *IEEE Journal of Solid-State Circuits*, 43, 243–246.

[32] Doms, I., Merken, P., Mertens, R. and van Hoof, C. (2008). Capacitive power-management circuit for micropower thermoelectric generators with a 2.1 µW controller. In *ISSCC 2008 Digest of Technical Papers*, 300–301.

[33] Doms, I., Merken, P., Mertens, R. and van Hoof, C. (2009) "Integrated capacitive power-management circuit for thermal harvesters with output power 10 to 1000 µW", In *ISSCC 2008 Digest of Technical Papers*, 300–301.

[34] http://www.seikowatches.com/technology/kinetic/

[35] Jones, G., Tudor, M.J., Beeby, S.P. and White, N.M. (2004). An electromagnetic vibration-powered generator for intelligent sensor systems. *Sensors and Actuators A*, 110, 344–349.

[36] Cheng, S., Wang, N. and Arnold, D. (2007). Modeling of magnetic vibrational energy harvesters using equivalent circuit representations. *Journal of Micromechanics and Microengineering*, 17, 2328–2335.

[37] Sterken, T., Fiorini, P., Baert, K., Puers, R. and Borghs, G. (2003). An electret based electrostatic microgenerator. In the *Proceedings of the 12th International Conference on Solid-State Sensors, Actuators and Transducers*, Boston, US, pp. 1291–1294.

[38] Torres, E.O. and Rincon-Mora, G.A. (2006). Electrostatic energy harvester and Li-ion charger circuit for micro-scale applications. In the *Proceedings of the IEEE Symposium on Circuits and Systems*, Kos, Greece, pp. 65–69.

[39] Despesse, G., Jager, T., Chaillout, J.J., Lger, J.M., Vassilev, A., Basrour, S.B. and Charlot, B. (2005). Fabrication and characterization of high damping electrostatic micro devices for vibration energy scavenging. In the *Proceedings of DTIP*, Montreux, Switzerland, pp. 386–390.

[40] Renaud, M., Karakaya, K., Sterken, T., Fiorini, P., van Hoof, C. and Puers, R. (2008). Fabrication, modeling and characterization of MEMS piezoelectric vibration harvesters. *Sensors and Actuators A*, 145–146, 380–386.

[41] Fanga, H.B., Liua, J.Q., Xub, Z.Y., Donga, L., Wang, L., Chena, D., Caia, B.C. and Liub, Y. (2008). A MEMS-based piezoelectric power generator array for vibration energy harvesting. *Journal of Microelectronics*, 39, 802–806.

[42] Keawboonchuay, C. and Engel, T.G. (2007). Design, modeling, and implementation of a 30-kW piezoelectric pulse generator. *IEEE Transactions on Plasma Science*, 30, 679–686.

[43] Kok, S.L., White, N.M. and Harris, N.R. (2008). A free-standing, thick film piezoelectric energy harvester. In the *Proceedings of IEEE Sensors*, Leece, Italy, pp. 589–592.

[44] Roundy, S. (2003). Energy Scavenging for Wireless Sensor Networks, Ph.D. Thesis, Southampton University.

[45] Bayrashev, A., Robbins, W.P. and Ziaie, B. (2004). Low frequency wireless powering of microsystems using piezoelectric-magnetostrictive laminate composites. *Sensors and Actuators A*, 114, 244–249.

[46] Bayrashev, A., Robbins, W.P. and Ziaie, B. (2001). Energy scavenging with shoe-mounted piezoelectrics. *IEEE Micro*, 21, 30–42.

[47] Kymissis, J., Kendall, C., Paradiso, J. and Gershenfeld, N. (1998). Parasitic power harvesting in shoes. In the *Proceedings of the Symposium on Wearable Computers*, Pittsburgh, US, pp. 132–139.

[48] Flatscher, M., Dielacher, M., Herndl, T., Lentsch, T., Matischek, R. and Prainsack, J. (2009). A robust wireless sensor node for in-tire-pressure monitoring. In the *Proceedings of IEEE Solid-State Circuits Conference*, San Francisco, US, pp. 286–287.

[49] Yang, G.Z. (2006). *Body Sensor Networks*, Springer-Verlag, Germany.

[50] Roundy, S.J., Wright, P.K. and Rabaey, J. (2003). A study of low level vibrations as a power source for wireless sensor nodes. *Computer Communications*, 26, 1131–1144.

[51] Von Büren, T. (2006). Body-Worn Inertial Electromagnetic Micro Generators, Ph.D. Thesis, Swiss Federal Institute of Technology.

[52] Perpetuum, http://www.perpetuum.com

[53] EnOcean, http://www.enocean.com

[54] Beeby, S.P., Tudor, M.J. and White, N.M. (2006). Energy harvesting vibration sources for microsystems applications. *Measurements Science and Technologies*, 17, 175–195.

[55] Mitcheson, P.D., Yeatman, E.M., Rao, G.K., Holmes, A.S. and Green, T.C. (2008). Energy harvesting from human and machine motion for wireless electronic devices. *Proceedings of the IEEE*, 96(9), 1457–1486.

[56] Elfrink, R., Kamel, T.M., Goedbloed, M., Matova, S., Hohlfeld, D., van Andel, Y. and van Schaijk, R. (2009). Vibration energy harvesting with aluminum nitride based piezoelectric devices. *Journal of Micromechanics and Microengineering*, 19, 094005.

[57] Beeby, S.P., Torah, R.N., Tudor, M.J., Glynne-Jones, P., O'Donnell, T., Saha, C.R. and Roy, S. (2007). A micro electromagnetic generator for vibration energy harvesting. *Journal of Micromechanics and Microengineering*, 17, 12–57.

[58] Sterken, T. (2009). Micro-Electromechanical Energy Harvesters, Ph.D. Thesis, Katholieke Universiteit, Leuven.

[59] Elfrink, R., Renaud, M., Kamel, T.M., de Nooijer, C., Jambunathan, M., Goedbloed, M., Hohlfeld, D., Matova, S., Pop, V., Caballero, L. and van Schaijk, R. (2010). Vacuum-packaged piezoelectric vibration energy harvesters: damping contributions and autonomy for a wireless sensor system. *Journal of Micromechanics and Microengineering*, 20, 104001.

[60] Renaud, M., Fiorini, P. and van Hoof, C. (2007). Optimization of a piezoelectric unimorph for shock and impact energy harvesting. *Smart Materials and Structures*, 16, 1125–1135.

[61] Renaud, M. (2009). Piezoelectric Energy Harvesters for Wireless Sensor Networks, Ph.D. Thesis, Katholieke Universiteit, Leuven.

[62] Yeatman, E.M. (2008). Energy harvesting from motion using rotating and gyroscopic proof masses. *Journal of Mechanical and Engineering Sciences*, 222, 27–36.

[63] Miao, P., Mitcheson, P.D. and Holmes, A.S. (2006). MEMS inertial power generators for biomedical applications. *Microsystems Technologies*, 12, 1079–1083.

[64] Kulah, H. and Najafi, K. (2004). An electromagnetic micro power generator for low-frequency environmental vibrations. In the *Proceedings of the IEEE Conference on MEMS*, Maastricht, the Netherlands, pp. 237–240.

[65] Sari, I., Balkan, T. and Külah, H. (2010). An electromagnetic micro power generator for low-frequency environmental vibrations based on the frequency up conversion technique. *Journal of Microelectromechanical Systems*, 19, 1075–1078.

[66] Renaud, M., Fiorini, P., van Schaijk, R. and van Hoof, C. (2009). Harvesting energy from the motion of human limbs: the design and analysis of an impact-based piezoelectric generator. *Smart Materials and Structures*, 18, 035001.

[67] Shenck, N.S. and Paradiso, J.A. (2001). Energy harvesting with shoe-mounted piezoelectrics. *IEEE Micro*, 21, 30–42.

[68] Guyomar, D., Magnet, C., Lefeuvre, E. and Richard, C. (2006). Nonlinear processing of the output voltage of a piezoelectric transformer. *IEEE Transactions on Ultrasonics Ferroelectrics and Frequency Control*, 53(7), 1362–1375.

[69] Dicken, J., Mitcheson, P.D., Stoianov, I. and Yeatman, E.M. (2012) Power-extraction circuits for piezoelectric energy harvesters in miniature and low-power applications. *IEEE Transactions on Power Electronics*, 27(11), 4514–4529.

[70] Brown, W. (1984). The history of power transmission by radio waves. *IEEE Transactions on Microwave Theory and Techniques*, MTT-32(9), 1230–1242.

[71] Brown, W. (1969). Experiments involving a microwave beam to power and position a helicopter. *IEEE Transactions on Aerospace and Electronic Systems*, AES-5(5), 692–702.

[72] Glaser, P. (1968). Power from the sun: It's future. *Science*, 162, 957–961.

[73] Nansen, R. (1996). Wireless power transmission: the key to solar power satellites. *IEEE AES Systems Magazine*, 11(1), 33–39.

[74] US Department of Energy (1979). Satellite Power Systems (SPS), Concept Development and Evaluation Program, Preliminary Assessment, Report DOE/ER 0041.

[75] Glaser, P. (1995). Energy for the global village. In the *Proceedings Canadian Conference on Electrical and Computer Engineering*, pp. 1–12.

[76] McSpadden, J. and Mankins, J. (2002). Space solar power programs and microwave wireless power transmission technology. *IEEE Microwave Magazine*, 3(4), 46–57.

[77] Sood, A., Kullaqnthasamy, S. and Shahidehpour, M. (2005). Solar power transmission: from space to earth. In the Proceedings IEEE Power Engineering General Meeting, pp. 605–610.

[78] Visser, H. (2005). *Array and Phased Array Antenna Basics*, John Wiley & Sons Ltd, Chichester, UK.

[79] Visser, H., Theeuwes, J., van Beurden, M. and Doodeman, G. (2007). High-Efficiency Rectenna Design, *EDN* 52(14), 34.

[80] Visser, H. (2009). *Approximate Antenna Analysis for CAD*, John Wiley & Sons Ltd, Chichester, UK.

[81] Press, W.H., Flannery, B.P., Teukolsky, S.A. and Vetterling, W.T. (1998). *Numerical Recipes: The Art of Scientifc Computing*, Cambridge University Press, Cambridge, UK.

[82] Harrison, R. and Le Polozec, X. (1994). Nonsquarelaw behavior of diode detectors analyzed by the Ritz-Galérkin method. *IEEE Transactions on Microwave Theory and Techniques*, 42(5), 840–846.

[83] Green, M.A. (2004). *Third Generation Photovoltaics: Advanced Solar Energy Conversion*, Springer-Verlag .

[84] Randall, J.F. (2005). *Designing Indoor Solar Products*, John Wiley & Sons Ltd.

[85] Wang, W.S., O'Donnell, T., Ribetto, L., *et al.* (2009). Energy harvesting embedded wireless sensor system for building environment applications. International Conference on Wireless VITAE, pp. 36–41, May 2009.

[86] Shmilovitz, D. (2005). On the control of photovoltaic maximum power point tracker via output parameters. *IEE Proceedings-Electric Power Applications*, 152(2), 239–248.

[87] Qiu, Y., Van Liempd, C., Op het Veld, B., Blanken, P. and van Hoof, C. (2011) 5 µW-to-10 µW input power range inductive boost converter for indoor photovoltaic energy harvesting with integrated maximum power point tracking algorithm. ISSCC 2011, San Francisco, USA, pp. 24–26.

[88] D'Hulst, R. (2009). Power processing circuits for vibration based energy harvesters, Ph.D. Thesis, Katholieke Universiteit, Leuven.

Index

Smart Sensor Systems: Emerging Technologies and Applications, First Edition.
Edited by Gerard Meijer, Michiel Pertijs and Kofi Makinwa.
© 2014 John Wiley & Sons, Ltd. Published 2014 by John Wiley & Sons, Ltd.

www.ingramcontent.com/pod-product-compliance
Lightning Source LLC
Chambersburg PA
CBHW082045280125
20788CB00044B/49